ROBERTSON'S INTRODUCTION TO FIRE PREVENTION

8th Edition

Michael T. Love
James C. Robertson, MIFireE

PEARSON

Boston Columbus Indianapolis New York San Francisco Upper Saddle River Amsterdam
Cape Town Dubai London Madrid Milan Munich Paris Montreal Toronto
Delhi Mexico City São Paulo Sydney Hong Kong Seoul Singapore Taipei Tokyo

Publisher: Julie Levin Alexander
Publisher's Assistant: Regina Bruno
Editor-in-Chief: Marlene McHugh Pratt
Product Manager: Sladjana Repic
Program Manager: Monica Moosang
Development Editor: Jo Cepeda
Editorial Assistant: Daria Galbo Ballard
Director of Marketing: David Gesell
Executive Marketing Manager: Brian Hoehl
Marketing Specialist: Michael Sirinides
Project Management Lead: Cynthia Zonneveld
Project Manager: Julie Boddorf
Full-Service Project Manager: Munesh Kumar, Aptara®, Inc.
Editorial Media Manager: Amy Peltier
Media Project Manager: Ellen Martino
Creative Director: Jayne Conte
Cover Designer: Nesbitt Graphics
Cover Image: William (Billy) Morris
Composition: Aptara®, Inc.
Printer/Binder: LSC Communications
Cover Printer: LSC Communications

Credits and acknowledgments of material borrowed from other sources and reproduced, with permission, in this textbook appear on appropriate pages with text.

Library of Congress Cataloging-in-Publication Data
Robertson, James C. (James Cole), 1929- author.
[Introduction to fire prevention]
Robertson's Introduction to fire prevention. — 8th edition/James C. Robertson, MIFireE, Michael Love.
pages cm
Revision of: Introduction to fire prevention. — 7th ed. — c2010.
ISBN 978-0-13-384327-9
ISBN 0-13-384327-0
1. Fire prevention. I. Love, Michael, 1954- author. II. Title. III. Title: Introduction to fire prevention.
TH9145.R55 2015
363.37'7—dc23 2014012285

ISBN-13: 978-0-13-384327-9
ISBN-10: 0-13-384327-0

CONTENTS

MEMORIAL TRIBUTE

Over the last year, as this edition was being finished, co-author James C. "Robbie" Robertson was slowly ebbing away from us. It was painful to observe and yet, right up until the end of his 84 years on this earth, Robbie remained passionate about the fire services, the fire protection disciplines, and his country. All across North America, he was recognized for his keen sense of fairness and willingness to help anyone committed to public service and safety.

It seemed that in our "world of fire," all those he knew were embraced as members of his family. Few in U.S. fire protection history ever had the far-reaching impact on persons of all ages and from a broad range of diverse backgrounds that Robbie Robertson did. Just like the seven earlier editions of this text, which for decades was *the* standard for fire prevention in college-level fire science and fire technology programs and was often referenced in fire officer promotional exams, this edition brings forward the latest insights, happenings, and techniques in all aspects of fire prevention.

Robbie was firm in his convictions about this nation and his sense of history, graciously offering to share what he knew. He was an avid collector of fire texts from near and far, even overseas, which he described as important in producing the various editions of this text. Additionally, his work on so many levels took him from coast to coast many times, all the while observing and making both mental and written notes, from which he affirmed and confirmed how groups dealt with service delivery expectations, wildly differing resources, and, of course, change.

The word *change* often engenders resistance and, when used with the term *new fire prevention,* initiatives similarly included push-back. It is intriguing to see prevention methodologies and advocacy now gaining acceptance in the hearts and minds of our colleagues who may have been criticized for such approaches only a few short years ago. Robbie's texts and constant promotion of prevention were significant in this evolution.

Whether as a volunteer or career firefighter, state fire service training instructor, state fire marshal, college professor, Commander in the U.S. Coast Guard Reserve, international representative of a prestigious fire safety association, or a member of numerous committees and study groups, Robbie's involvement made for great strides that continue forward today. He loved to travel and interact with others, often reporting in the following days and weeks that he "just had a great visit with . . .". The blank was always a who's who in our respective fields. He seemed to be able to recall everything and helped remind us of the need to be comprehensive and thoughtful in our work.

In his passing, one thing that has surprised even those who knew him well is just how involved he was in so many different yet interrelated parts of the fire safety puzzle. Be it on matters of public policy or advancing the cause of training and education, he could be counted on for important reminders drawn from what he had seen firsthand and had also done during his many travels and in his many multifaceted roles. Robbie's friendship and influence will be missed, but his work was important and lives on. Thank you, Robbie.

R. Wayne Powell, Executive Director, National Fire Heritage Center, Emmitsburg, Maryland

PREFACE

This Eighth Edition of *Robertson's Introduction to Fire Prevention* represents over 38 years of service to the students of fire science. Despite the passing of author James C. "Robbie" Robertson in 2013, his legacy of introducing students to an awareness of and general familiarity with fire prevention lives on in this edition. *Robertson's Introduction to Fire Prevention,* Eighth Edition, is updated to sustain a strong core of fire prevention history and some of the latest information in fire prevention research, statistics, and progress in fighting the fire problem. It is equally useful to the first-year fire science student, as a midcareer reference for fire officer promotional exams, and as a go-to reference on the shelf of fire chiefs.

Robertson's Introduction to Fire Prevention is a general survey of the many concepts associated with the fire prevention industry. Other publications provide more specifics on the technical side of managing fire prevention programs. Along with a historical perspective, the book offers insight into the philosophy behind efforts to prevent fires or to reduce their impact and then narrows its focus to subsets of elements of the fire prevention discipline. Elements covered in the book include, for example, the status of the "Three Es" of fire prevention and the addition of two new "Es"; the organization and administration of local-government fire prevention units; public fire education programs; fire investigation and arson investigation; and the means of proving when fire prevention efforts work.

The reader will find in this edition of *Robertson's Introduction to Fire Prevention* the most current information available relating to fire prevention. In addition, the chapter on international fire prevention perspectives has been moved out of the book to a new online resource that will include much more in the way of supplemental material, links to key resources, and articles that offer a bit more depth. More information will be available on this new feature in the future.

Readers will find some useful tools in *Robertson's Introduction to Fire Prevention* that will help them to navigate through the book and to readily recognize key information by way of boldface and defined terms. The Table of Contents offers an intuitive design to make clear the main sections of each chapter. Within each chapter, readers will recognize immediately what they should learn from that chapter by reading the list of objectives. When they have finished reading, they can test their knowledge with the summary questions at the end of the chapter. Each chapter also includes end notes that offer the student an opportunity to find more detailed information about notable areas of study. Many of the end notes have online World Wide Web addresses that allow readers to find the original sources of quotes and critical information, as well as more discussion of the subject. A glossary at the end of the book lists the critical terms used throughout the book, with definitions. The index identifies one or more pages where key concepts appear.

In addition, readers will find updates reflecting current conditions and statistics, as well as new and emerging concepts in fire prevention. Some examples are an introduction to the evolving concept of Community Risk Reduction (CRR). This integrated approach to managing community safety is just starting to show positive outcomes. Readers will learn how communities have taken fire safety directly to people, with home safety visits, workshops in assembly occupancies, and public information. They will also read about the latest efforts to increase the acceptance of fire safety technology by creating incentives for installing residential fire sprinklers.

Readers will find many examples of innovation, sometimes resulting from shortages in funding and resources and possibly related to a dangerous fire trend emerging in their own communities. Asheville, North Carolina, is one of these examples, where funding pressure led to a reengineering of the inspections process and created more opportunity for improved customer service. Another example is the research work of Fire Chief Brian Crawford, who brought about changes in Shreveport, Louisiana, that address some of the inherent fire risk associated with poverty.

Readers will discover what researchers have learned about typical human reactions to being threatened by fire. The discussion of the latest information on fire and arson investigation offers up-to-date statistics from the FBI and describes new approaches to arson-related investigative techniques, as well as a view of expert testimony in arson cases.

Readers will also learn of the continuing advancement in fire science and fire safety, from the urban residential areas of London, England, to the remote wildland areas of the United States. They will explore some of the principles of evaluating fire prevention programs and will discover just some of the positive outcomes available from fire safety organizations proving that fire prevention works.

ACKNOWLEDGMENTS

The contribution of a number of people who reviewed certain chapters and suggested revised wording is greatly appreciated. They include Marty Ahrens, Meri-K Appi, Charles Burkell, Brian Crawford, Michael A. Donahue, Cheryl Edwards, Rita Fahy, Wayne Hamilton, J. Frank Hodges Jr., Dean Hunt, David Icove, Chris Jelenewicz, Tom Lia, Lorne MacLean, Daniel Madrzykowski, Mary Marchone, Brian Meurer, Lori Moon, James G. Munger, Robert Neale, Jon Nisja, Joe Pierce, Fred Prather, Vincent Quinterno, Tammy Roddey, Steve Sawyer, Phil Schaenman, Edwina Scott, L. Charles Smeby Jr., Mark V. Smith, and Ken Tennant. Other contributors are recognized in endnotes. A big thanks to Billy Morris who stepped in late in the process to help acquire and organize photos. Special recognition is due to Kim Borofka, whose experience from the 7th Edition and early work on the 8th Edition helped us get moving and kept us on track. I would also like to thank Monica Moosang and Stephen Smith from Pearson. Through her editorial excellence, Development Editor Jo Cepeda helped us create a manuscript that is clear, readable, and correctly composed.

Thanks also to the following reviewers:

Dave Sauerbrey
Portland Community College, OR

Nathan Sivils, B.S.
Blinn College, TX

Kevin J. Abbott
Columbia Southern University, AL

Thomas Y. Smith Sr., M.S.
West Georgia Technical College, GA

Mike Kennamer, Ed.D.
Northeast Alabama Community College, AL

Matthew Marcarelli, A.A.S., B.S.
New Haven Regional Fire Training Academy, CT

Captain Michael J. Wheeler, CFI, CFPE
Prairie State College, IL

ABOUT THE AUTHORS

James C. Robertson, MIFireE, was a consultant on fire protection for various cities and countries. He received an A.S. in Fire Protection at Oklahoma State University and a B.S. with specialization in Fire Administration from the University of Southern California. He served in the U.S. Coast Guard Reserve, retiring as Commander. He served as both a career and a volunteer firefighter for 15 years. Robertson was an instructor for the University of Maryland's Fire Service program and an Assistant Fire Chief in Gainesville, Florida, and he represented the National Fire Protection Association (NFPA) in the southern and midwestern states. Robertson served as Maryland's State Fire Marshal for 18 years.

Michael T. Love worked as a firefighter for Montgomery County, Maryland, for over 33 years. He served in a diverse range of positions that included extensive experience as an operational company officer, a firefighter, and an instructor; Field Battalion Chief; Assistant Chief, commanding one of three shifts of countywide operations; Chief of Emergency Medical Services; and Special Operations Chief, Supervisor of Dispatch. Most recently, he commanded a division that included the Fire Marshal's Office, Public Information, Community Outreach, Recruiting, and Planning and Research. As Division Chief, Love served as the Montgomery County Fire Marshal. He is a graduate of the University of Maryland with a bachelor's degree in Fire Science, a graduate of the Executive Fire Officer Program, and a Certified Public Manager. Love is currently a freelance planning consultant and technical writer and is a member of the Maryland Fire and Burn Safety Coalition.

The following grid outlines Fire Prevention course requirements and where specific content can be located within this text:

Course Requirements	1	2	3	4	5	6	7	8	9	10	11	12	13
Define the national fire problem and main issues relating thereto.	X	X	X		X			X			X	X	X
Recognize the need for, responsibilities of, and importance of fire prevention as part of an overall mix of fire protection.		X					X						
Recognize the need for, responsibilities of, and importance of fire prevention organizations.		X							X	X			
Review minimum professional qualifications at the state and national level for Fire Inspector, Fire Investigator, and Public Educator.						X					X		
Define the elements of a plan-review program.				X									
Identify the laws, rules, codes, and other regulations relevant to fire protection of the authority having jurisdiction.	X	X		X	X						X		
Discuss training programs for fire prevention.			X	X		X					X		
Design media programs.			X						X				
Discuss the major programs for public education.		X	X			X						X	X

CHAPTER

1 History and Philosophy of Fire Prevention

Tom Reel/Thomson Reuters (Markets) LLC

KEY TERMS

conflagration, *p. 15*
curfew, *p. 2*
fire-resistive, *p. 5*
fire exits, *p. 6*
fire prevention, *p. 17*
fireproof, *p. 12*
fire safety, *p. 3*
public assembly, *p. 7*

OBJECTIVES

After reading this chapter, you should be able to:

- Identify tragedies that have led to regulations in specific types of occupancies.
- Identify contributing factors leading to fires of historical impact.
- List five contributing factors to loss of life in fires.
- Identify ways that arsonists have been punished for this crime in the past.
- Describe ways that early governments managed the fire risk of commonly found fuel.
- Identify early fire prevention measures and make the connection to modern fire prevention practices in some cases.
- Identify the major classifications of occupancies.

Fire's Early Impact

A great deal can be learned by studying the historical development of fire prevention. The brief review given here will assist the reader in recognizing and understanding the reasons for certain procedures that are followed in the field today.[1]

As early as 300 BCE, the Romans established a "fire department," which was composed primarily of slaves. The response of those individuals is reported to have been quite slow. Little else is known about their procedures. However, the program apparently was so unsuccessful that it was necessary to convert the department into a paid force in the year 6 CE.

This conversion apparently proved successful, and by 26 CE the full-time fire force in Rome had grown to approximately 7,000. Those individuals were charged primarily with a responsibility for maintaining fire prevention safeguards. The population of Rome at the time was just under 1 million.

The fire brigades of Rome patrolled the streets to make sure people followed proper fire prevention procedures. Granted the authority to administer corporal punishment to violators of fire codes, they were provided with rods. Records indicate that "most fires are the fault of the inhabitants."[2] As an interesting sideline, in addition to their fire prevention duties, the fire brigades of Rome had the responsibility of keeping a watchful eye on the clothing of individuals who were using the public baths, and they were required to make inspections of the baths on a regularly scheduled basis to prevent theft.

In 872, according to history, a bell was used in Oxford, England, to signal the time to extinguish all fires. The Anglo-Norman word *covrefeu,* which means "cover fire," evolved into the English word **curfew**. Records from 1066 indicate that during their occupation of England, the Normans strongly enforced the requirements for extinguishing fires at an early hour in the evening. Because construction at the time allowed dwelling fires to spread easily, this preventive measure was an effective safeguard.

curfew
▪ an official action intended to restrict some activity; derived from the Anglo-Norman word *covrefeu,* which means "cover fire."

As an added means of fire prevention, certain building code requirements were imposed. Fitzstephen, writing in 1189 during the reign of Henry II, stated: "The only plagues of London were immoderate drinking by idle fellows and often fires."[3] This comment suggests London had a severe fire problem in those early days. In an effort to control the fire situation, the lord mayor of London issued an order in 1189 to the effect that "no house should be built in the city but of stone and they must be covered with slate or tiled."[4] This requirement was apparently vigorously enforced in structures built from that day on.

In 1190, Oxford imposed a requirement for firewalls to be placed between every six houses. This is another early example of a major city's efforts to rapidly control a fire and limit its spread.

The Scottish Act of 1426 emphasized fire prevention. For example, it ordered "no hemp, lint, straw, hay or heather, or broom be stored near a fire."[5] Edinburgh merchants selling such wares were permitted to use lanterns, but not candles, and citizens in general were forbidden to carry open flames from house to house.

Fire precautions figured prominently in the Edinburgh Improvement Act of 1621. It ordered noncombustible roofs and required tradespeople who kept "heather, broom, whins, and other fuel"[6] in the center of town to remove the material to remote areas.

High-rise buildings were a problem in 17th-century Edinburgh. In those early times the use of many-storied buildings introduced considerable fire risk because there were limited means of egress and no limits to fire spread, not to mention the incredible increase in combustibility of the high buildings, which enabled fire spread and conflagration. Some buildings in Edinburgh reached 14 stories. As a result, the Scottish Parliament issued a regulation in 1698 restricting new buildings to a height of 5 stories. The regulation did not, however, affect existing buildings.[7]

Arson was a problem during riots in the early days of England. In 1272 in Norwich, for example, 34 rioters involved in arson and looting were captured. Their punishment consisted of being dragged about town until dead. One woman arsonist was burned alive as punishment for her act. Arson laws were just as severe 300 years later: In 1585 a 15-year-old boy in Edinburgh was judged responsible for setting fire to peat stacks and was burned alive as punishment for his act.

Specific punishments for fire prevention violations are also noted in historical documents. The city records of Southampton, England, contain a late-1500s case in which a baker was fined 2 shillings for having combustibles too close to an oven. A 1566 law forbade Manchester bakers to keep gorse (barley) "within two bays of the ovens."[8]

Charles II in 1664 gave authority for imprisoning those who contravened building regulations. The regulations related, then as now, to **fire safety**.[9] A 1763 act prohibited the piercing of fire walls.[10]

fire safety
▪ the concept of actions planned and taken that reduce the risk of human exposure to fire.

Among fire prevention recommendations issued to the public in England was one in 1643 that suggested candles be placed in water-based holders. The thought was that an unattended candle would burn down and go out before causing trouble. Before that, an act of Parliament in 1556 had required bellmen to patrol the streets and cry out, "Take care of your fire and candle."[11]

In 1212 a fire in London caused 3,000 deaths. No fire recorded before then had caused such a great loss of life. More than 400 years later, in 1666, another major fire struck the city. Referred to as the Great Fire of London, it burned for four days and destroyed five-sixths of the city. Amazingly, only six deaths occurred. The effectiveness of the previously imposed fire prevention requirements undoubtedly had a bearing on the reduced number of deaths. Although thousands of structures were destroyed, the progress of the fire was retarded long enough to allow the occupants to vacate their premises.

As a further indication of efforts in the fire prevention field, in 1722 an English citizen named David Hartley secured a patent for a fire prevention invention. The invention consisted of steel plates with dry sand between them, meant as a means of reducing fire spread from one floor to another. Hartley's construction innovation was used successfully in some buildings prior to his securing a patent. Mr. Hartley's invention was considered noteworthy enough that a statue was erected in his honor.[12]

In 1794, theater fire protection was given a boost by the placement of a water tank on the roof of a theater in England. The tank provided a curtain of water in the event of a fire. In addition, an iron safety curtain was provided to separate the theater patrons from fire on the stage.[13]

A February 1849 fire involving a burning piece of paper and a small gas leak caused 65 fatalities, mainly of young people in a Glasgow, Scotland, theater. Many tripped and fell as they tried to escape.[14]

The Birmingham, England, fire brigade issued a requirement in 1884 for inhabited tall buildings to have two staircases. This requirement was considered a progressive fire protection measure.[15]

Early Fire Prevention Measures in North America

During the 1600s, America's colonists generally reacted to fire danger by implementing stringent regulations.[16] For example, chimneys were a major fire problem then. So, in 1631, as a result of a serious fire in Boston, Massachusetts, Governor John Winthrop issued an order that simply prohibited wooden chimneys and thatched roofs.[17]

Fire inspections in the New World probably began in 1648 when the New York governor, Peter Stuyvesant, appointed four fire wardens to inspect wooden chimneys of thatched-roof houses in New Amsterdam (later New York City). The fire wardens were also empowered to impose fines for chimneys that were improperly swept.[18]

The following statement from *WNYF*, New York's official training magazine, discusses early fire prevention practices in the United States, including inspecting chimneys and punishing offenders, in an effort to avoid the dire consequences of even small fires, as follows:

> Far from being a new concept, the principle of fire prevention in this country dates back to the days of our earliest settlers. As we scan the aged and yellow pages of books dealing with fires in Olde New York, we note that mention is often made of men assigned to inspect chimneys and hearths, and report if they were inadequately constructed.
>
> There seems to be little doubt that taking precautions against fire received high priority as far back as the early 1600s. The records indicate that even when a small fire occurred, it usually resulted in the destruction of many buildings before being brought under control. In an attempt to curb the problem, lists were published naming persons who maintained faulty chimneys and hearths. If the owner failed to correct the condition leading to his violation order, a heavy fine was levied.[19]

Other examples of early American attempts to prevent fires related to chimneys include the following:

- In 1663, Salem, Massachusetts, imposed a fire safety ordinance requiring that chimneys be swept each year.[20]
- In 1696 Philadelphia found it necessary to prohibit burning out chimneys in order to clean them. In addition, colonists were not allowed to smoke on the street at any time, and the possession of more than 6 pounds of gunpowder within "forty paces of any building or dwelling" was prohibited.[21]
- In 1731, Norfolk, Virginia, prohibited wooden chimneys.[22]
- In 1791, Easton, Maryland, required chimneys to be built of brick or stone.[23]
- In 1796, New Orleans, then a Spanish province, passed an ordinance against the use of wooden roofs.[24]

Rhode Island's first fire prevention law was enacted in 1704. It banned the setting of fire "in the woods in any part of this colony on any time of the year, save between the tenth of March and the tenth of May annually nor on the first or seventh day of any week." A subsequent measure enacted in 1731 prohibited unauthorized bonfires.[25]

Fire prevention enforcement measures were initiated in many communities during the early days of our country. As an example, in 1785 a city ordinance in

Reading, Pennsylvania, imposed a fine of 15 shillings for each chimney fire that occurred in the city.[26] The fine was collected by the city and turned over to the fire company that had responded to the alarm. This ordinance was later repealed. Another requirement in Reading was the alteration of chimneys in blacksmith shops to make them **fire-resistive**, with a fine of $20 for violation.[27]

fire-resistive
▪ a term currently used for some buildings that can resist and even prevent the spread of fire and products of combustion.

In addition, an 1807 ordinance in Reading prohibited the smoking of cigars on the street after sunset.[28] It also forbade people to sit on porches or in the doorway of any house with a lighted cigar or pipe without the consent of the owner. A $1 fine was imposed for violations of this ordinance. The use of firecrackers was also prohibited, with a fine of $1 or 12 hours in jail for violators. A duty was imposed on the citizens of Reading to confiscate and destroy fireworks found in the possession of a child.

The Board of Aldermen in Pensacola, Florida, passed an ordinance in 1821 requiring chimneys to be kept swept. A $10 fine was levied against the owner of any house whose roof caught fire.[29]

Jamestown, New York, imposed fire prevention regulations in 1827. Fire wardens were required to examine all chimneys, stoves, and other fireplaces used within Jamestown and to direct "such reasonable repairs, cleansings, removals, or alterations as shall be in his or their opinion best calculated to guard against injury by fire." Fines were imposed for failure to comply or for refusal of entry to the warden. Occupants of shops or other places in Jamestown where rubbish might accumulate were required to remove accumulations as often as the warden saw fit. Fines were imposed for each day the violation continued.[30]

The first fire safety ordinance in Greensboro, North Carolina, enacted in 1833, required each household to have two ladders on its premises to remove accumulations of combustible materials from the roof, "one which shall reach from the ground to the eaves of the house, the other to rest on top of the house, to reach from the comb to the eaves." Two inspectors were appointed to enforce this requirement and to ensure that all rubbish and nuisances were cleared from backyards. A $5 fine was imposed for each violation.[31]

In most newly formed towns, fire suppression forces were organized before the advent of fire prevention efforts. However, in 1860 in Auraria, a section of what is now Denver, Colorado, the legislative council appointed six fire wardens "to inspect buildings and their chimneys and to prevent the accumulation of rubbish" as the result of a large livery barn fire. The first firefighting company was formed there in 1866.[32]

More comprehensive fire prevention regulations were imposed in New York City in 1860 subsequent to a tenement building fire in which 20 people were killed. The ordinance required all residential buildings built for more than eight families to be equipped with fireproof stairs and fire escapes.[33]

Several major fires occurred in the early 1800s in Montpelier, Vermont. As a result, "the village appointed a committee of three to report a code of by-laws for the preservation of buildings from fire. The bailiffs were required to inspect every house in their ward to see that there was no fire hazard and that each place had, as the by-laws required, a fire bucket and ladder." Another by-law required that no fire be left burning in a house unoccupied between the hours of 11:00 P.M. and 4:00 A.M., if adjacent to another.[34]

Pierre, South Dakota, had its first major fire in 1884. Thirty buildings were destroyed. The city council immediately passed an ordinance creating a fire district covering much of the downtown area. New buildings were to have 8-inch-thick brick walls. Roofs were to be "fireproof."[35]

Fire escapes and exits attracted the attention of the Boise, Idaho, city council in 1887, when it imposed a requirement that doors on halls in theaters be made to swing outward. The council was concerned about the possibility of a disaster at a performance in one of the city's places of assembly.[36]

Fire alarms and fire escapes, of course, had been invented, but they were not yet generally accepted. In fact, in 1897 the Illinois legislature attempted to enact a fire escape law, one that would have replaced earlier, ineffective legislation. The 1897 act required fire escapes in all buildings more than four stories high and in all buildings higher than two stories if the structures were used as manufacturing places, hotels, dormitories, schools, or asylums. According to the *Centennial History of Illinois*, this act was bitterly fought by the Manufacturers' Association of Illinois. When passed, it proved impossible to enforce and was repealed in 1899. As late as 1912, a total of 308 fire deaths were reported in Illinois, with slightly fewer than half occurring in Cook County alone. Most victims were trapped in burning buildings. This entrapment suggests a continuing problem with **fire exits** and escapes, although the circumstances of the deaths were not individually reported.[37]

fire exits
■ exits specifically identified and maintained as a means of egress, often a door or other opening that provides occupants a safe way out of a building or other structure.

An 1896 fire that destroyed a saloon and hotel brought about the first fire prevention code in West Palm Beach, Florida. The ordinance established a fire district in which no building could be erected unless it was of brick, brick veneer, or stone construction.[38]

As early as 1900 captains of steam fire engine companies in Memphis, Tennessee, performed inspections to locate and correct rubbish conditions in buildings, dangerous stovepipes, obstructed fire escapes, and defective chimneys and flues. The great amounts of cotton stored in vacant lots and on streets further contributed to the fire problem.[39]

Formal fire prevention measures in Tulsa, Oklahoma, apparently began with a 1906 requirement that owners of all buildings with three or more stories install fire escapes. Failure to comply by a set date resulted in a fine of $15 per day. Storeowners were prohibited from using rubber tubing for gas connections. Failure to comply resulted in the installation of steel piping at the owner's expense.[40]

Fire chiefs in the United States have long had an interest in fire prevention. Conflagrations in Chicago, 1871 (approximately four square miles of buildings destroyed); Boston, 1872 (65 acres of the central business district destroyed) and again in Chicago, 1874 (approximately 16 acres with 800 buildings destroyed) peaked the fire chiefs' interest and elevated fire prevention as a high priority agenda discussion within their trade. In Boston in 1873, at the First Annual Conference of the National Association of Fire Engineers (predecessor to the International Association of Fire Chiefs), the first general topic on the agenda was fire prevention. The association considered a number of subjects related to fire prevention as they became more aware of the threat of fire in the growing urban areas of the United States. It was becoming evident that fire safety was more than just preventing a single fire from starting.[41]

The National Association of Fire Engineers' (NAFE) considered a number of subjects within their general fire prevention discussion that included for example fire protection systems for buildings, flammability of structural components, the increasing flammable building contents of 19th century industry, concern for fire safety in high-rise buildings, design of passive resistance to fire movement in buildings, improved construction of heating equipment, the need earlier detection and fire department notification through human surveillance, improved emergency egress from buildings and the need to investigate cause of all fires and prosecution of maliciously set fires.[42]

The increasing density of urban business districts and increasing height of combustible buildings concerned fire chiefs. Impact from recent urban conflagrations in Chicago and Boston increased concern that fires were beyond their capacity to stop them. Essentially they were beginning to identify the first outlines of building codes to increase fire safety. This discussion included that tall buildings be equipped with vertical waterways that were large enough to support firefighting hoses and have reliable valves on each floor landing. It also included discussion of improved egress from these tall buildings, as it had been observed that fire escapes of early design were not serviceable for people that were not ambulatory or not of the best level of fitness or physical strength. This discussion also considered an interest into segregating residential units in multi story buildings to self-contained compartments that would restrict the spread of fire within the building. The fire chiefs also were interested in pursuing a detailed system of immediate and objective investigation of every fire not only to determine the fire's cause but also to discover ways that the fire could be avoided in the future.[43]

The use of fire suppression personnel for prefire planning inspections was discussed by the Salt Lake City fire chief at the 1901 conference. At the 1902 conference, fire chiefs discussed developments in fire-retardant paint and slow-burning wood.[44]

In Milwaukee, Wisconsin, in 1888, the first fire prevention requirements were imposed on places of **public assembly**. Apparently, they were the only regulations of a fire prevention nature in effect. Violations of the regulations carried fines of $5 to $100. By 1913, Milwaukee had a force of 30 men, strictly devoted to fire prevention duties in the city's 90,000 buildings and paid entirely through the returns from an insurance premium tax. By 1919, more than 250,000 inspections were being conducted each year by this fire prevention bureau.[45]

public assembly
▪ a type of area where at least fifty people tend to congregate, such as theaters, churches, auditoriums, dance halls, nightclubs, and restaurants.

Fire prevention bureaus were started after 1900 in a number of larger cities. Long Beach, California, established such a bureau in 1917, and Phoenix, Arizona, started its in 1935. At that time Phoenix had a population of 46,500.[46, 47]

The development of water distribution systems has played a major role in community fire defense. In Houston, Texas, the first fire engine arrived in 1839. However, a public waterworks did not come about for many years. By the mid-1870s, most businesses had cisterns for fire protection. In late 1878 the city of Houston signed a contract for the development of a water distribution system, which was in service by the following summer. This pattern of water system development is typical of North American cities.[48]

Unfortunately, some fire safety provisions were not effective, as noted in the following report from Evansville, Indiana:

> As time passed without a big fire, the city grew lax. In spite of the ordinance against frame buildings within the fire limits, the Council routinely allowed variances. Other builders simply violated the building codes. The tightly packed frame buildings were rightly perceived as a fire hazard. In June 1850 the Council required the city marshal to begin investigating all building code violations within the fire limits. They also asked the city attorney to determine whether they could prosecute carpenters, brick and stone masons, and "other mechanics" who violated the codes.[49]

Tragedy: A Spur to Regulations

It has been said that "in the realm of fire 'the law' is a thing mothered by necessity and sired by great tragedy."[50] The truth of this statement becomes clearer in a review of some of the major fires that have occurred through the years within the context of the development of fire safety regulations and procedures in the United States.

PUBLIC ASSEMBLY

On December 5, 1876, a major fire consumed the Brooklyn Theater in New York. In this fire a stage backdrop was ignited and 295 people were killed under conditions similar to those in Chicago's Iroquois Theater fire 27 years later.[51]

The Iroquois fire, notorious among public assembly fires, occurred in 1903 during a Saturday matinee of a new play, *Mr. Bluebeard.* There were 2,000 people present for the performance. The Iroquois was Chicago's newest theater and was also considered its safest; in fact, it was advertised as being "Absolutely Fireproof." Arc lamps were used in the theater. A light set a curtain on fire, and flames and smoke rapidly made the structure untenable. Despite heroic efforts, panic ensued, and human logjams developed at each of the doors. No fire extinguishers were provided. The curtains were combustible, and exits were improperly marked and swung inward. No venting was provided for the stage area and there was no way to immediately remove hot gases and smoke. This tragic fire took 603 lives and provided a great impetus to the fire prevention movement, especially in the field of public assembly occupancies.

On an earlier date, the day after Christmas in 1811, some 600 people were in Virginia's Richmond Theater when scenery caught fire and 72 perished, including the governor of the state. The 200th anniversary of this disaster was commemorated in Richmond in 2011.

In 1940 in Natchez, Mississippi, a fire in a small dance hall, the Rhythm Club, took 207 lives and caused injuries to 200 more. Combustible decorations and one exit with the door opening inward were the factors responsible for the tragedy. More than 700 patrons had been packed into the one-story building, which measured only 120 feet by 38 feet.

During the early days of World War II a major fire struck the Cocoanut Grove nightclub in Boston, Massachusetts. On the night of the fire, November 28, 1942, the club had approximately 1,000 occupants, many of whom were people preparing to go overseas on military duty. A lighted match used by an employee changing a light bulb was considered a likely cause of this tragic fire, which took 492 lives. Almost half of the occupants were killed and many were seriously injured. Flammable decorations spread the fire rapidly. Men and women were reported to have clawed inhumanly in an effort to get out of the building. The two revolving doors at the main entrance had bodies stacked four and five deep after the fire was brought under control. Authorities estimated that 300 of those killed might have been saved had the doors swung outward. It should be noted that the capacity of the structure had also been exceeded.

The Cocoanut Grove fire prompted major efforts in the field of fire prevention and control for nightclubs and other related places of assembly. Immediate steps were taken to provide for emergency lighting and occupant capacity placards in places of assembly. Exit lights were also required as a result of the concern generated by this fire.

On July 6, 1944, fire protection under the big top received attention as the result of the fire that struck the Ringling Bros. and Barnum & Bailey Circus. The circus was playing in Hartford, Connecticut. Seven thousand people attended the daytime performance. The circus tent, which measured 425 feet by 180 feet, was apparently not properly flame-retardant, and the fire caused 163 deaths and 261 injuries. After the fire, many states and municipalities gave more attention to circus fire safety requirements. It is ironic that the fire occurred in Hartford, a city that had had an outstanding fire prevention program for many years.

On May 28, 1977, a tragic fire struck the Beverly Hills Supper Club in Southgate, Kentucky. At the time of the fire, which took 165 lives, the club was occupied by 3,000 to 3,400 people. The building, which had an area of 54,000 square feet, was of unprotected, noncombustible construction. Fire separations, automatic sprinklers, and other safeguards were lacking. Exits were insufficient for the capacity crowd. Interior furnishings were made of combustible materials.

The Beverly Hills fire spurred new demands for improved fire safety measures, including inspection improvements. Many of the patrons in the club at the time of the fire were from other jurisdictions that strongly enforced codes for public assembly occupancies. National political leaders raised the question of the propriety of citizens of one jurisdiction being exposed to fire danger when visiting an area where code enforcement is not as stringent. The impact of this fire would be felt for many years to come. Destruction was thorough and the site of the club remains unused over 35 years later (Figure 1-1). The cause remains a subject of conjecture.

On February 17, 2003, Chicago was again in the national news, when 21 people were killed and 57 injured as they attempted to leave a nightclub where

AP images

FIGURE 1-1 A fire in the Beverly Hills Supper Club in 1977 killed 165 people due to inadequate exits for the number of people in the club. The site of the club remains unused at the time of publication.

pepper spray had been used to quell a fight. Three days later 100 died in a West Warwick, Rhode Island, nightclub fire. You might think that these and similar fires would result in common lessons learned. Unfortunately, public assembly fires continue to occur with predictable results.

At approximately 2:00 A.M. on January 27, 2013, a musical band's pyrotechnics ignited acoustical insulation in the stage area of the Kiss Nightclub in Santa Maria, Brazil, killing more than 230 people. It was reported that the club had over twice its maximum occupant load at the time of the fire and that occupants were blocked from exits by security guards trying to prevent people from leaving before they paid their bills. Many occupants of the club also reported not being able to see in the dark conditions and the absence of exit signs.[52]

VESSELS

Another fire that had a major impact in a fairly limited occupancy arena was the disaster that struck the excursion steamer *General Slocum* on June 15, 1904. The vessel, which had been constructed primarily of wood, steamed down the East River in New York with 1,400 passengers on board. Within half an hour, fire was discovered on the forward deck. Efforts by the untrained crew to control the fire were futile because the hose burst upon being pressurized. Life preservers were faulty and lifeboats were entirely inadequate. The vessel was eventually beached, and 1,030 persons perished either from the effects of the fire or by drowning.

This tragedy led President Theodore Roosevelt to appoint a commission to study the disaster and to make recommendations for future action. The investigation found that officers of the Steamboat Inspection Service had been negligent in their duties. The president ordered the dismissal of those individuals, and Congress soon passed legislation expanding the duties of the Steamboat Inspection Service and giving its personnel more authority to address problems. In 1942, all duties and responsibilities relating to vessel inspection and certification of shipboard personnel were transferred by executive order to the U.S. Coast Guard.[53]

On April 16, 1947, a major disaster struck the waterfront of Texas City, Texas. The S.S. *Grandcamp* was taking on a shipment of ammonium nitrate fertilizer at the pier of the Monsanto Chemical Company. A small fire was discovered aboard, but before it could be brought under control, the ship blew up. The explosion instantly killed all but seven of the ship's crew and almost the entire fire department of Texas City, which had responded to the first alarm of fire. Fire and other explosions that followed during the ensuing hours in the waterfront industrial area resulted in the deaths of 468 people, more than 2,000 injuries, and property loss of more than $67 million. This disaster emphasized the need for regulations in the control of fertilizer-grade ammonium nitrate.

INDUSTRIAL FACILITIES

Another tragedy that had an impact on structural fire safety regulations was the fire that occurred in the Triangle Shirtwaist Factory in New York City on March 25, 1911. More than 600 women, most of them young, were working on the 8th, 9th, and 10th floors of this loft building. To prevent unauthorized removal of products, the factory management had made a practice of checking the purses and bags of all employees as they left the premises. During this procedure, the exit doors to the stairs were locked.

The fire started from an unknown cause on the 8th floor at approximately 4:45 P.M. The interior standpipe hoses were rotten and completely ineffective. The fire spread out the windows and onto the floors above, and 60 people jumped to the ground from the 8th, 9th, and 10th floors. A total of 145 people were killed in this disaster and 70 people were seriously injured. The owner of the factory and his family, who happened to be in the building, escaped by going to the floor above. Charges were brought against the owner as a result of the investigation. He was acquitted, however, because he apparently proved that he was unaware of the practice of locking the stairway exit doors during employee searches.

This fire focused attention on the need for fire safety measures and the safeguarding of occupants in similar buildings. The nation was shaken by this incident and regulations were established to preclude the possibility of a recurrence. The fire led to enactment of New York State's Labor Law and establishment of the Fire Department's Bureau of Fire Prevention.[54]

Mine safety, incidentally, followed a somewhat similar pattern. Pennsylvania adopted the first state mine safety regulations in 1870 as the result of the deaths of 110 miners at the Avondale Colliery disaster at Plymouth the preceding year. In 1910, the U.S. Bureau of Mines was created with improved safety as a goal.

On April 17, 2013, a fire occurred at the West Fertilizer Company in West, Texas. The company's facility in West was described as a distributer of chemicals for the agricultural industry in the region, where chemicals such as anhydrous ammonia and ammonium nitrate were prepared for sale.[55] Around 8:00 P.M., as firefighters were attempting to extinguish the fire, an explosion occurred that killed at least 14 people and injured approximately 160. Firefighters and other responders were among the dead. The explosion in this city, which is located just to the north of Waco, Texas, was recorded by the U.S. Geological Survey as a 2.1 magnitude tremor, and it damaged or destroyed approximately 150 buildings, with reports of 50 homes destroyed.[56] A close-up of investigators making precise forensic measurements of the crater also shows the debris and devastation left after the explosions (Figure 1-2).

The West Fertilizer Company was reported to have had as much as 270 tons of ammonium nitrate used as agricultural fertilizer in a February 2013 disclosure to the state, but the company's general manager believed only 50 tons had been present at the time of the explosion.

The impact of any future changes as a result of the explosion is unknown, but news coverage reported the city may not have been fully aware of the risk of ammonium nitrate's explosive nature. One news story described the reaction of West's mayor: "He never imagined the plant could explode."[57]

A State of Texas legislative panel conducted a special hearing shortly after the explosion and became aware that there was little in the way of top-down oversight in the facility. The panel was seeking answers on the level of risk compared to the level in other, similar facilities in the state.[58]

Fire struck the Imperial Foods Processing Plant in Hamlet, North Carolina, on September 3, 1991. The fire's intensity, coupled with several inoperable exits, resulted in 25 fatalities and 54 injuries in the unsprinklered one-story, windowless structure. This fire emphasized the need for adequate means of egress in all industrial plants, including food processors.

FIGURE 1-2 Investigators take precise measurements of the crater left after a fire and explosion in a Texas chemical plant containing ammonium nitrate. Some of the devastation is apparent in this photo.

Tom Reel/Thomson Reuters (Markets) LLC

HOTELS

Nationwide interest in hotel fire safety was awakened by the fire that swept the Winecoff Hotel in Atlanta, Georgia, on December 7, 1946. The building was widely advertised as a "fireproof" structure and was, in fact, of fire-resistive construction. An open stairway, however, permitted the rapid spread of smoke and heat up the stairs from floor to floor. There were 304 guests in the 15-story building the night of the fire; 119 people were killed and 168 injured. A number of occupants jumped from windows.

fireproof
■ a term used, usually in the construction of buildings, to describe materials that do not demonstrate the characteristic of combustibility. This term was found to serve little value since the structure was often filled with other combustible materials that still presented a hazard to the occupants of the building.

A definition of the term **fireproof** from a 1923 publication is as follows:

> A so-called fireproof building bears about the same relation to its contents that a furnace or other stove does to the material put into it to burn. As a rule, the fireproof building will prevent the spread of fire to other buildings just as a fire will not spread from one stove to another placed near it; but the contents of a fireproof building will be consumed once the fire is well under way just as thoroughly as the coal and wood in the stove. Further, the heat will be retained in the fireproof building and human beings, if they fail to get out quickly, will be killed.[59]

Reports on the Winecoff Hotel fire indicate that some people who stayed in their rooms were not injured by the fire. They were protected by room doors and managed to obtain air for breathing by opening the windows. Great strides in hotel fire safety came about as a result of this fire. Fire safety improvements were made within the entire state of Georgia, which adopted a fire code soon after the incident.[60]

During the same year, a fire at the La Salle Hotel in Chicago took 61 lives. In Dubuque, Iowa, 19 more lives were lost in a fire in the Canfield Hotel. Many similarities existed in the fires in the three hotels.

A 1986 midafternoon New Year's Eve fire struck the DuPont Plaza Hotel and Casino in San Juan, Puerto Rico. The fire resulted in 96 fatalities and more than 140 injuries. This 20-story, nonsprinklered hotel contained a first-floor ballroom, a second-floor casino, and various mercantile shops, restaurants, and conference rooms. Fire consumed the contents of the first-floor areas and was primarily confined to that level. Investigation revealed that the fire had been deliberately set in a large stack of recently delivered furniture that had been temporarily stored in the ballroom.

The government of Puerto Rico convened a commission to study the need for improvements in its fire safety code. The commission report brought about improved fire safety standards.

NURSING HOMES, HOSPITALS, AND HOUSING FOR OLDER ADULTS

A number of serious fires have occurred in nursing homes throughout the country. One that focused attention on the field of fire protection and prevention was the Katie Jane Nursing Home fire in Warrenton, Missouri, in 1957. The fire occurred on a Sunday afternoon when many visitors were in the home. It took 72 lives. Construction deficiencies, the lack of an automatic sprinkler system, and a number of other factors were paramount in the disaster. The fire and a 1963 fire at a Fitchville, Ohio, home for the aged, which took 63 lives, did result in the improvement of nursing home fire safety regulations and procedures in a number of jurisdictions. There was, however, a continuing pattern of disastrous fires in nursing homes and related institutions until corrective action was initiated at the federal level early in the 1970s.

Fire safety problems in hospitals were brought to national attention by the St. Anthony's Hospital fire in Effingham, Illinois, on the night of April 4, 1949; 74 lives were lost. Combustible interior finish, the lack of automatic sprinkler protection, and general construction were factors in this fire.

A fire that occurred in 1961 in the Hartford Hospital in Hartford, Connecticut, was responsible for 16 deaths. Combustible ceiling tile contributed to the fire's spread, as did the fact that trash and linen chutes in the building opened directly onto the corridors. The structure was a modern fire-resistive building and proved that even such structures are not immune to fires.

In just 13 days in December 1989, fires killed 23 residents in three different housing facilities for the elderly. Not one of them was licensed or operated as a nursing home.

The first occurred in a six-story, fire-resistive, unsprinklered apartment house for older adults in Roanoke County, Virginia. Four residents died on the third floor, although the fire was confined to the room of origin. The investigation revealed that some of the occupants were difficult to awaken and reluctant to leave their apartments.

Watertown, New York, was the scene of the second fire. Three fatalities occurred, all on the upper floors of the nonsprinklered, fire-resistive apartment building, which had been constructed specifically for housing older adults. Smoke movement from the first-floor point of origin was responsible for the fatalities in this December 15 late-night fire.

On December 24, the third—a late-afternoon fire—started on the first floor of an 11-story fire-resistive apartment house in Johnson City, Tennessee. The nonsprinklered structure had served as a hotel for many years before being converted to housing for older adults. The toll was 16 fatalities and 40 injuries. As in the other two fires, combustible interior finish was a factor in the incident. Most of the fatalities were on upper floors.

Generally, fire safety requirements in apartments for older adults are no more stringent than those for any other apartment house. Special licensure is usually not required.[61]

SCHOOLS

A six-year-old three-story school in Colinwood, Ohio, was the scene of tragedy in 1908. The Lakeview School was occupied by more than 300 students at the time of the fire. Of those students, 175 were killed. Open stairways and combustible materials contributed to the fire spread.

During the commencement ceremony at the Cleveland School near Camden, South Carolina, an oil lamp fell on the stage of the upstairs auditorium, causing the death of 77. The school was slated to be replaced by a new building the next school year, so the fatal fire on May 17, 1923, marked the building's last scheduled day as a school. A move by fire chiefs in the state brought about improvements in school fire safety.[62]

Another school fire brought with it a realization that fire safety should be considered in even the smallest of schools. The fire occurred in a two-room elementary school at Babbs Switch, Oklahoma, on December 24, 1924. The school was being used for a Christmas event attended by more than 200 community residents. A Christmas tree had been placed in the opening between the two rooms. The windows in the school had previously been barred, and there was only one exit. The fire was started by a toppled candle on the tree. As a result of the unsafe conditions, 36 individuals died.

In March 1937, a major explosion struck a New London, Texas, school. The disaster, which killed 297 people, was caused by the improper use of gas. Again, school fire safety became of immediate paramount concern in communities throughout the country.

School fire prevention and fire protection efforts were greatly enhanced by the tragic fire at Our Lady of Angels School in Chicago on December 1, 1958. The fire, in which 95 lives were lost, was reported to be of incendiary origin. The old multistory structure was of ordinary construction. A delay in turning in an alarm within the building and in contacting the fire department compounded the tragedy. As a result, tremendous efforts were made to upgrade school fire prevention regulations throughout the country. Many schools were provided automatic sprinkler protection. In addition, a number of states started vigorous fire inspection programs and took other measures to reduce the possibility that fires would take lives because of lack of safeguards.

PRISONS

The Ohio State Penitentiary fire of April 21, 1930, in Columbus aroused interest in problems relating to fire prevention in prisons. The penitentiary had 4,300 inmates at that time and 320 were killed in the fire. Most killed were trapped in their cells. The fire brought about immediate demands for improvements in fire safety in penal institutions.

Three major prison fires killed a total of 68 people during 1977. There were 5 fatalities in the federal prison in Danbury, Connecticut; 42 in the county jail in Maury County, Tennessee; and 21 in a St. John, New Brunswick, Canada, penal institution. The fires led to new demands for tighter regulation of fire safety within places of incarceration. Interior finishes had become a major contributory factor to fires in such places.

CONFLAGRATIONS

Until the 20th century, communities had been built to burn while the country expanded rapidly across the continent. Building conditions in the United States were long marked by excessive use of combustible materials without much regard for protection from fire. Unsafe factors that contributed to the rapid spread of fires were characteristic of the American scene and included large individual buildings housing vast amounts of combustible stocks under one roof, lack of fire walls and vertical cutoffs, and wood shingle roofs. In many parts of the United States, seasonal droughts and high winds aggravated fire conditions and resulted in area conflagrations that contributed substantially to the high annual record of United States fire losses (Figure 1-3). The term **conflagration** usually refers to a major destructive fire.

conflagration
▪ a destructive fire covering a large area, often able to cross natural barriers and frequently causing major life and property loss.

In a conflagration on July 4, 1866, in Portland, Maine, 1,500 buildings were destroyed by a fire believed to have been started by fireworks. Two fatalities occurred and 10,000 persons were left homeless.

October 9, 1871, was the day of the two greatest fires in the history of the Midwest. Both started at about the same hour, were compounded by tinder-dry wood resulting from a prolonged drought, and involved human carelessness. The better known of the two struck Chicago, lasted two days, took 300 lives and consumed structures in an area 4 miles long and 2/3 mile wide.

Steven F. Mullensky/Corbis

FIGURE 1-3
Conflagrations often spread by wind-driven burning embers and can destroy homes and other structures before the flaming fire front even reaches residential developments. As the embers fall the wind drives them into receptive fuels on structures where they ignite the structure. This process repeats itself until the fire is controlled.

The lesser-known fire ravaged 2,400 square miles and took 1,152 lives in a forested area of northeastern Wisconsin. This fire is recorded as the Great Peshtigo fire, because the majority of fatalities occurred in that Wisconsin town.

On June 6, 1889, the Great Seattle fire swept through what was then downtown Seattle, wiping out 66 square blocks and paving the way for "The Forgotten City" lying beneath the city's modern streets. Because much of the downtown area was on stilts, the sidewalks and streets were great fire carriers. In one instance, firefighters were able to stem the fire north of University Street simply by tearing up the streets and sidewalk and tossing them over the cliff into the bay. To fight the fire, a 200-person bucket brigade was formed along the river that ran past the Olympic Hotel. Water was hauled up from the river and sloshed against the buildings.[63]

Major conflagrations occurring in U.S. cities have brought about reforms in building and fire prevention codes. In 1904, Baltimore was swept by a conflagration that destroyed 80 blocks in the downtown business center. The effectiveness of incoming firefighters from other cities, including New York, was hampered by incompatible hose threads. This fire led to the improvement of construction standards and the development of new procedures in fire prevention.

Paterson, New Jersey, lost nearly 500 buildings in a 1902 fire. Other North American cities and towns have had somewhat similar occurrences. For example, an 1861 fire destroyed most of the town of Lindsay, Ontario. Four hotels, two mills, the post office, the customs office, and 83 other buildings were destroyed in the town of 2,000. A history of Lindsay states: "The fire spurred the construction of many fine brick buildings to replace the wooden structures which had been consumed."[64]

The City of Chelsea, Massachusetts experienced two catastrophic fires that destroyed significant numbers of buildings. The two fires had remarkable similarities that enabled the widespread destruction, including extremely dry conditions, sustained winds, and buildings that were old and constructed mostly of wood, so they were readily ignitable. A fire in 1908 destroyed 1,500 buildings and one in 1973 destroyed 18 city blocks.[65]

In 1923, Berkeley, California, was ravaged by a fire that destroyed 640 structures. In 1961, a conflagration in Los Angeles spread through wooded lands and resulted in the destruction of more than 500 dwellings. Some large communities permitted the use of wood-shingle roofing. The Los Angeles fire clarified the need to return to more restrictive code provisions regarding such roofing.

Oakland, California, was the site of a devastating fire that took 26 lives and resulted in the loss of 3,469 housing units on October 20, 1991. Property losses were estimated at $1.5 billion. Dry conditions and steep terrain contributed to this loss.

EARTHQUAKES AND FIRE

Although it was not directly controllable under normal fire safety procedures, the 1906 earthquake and fire in San Francisco must be mentioned in any recounting of U.S. fires of major significance. The severity of the fires was much greater because of the earthquake. Because water mains were broken, the amount of water available for firefighting purposes was greatly reduced. In addition, the earthquake made response by firefighting equipment most difficult. A number of fires were uncontrolled.

In that disaster, 422 lives were lost, 28,000 buildings were destroyed, and property loss was estimated at $350 million—over $7 billion today. This devastating event prompted reconsideration of fire prevention procedures in all areas of the country where earthquakes might be anticipated. The Baltimore and San Francisco conflagrations also led the insurance industry to develop advanced methods of analyzing risks.

Variables in the Philosophy of Fire Prevention

The philosophy of fire prevention includes many variables. The term *fire prevention* varies in interpretation within the fire protection field. There has been a tendency to include some closely related activities under fire prevention responsibilities. Some are not truly preventive in nature. Activities that are actually *fire reactions* often are considered fire prevention practices by both lay and professional people. For example, the practice of home and office fire drills is usually associated with fire prevention, even though it is actually a fire reaction program. A reason for this confusion may be that the same individuals generally promote both programs.

Prevention is any act of intervention that stops something from happening. The term **fire prevention** came into use to identify the actions that eliminate the hazardous conditions that can lead to a fire. Measures such as better control or protection of open flames and elimination of the accumulation of combustible materials such as leaves and solid waste materials were some of the very earliest work involved in preventing fires. But as humans became more aware of the science of fire and how fire behaves, it became necessary to expand the scope of fire prevention. Fire prevention now includes, for example, human behavior research, fire education and awareness, fire science research, fire and building code development and management, standards, community risk reduction campaigns, and fire protection systems. Prevention activities described in some of the oldest historical references and right up through our current time show a constant broadening of the scope of concern regarding fire safety. The fire experience drives the development of fire prevention activities. Each fire that occurs, especially one involving a tremendous waste of life and property, leaves some lasting legacy of improved understanding of fire, which in turn often results in some new way of trying to prevent the next wasteful loss.

fire prevention
■ a level of effort to decrease the chances of unwanted fire ignition; the philosophy and practice of reducing the hazards and risk of fire with the goal of decreasing the loss of life and property.

Certain fire prevention concepts are not strictly related to the prevention of fire and are more closely related to the prevention of the spread of fire. For example, wearing noncombustible clothing will not prevent the ignition of a match, but wearing such clothing does retard the possible spread of a fire that might be started on the clothing by the match. The personal steps taken to reduce fire spread possibilities on clothing are one aspect of fire prevention; however, measures taken to preclude the possibility that a fire will start are more truly the essence of fire prevention.

Another fire prevention measure is to ensure that a fire that occurs in a structure will not entrap individuals. People who happen to be in the structure at the time of the fire should have every opportunity to leave the building safely. An important factor in such a situation is an individual's conditioned reaction to fire, in which health, physical abilities, past exposure to fire, and many other variables play a part. The term *fire safety* is seen by many as encompassing fire prevention, fire reaction, and prevention of fire spread.

CHAPTER REVIEW

Summary

History teaches us much about fire prevention. It shows us that virtually all fire prevention programs have been the result of disastrous fires. This has not changed through the years. In spite of improved technology, science, and the efforts of many fire prevention professionals, significant changes in fire prevention practices almost always require a large loss of life or property.

A review is necessary of some of the significant factors that contribute to fire disasters, whether they be blocked exits, doors that open inward, flammable or combustible finishes, absence of alarms, or lack of automatic sprinklers. We can also note that politics and business (profits and the bottom line) greatly affect fire prevention.

The ability to find a solution to fire risks is not enough. It takes leadership and a willingness to influence those who have the power to make the necessary changes.

The danger of fire to life and property has always been present. Due diligence of fire prevention professionals must be constant.

Review Questions

1. The term *curfew* originally meant:
   a. setting a time to vacate public areas.
   b. the time when fires had to be extinguished.
   c. getting children off the streets.
   d. establishing corporal punishment for fire prevention violations.
2. Early fire prevention efforts included:
   a. construction requirements.
   b. maintenance of fire safety devices.
   c. punishment for violators.
   d. all the above.
3. Fire inspections in the New World probably began in:
   a. New Amsterdam.
   b. Boston.
   c. Williamsburg.
   d. Philadelphia.
4. Which one of the following was not a topic of discussion in the first annual conference of the National Association of Fire Engineers in 1873?
   a. Repression of incendiarism
   b. Automatic sprinklers
   c. Reduction of excessive height of buildings
   d. Fire investigations for cause
5. In 1913, how many employees did Milwaukee devote to fire prevention?
   a. 5
   b. 10
   c. 20
   d. 30
6. Common contributing factors to the disastrous fires at the Iroquois Theater, Rhythm Club, and Cocoanut Grove included:
   a. inadequate exits.
   b. inward-opening doors.
   c. flammable decorations.
   d. all the above.
7. Which fire contributed to the beginning of labor laws for the safety of workers?
   a. *General Slocum*
   b. Triangle Shirtwaist Factory
   c. General Motors Transmission Plant
   d. MGM Grand Hotel
8. A school fire in which city contributed the most to today's fire prevention efforts in schools?
   a. Colinwood, Ohio
   b. Babbs Switch, Oklahoma
   c. New London, Texas
   d. Chicago, Illinois

9. Which two great conflagrations in the United States started on the same day?
   a. Chicago, Illinois, and Peshtigo, Wisconsin
   b. Chicago and Seattle
   c. Seattle and Baltimore
   d. Baltimore and San Francisco
10. Fire prevention or fire safety includes:
   a. fire reaction.
   b. fire prevention.
   c. prevention of the spread of fire.
   d. all the above.
11. In 1961 a fire at the Hartford Hospital claimed 16 lives. The building was of fire-resistive construction. What was learned about fire-resistive construction?
   a. Vertical openings automatically transmit smoke throughout the building.
   b. Arson is a leading cause of death in health-care facility fires.
   c. Fire-resistive buildings are not immune to fires.
   d. The improper use of anesthetizing gas precipitates most hospital fires.
12. What is considered a major contributing factor to fatal fires in places of incarceration?
   a. Interior finish
   b. Vertical openings
   c. Blocked exits
   d. Unmaintained sprinklers

End Notes

1. G. V. Blackstone, *A History of the British Fire Service* (London: Routledge & Kegan Paul, 1957).
2. Ibid.
3. Ibid., p. 10.
4. Ibid.
5. Ibid.
6. Ibid.
7. Alexander Reid, *"Aye Ready!"* (Edinburgh: Geo. Stewart & Co., 1974), p. 5.
8. Blackstone.
9. Ibid., p. 33.
10. Ibid., p. 87.
11. *Project 9: The Story of Fire Fighting* (London: The Home Office and the Central Office of Information, 1976), p. 1.
12. Blackstone.
13. Ibid., p. 87.
14. Campbell Steven, *Proud Record: The Story of the Glasgow Fire Service* (Glasgow: Holmes McDougall Ltd. Glasgow, 1975), p. 33.
15. Harry Klopper, *The Fight Against Fire* (Birmingham, England: Birmingham Fire and Ambulance Service, 1955), p. 36.
16. *Arizona Daily Star*, September 24, 2006.
17. Paul R. Lyons, *Fire in America!* (Boston: National Fire Protection Association, 1976), p. 2.
18. Charles L. Radzinsky, *100 Years of Service 1872–1972* (Rensselaer, NY: Hamilton Printing Company, 1972), p. 9.
19. Bernard Neer, *WNYF (With New York Firefighters),* the official training publication of the New York City Fire Department, 3rd issue, 1980, p. 20.
20. Edward R. Tufts, *A History of the Salem Fire Department* (Salem, MA.: Holyoke Mutual Insurance Company in Salem, 1975), p. 3.
21. *History of the Fire Department, Norfolk, Virginia* (Norfolk, VA: Norfolk Firefighters' Association, 1975), p. 32.
22. Ibid., p. 8.
23. James C. Mullikin, *A History of the Easton Volunteer Fire Department* (Easton, MD: Easton Volunteer Fire Department, 1962), p. 9.
24. National Commission on Fire Prevention and Control, *America Burning* (Washington, DC: Author, 1973), p. 79.
25. Patrick T. Conley and Paul R. Campbell, *Firefighters and Fires in Providence* (Providence, RI: Rhode Island Publications Society, 1985), p. 3.

26. *Reading's Volunteer Fire Department,* comp. Federal Writers Project of Works Progress Administration in the Commonwealth of Pennsylvania, Berks County Unit (Philadelphia: William Penn Association, 1938), p. 6.
27. Ibid., p. 7.
28. Ibid.
29. Arnold Rosenbleeth, *Firefighting in Pensacola* (Pensacola, FL: Pensacola Historical Society Quarterly, Winter 1980), p. 2.
30. Leo E. Duliba, *A Transition in Red* (Merrick, NY: Richwood Publishing Co., 1976), pp. 12–13.
31. *Greensboro Fire Department, 1808–1984* (Dallas, TX: Taylor Publishing, 1984), p. 6.
32. "Past and Present History Denver Fire Dept.," unpublished paper, Denver, CO, p. 2.
33. Lyons, pp. 29–30.
34. Perry H. Merrill, *Montpelier, the Capital City's History* (Montpelier, VT: Published by author, 1976), p. 62.
35. Harold H. Schular, *A Bridge Apart: History of Early Pierre and Fort Pierre* (Pierre: South Dakota State Publishing Co., 1987), Ch. 1, p. 87.
36. Arthur A. Hart, *Fighting Fire on the Frontier* (Boise, ID: Boise Fire Department Association, 1976), p. 33.
37. Betty Richardson and Dennis Henson, *Serving Together: 150 Years of Firefighting in Madison County, Illinois* (Collinsville, IL: Madison County Firemen's Association, 1984), p. 45.
38. *The Fire History of the City of West Palm Beach* (West Palm Beach, FL: West Palm Beach Fire Department, 1980), p. 6.
39. Frank Myers and Gennie Myers, *Memphis Fire Department* (Marceline, MO: III Walsworth, 1975), p. 30.
40. *Tulsa Fire Department 1905–1973* (Tulsa, OK: Intercollegiate Press, 1973), p. 17.
41. Donald M. O'Brien, *The Centennial History of the International Association of Fire Chiefs* (n.p., 1973), p. 6.
42. Ibid., p. 6.
43. Ibid., p. 6.
44. *Tulsa Fire Department 1905–1973* (Tulsa, OK: Intercollegiate Press, 1973), p. 43.
45. R. L. Nailen and James S. Haight, *Beertown Blazes: A Century of Milwaukee Fire Fighting* (Milwaukee, WI: NAPCO Graphic Arts, 1971), pp. 21–22.
46. *Long Beach Fire Department* (Long Beach, CA: Long Beach Fire Department, 1976), p. 29.
47. *Phoenix Fire Fighters* (Phoenix, AZ: Phoenix Fire Department, 1983), p. 23.
48. Houston Fire Museum, Inc., *Houston Fire Department 1838–1988* (Dallas, TX: Taylor Publishing, 1988), p. 13.
49. Sharron K. Brace, *Fire on the River* (Inglefield, IN: APS Publishing, 1995), pp. 28–29.
50. Bob Considine, *Man Against Fire* (Garden City, NY: Doubleday and Co., 1955), p. 134.
51. *Great Fires of America* (Waukesha, WI: Country Beautiful Corporation, 1973), p. 134.
52. Associated Press, "Deadly Smoke, Lone Blocked Exit: 230 Die in Brazil, 2013." Accessed May 3, 2013, at http://bigstory.ap.org/article/more-90-dead-nightclub-fire-brazil
53. Walter C. Capron, *U.S. Coast Guard* (New York: Franklin Watts, Inc., 1965), pp. 43–44.
54. *WNYF* (New York: New York Fire Department, 3rd issue, 1980), p. 20.
55. DFWCBSLocal.com, "I-Team: What Went Wrong at West Fertilizer Plant," 2013. Accessed May 8, 2013, at http://dfw.cbslocal.com/2013/04/18/i-team-what-went-wrong-at-west-fertilizer-plant/
56. KWTX.com, "West Explosion Claims 14 Lives; 9 Were First Responders." Accessed May 3, 2013, at http://www.kwtx.com/home/headlines/Explosion-Injuries-Reported-At-West-Fertilizer-Plant-203505331.html
57. DFW.CBSLocal.com, "Fertilizer Plant Manager Believes 50 Tons of Ammonium Nitrate in Building That Exploded," 2013. Accessed May 8, 2013, at http://dfw.cbslocal.com/2013/05/06/fertilizer-plant-manager-

believes-50-tons-of-ammonium-nitrate-in-building-that-exploded/

58. Brandon Formby, DallasNews.com, UPDATE, "State Officials Tell Legislators That Educating West Residents About Plant Was Local Responsibility," 2013. Accessed May 8, 2013, at http://thescoopblog.dallas-news.com/2013/05/state-house-members-begin-questioning-experts-on-west-explosion.html/
59. C. C. Dominge and W. O. Lincoln, *Fire Insurance Inspection and Underwriting* (New York: Spectator Co., 1923), p. 323.
60. Sam Heys and Allen B. Goodwin, *The Winecoff Fire* (Marietta, GA: Longstreet Press, 1993).
61. "Alert" (Bulletin No. 90–1, Quincy, MA: National Fire Protection Association, 1990).
62. Visit to anniversary event in May 2012 by author Robertson.
63. "The Great Seattle Fire," *Wildlife News Notes*, NFPA, June 1998.
64. Alan R. Capon, *Historic Lindsay* (Belleville, Ontario: Mika Publishing, 1974), p. 18.
65. Celebrate Boston, "The Great Chelsea Fire." Accessed October 16, 2013, at http://www.celebrateboston.com/disasters/great-chelsea-fire-1908.htm

CHAPTER

2 Status of Education, Engineering, and Enforcement in the United States

Geoff Manasse/Photodisc/Getty Images

KEY TERMS

Community Risk Reduction (CRR), *p. 23*
compliance, *p. 32*
economic incentives, *p. 24*
education, *p. 25*
emergency response, *p. 24*
enforcement, *p. 28*
engineering, *p. 28*
fire investigation, *p. 25*
fire prevention, *p. 23*
International Building Code (IBC), *p. 31*
interventions, *p. 23*
mechanical codes, *p. 35*
mini–maxi codes, *p. 32*
performance-based design, *p. 32*
plan review, *p. 25*
wildland, *p. 36*
zoning codes, *p. 34*

OBJECTIVES

After reading this chapter, you should be able to:

- Identify the basic concepts and definitions of education, enforcement, and engineering and how each relates to fire prevention.
- Describe how the local jurisdiction applies code to new and existing occupancies and how the code is enforced.
- Describe the relationship between performance-based codes and performance-based designs.
- Describe the process by which codes are adopted and applied.
- Explain the concept of Community Risk Reduction.
- List the five interventions associated with Community Risk Reduction.

The proceedings of the *Official Record of the First American National Fire Prevention Conference* published in 1914 contain many **fire prevention** concepts that are just as applicable today as they were in the early 20th century.[1] Subsequent national gatherings related to fire prevention contain similar applicable suggestions. As years pass, fire prevention continues to evolve. Today, fire prevention is less likely to be a stand-alone effort. To accomplish its life safety mission more effectively, fire prevention is more often a function integrated with other agencies and stakeholders.

fire prevention ■ a level of effort to decrease the chances of unwanted fire ignition; the philosophy and practice of reducing the hazards and risk of fire with the goal of decreasing the loss of life and property.

In 1947 President Harry Truman announced a National Conference on Fire Prevention and expressed his deep concern about preventable fire deaths, which had averaged 10,000 a year in the previous decade and accounted for billions of dollars of property loss. This President's Conference focused on the nation's need to take responsibility for fire safety, and it laid a solid foundation for future efforts that are still producing results today. Major General Phillip G. Fleming, General Chairman for the Conference, provided a sustaining thought about the approach to fire safety when he wrote, "I believe the keys to the fire prevention problem are Education, Enforcement, and Engineering. We need widespread education in methods of fire prevention and control, more adequate laws and their rigid enforcement, and better engineering to make buildings fire resistant."[2]

Chairman Fleming's comments focused on three critical intervention areas already being implemented by the transportation industry to reduce the rate of casualties from automobile collisions. The "Three Es" as they are now called in the fire prevention field are relevant today as **interventions** to prevent fires, injuries, and property loss associated with fire.

interventions ■ fire prevention or Community Risk Reduction (CRR) actions to prevent or reduce loss; can include changing unsafe behavior, separating building occupants and combustible products, and installing fire protection or smoke detection.

The mission of the fire department has changed from fighting fires only to firefighting plus emergency medical services and hazardous materials and technical rescue. Fire prevention has also expanded its scope to include services such as injury prevention, blood pressure screening, and providing child passenger safety seats. One inclusive approach to risk reduction is the use of the **Community Risk Reduction (CRR)** model, which is currently being taught at the National Fire Academy in Emmitsburg, Maryland.

Community Risk Reduction (CRR) ■ local-level commitment to develop internal and external partnerships with the community to implement programs, initiatives, and services that promote an integrated approach to the risks of injury and loss through education, engineering, enforcement, economic incentives, and emergency response.

Community Risk Reduction

The purpose of Community Risk Reduction (CRR) is to reduce hazards and mitigate risks within any given community. CRR is a multifaceted and integrated planning process that involves three basic functions, including the identification of fire risk in the response area, prioritization of risks to be addressed by the CRR process, and the coordination of resources that perform preventive actions that mitigate risks. Fire is only one of many risks communities face, and efforts to reduce it must include all residents. CRR involves citizens, community organizations, community leaders, businesses, most government services, advocacy groups, schools, and public health agencies. All must be prepared to deal with incidents involving hazardous materials, severe weather, terrorism, floods, and transportation. The fire department generally plays a critical role.

In 2010, according to the Centers for Disease Control and Prevention (CDC), 44,743 unintentional injuries resulted in death among people between the ages of

1 and 44. Such injuries were the leading cause of death for this segment of the U.S. population.[3] A multidisciplined approach such as CRR is an effective way to plan for, prevent, mitigate, and respond to such incidents.

CRR's comprehensive approach complements any fire prevention program and readily integrates with it. For example, a CRR program may initiate an enhancement project that focuses on blighted and vacant buildings, thereby eliminating a potential for set fires and drug-related and other criminal activity.

Prevention Intervention

According to the NFPA's 2013 report *Home Structure Fires*, the annual average for fires in one- or two-family dwellings was 366,600 for the period 2007–2011.[4] Despite nearly a 50 percent drop from the number of home fires reported in 1980 and a parallel drop in home fire deaths, the home is still at risk for fire.[5] Between 2009 and 2011, the U.S. Fire Administration (USFA) reported that 76 percent of fire injuries occurred in residential buildings.[6] The USFA also reported that residential fire fatalities during the period 2008–2010 accounted for an estimated 2,560 deaths, 75 percent of the fire deaths in the United States, or about 5.5 deaths per 1,000 residential fires.[7] Residential fires and the resulting injuries and deaths remain a significant challenge in this country.

The need continues as fire and life safety professionals discover new ways to prevent fires and reduce the impact of those that do occur. While the traditional "Three Es" of fire prevention remain the core of intervention efforts, the CRR process has offered two additional interventions: **economic incentives** and **emergency response.** The primary toolbox for fire prevention now includes "Five Es": education, engineering, enforcement, economic incentives, and emergency response.

economic incentives ▪ an intervention that promotes a variety of savings to developers, builders, and building owners in exchange for employing desired but often not required fire protection systems, materials, or construction practices.

emergency response ▪ the systematic and immediate response of fire suppression or other emergency services with the goal of stopping the loss of life or property or otherwise reducing the risk to community by the deployment of personnel and materials.

The Traditional "Three Es"

The traditional "Three Es" might be expressed as follows:

Education + Engineering + Enforcement = Fire Prevention

Fire prevention is defined by the *Municipal Fire Service Workbook* as "the effort to decrease the chances of unwanted ignition and, to some extent, to limit the spread of fire by methods that are independent of actions taken after ignition occurs."[8] This definition is designed to include fire prevention activities carried out by municipal fire departments but not fire suppression, which involves the actions taken by the fire department to bring an unwanted fire under control.

Education is a core activity of community fire safety programs. It is increasingly followed in the CRR program modeled on the fire service in the United Kingdom. There, operational firefighters go door to door, often accompanied by a partnering member of the community, to ensure the presence of working smoke alarms. That effort sometimes involves the installation of a new smoke alarm or batteries, paired with a message of smoke alarm maintenance, awareness of fire hazards in the home, and facilitation of the skills of planning and conducting home fire drills.

Consider some of today's problems. Greater emphasis is still needed on fire prevention—as opposed to suppression—as a fire department function, but progress has been made. It is rare to find a community with a population of more than 10,000 that does not have at least one person in the fire department, whether career or volunteer, assigned to fire prevention. In a small community, that one person is usually attempting to address all of the "Three Es," while the remainder of the force devotes its time primarily to suppression duties.

In 2006 the U.S. Fire Administration commissioned a study to assess the needs of a broad spectrum of U.S. fire departments. Of approximately 15,000 invited to complete the survey, 30 percent, or approximately 4,700, responded. Out of 15 departments responding to the survey with a population of more than 1 million, only 5 had no fire prevention or code enforcement programs. In cities with populations less than 1 million, the percentage of departments lacking such programs increased substantially. The study concluded that an estimated 20.3 million people who live in small cities and towns had no established programs. Sadly, those departments protect about 29 percent of the U.S. population.[9]

Areas of responsibility for fire prevention–related duties have become more specialized through the years. For example, more fire departments, especially larger ones, have staff trained specifically in public fire-safety education. **Plan review** and **fire investigation**, which are often overlooked as fire prevention measures, are more often specialized assignments. Certainly, both contribute materially to the prevention of fire. Abatement of arson by way of thorough investigation is a major contributor to fire prevention. Likewise, plan review by a trained fire protection specialist can result in a decrease in the loss of life and property in a community.

plan review
▪ prior to new construction or building modifications, an evaluation by certified officials that ensures compliance of architectural or engineering plans with building and fire codes.

fire investigation
▪ a scientific inquiry conducted to determine the cause of fire so as to prevent a recurrence of the same scenario.

All of the "Three Es" have seen a higher degree of specialization in their administration. Having college-trained fire protection engineers involved in reviewing construction plans and specifications for both governments and the private sector was only a dream when the first college-level fire protection engineering program opened in the early 1900s. At that time, mass media concepts such as television and radio were not available to disseminate fire safety education concepts. Probably the strongest application of one of the "Three Es" was in the enforcement realm.

Education

Fire safety **education**, the first of the traditional "Three Es," has become much more widespread in recent years (Figure 2-1). Smoke alarms are now found in most homes. Increasingly, smoke alarms are being hardwired and interconnected as a result of requirements for new dwellings, thereby reducing the reliance on consumers to change batteries each year. Smoke alarms are estimated to be responsible for cutting home fire fatalities in half. However, thousands of smoke alarms are reaching their life expectancy, and replacement is becoming a major concern.

education
▪ actions that teach, promote awareness, present information, and conduct activities intended to change behaviors that increase people's safety from fire.

FIGURE 2-1 Educating residents about how to conduct home fire drills and plan their home escape routes is an essential component of their home safety. Exit drill education is effectively presented during fire department home safety visits.

Stewart Cohen/Blend Images/Getty Images

Residential sprinkler requirements are now included in all the U.S. model codes pertaining to residential occupancies, including the International Residential Code (ICC), NFPA codes dealing with building and life safety, and the International Building Code (IBC). Although residential sprinklers are slowly gaining acceptance, they have yet to make a major impact on home fire fatalities because they are installed in so few homes. There is no doubt they will eventually have a profound impact on fire safety in dwellings. Because sprinklers have proven to be lifesavers in a number of kinds of occupancies, private homes will someday be added to that list (Figure 2-2).

"Stop, drop, and roll" has become a well-practiced alternative to running with burning clothes. The advent of child-resistant lighters has also had a positive effect on home fire safety, although a newer challenge has emerged as lighters, made to look like toys, have become a popular collector's item. Many jurisdictions have implemented regulations banning the novelty lighters. Maine was the first state to actually ban them.

Smoking, coupled with alcohol, continues to be a major contributor to home fire fatalities. New York State leads the way as the first state to adopt a model standard for reduced-ignition-propensity cigarettes, which extinguish more quickly than standard cigarettes if ignored. All states currently have completed implementation or are in the process of implementing the model standards for reduced-ignition-propensity cigarettes. Measures to reduce the combustibility of mattresses and other bedding material have been helpful, especially in institutional occupancies.

Although fire safety in nursing homes has been greatly improved, 15 people were killed in a Hartford, Connecticut, facility in February 2003, followed by 15 fatalities in a similar facility in Nashville, Tennessee, in September 2003, and 5 in a Maryville, Tennessee, nursing home in January 2004.[10]

FIGURE 2-2 Residential fire sprinklers have been referred to as firefighters living in your home with you.

Geoff Manasse/Photodisc/Getty Images

The 20th century saw the introduction of hundreds of new chemicals and related hazardous materials, many of which created code enforcement and suppression problems for the fire service. Specialized hazardous-materials response units abound in the fire service. The everyday use of many hazardous materials has become a necessity of modern life, and fire codes have had to adjust to the change.

Greater examination of specific needs in the fire safety education field is another trend that developed in the latter part of the 20th century and continues in the early part of the 21st century. Individual groups are now being spotlighted for special attention geared to their needs. This trend is replacing the broad-brush approach used in the past]. For many years public fire safety education emphasized only the elimination of ignition sources. Now, palliative measures such as smoke alarm installation take precedence.

This approach is exemplified by the 2002 U.S. Fire Administration report titled *Babies and Toddlers Fire Death Data Report.*[11] The report concludes that children under 5 years of age are twice as likely to die from fire as the rest of the population. The document includes data from all state fire marshals indicating the trends in their individual states. The publication shows the ranking of fire deaths for children under age 5 compared with other causes of death for this population group in each state. It also includes comparisons of these fatalities with those of several other countries. Public fire safety educators and others interested in the field are strongly encouraged to develop and carry out programs that will make parents and guardians more aware of how they can prevent young children's death by fire.

There is no doubt that improved application of building and life safety codes, increased use of fire safety education, installation of smoke alarms and sprinklers in residences, and other measures initiated by the fire service and allied organizations did much to reduce fire occurrences and most particularly fire fatalities during the last 10–15 years of the 20th century. The advent of the U.S. Fire Administration and its National Fire Academy has also contributed to improved fire safety in the United States.

A review of fire fatality statistics of the early years of the 20th century provides a striking comparison with the statistics for the present era. In 1913 the U.S. fire fatality death rate was 9.1 persons per 100,000 population; by 2002, just 90 years later, it had dipped to 1.4 per 100,000 population. Canadian fatality rates were slightly higher in both year groupings. Although the data represent a drop of nearly 400 percent in less than 100 years, it must be recognized that the compilation of fire fatality statistics is more accurate today than in 1913. The effect of this greater accuracy on the death rate statistics is unclear.[12] However, the downward trend of deaths by fire continues, and the most recent NFPA report shows fatalities down to 2,380 in 2012, a 35-year low since 1977, when 5,865 fire deaths were recorded. Unfortunately, the report points out a troubling trend in the 5 years between 2008 and 2012: Communities of less than 25,000 population had a noticeably higher fire death rate per million population. The fire death rates of all communities in the analysis between 25,000 and greater than 1 million were below 9 deaths per million, but communities under 25,000 had more than 10 deaths per million, and communities under 2,500 had almost 20 deaths per million.[13]

A National Fire Protection Association report issued in 2010 indicates that fire deaths in rural communities (21.8 deaths per million) are more than twice the rate for nonrural areas (9.8 deaths per million).[14] Fire death rates by race show that African Americans in rural areas are twice as likely to die in a fire as rural whites. Children under the age of 5 and adults 65 years and older also face a higher risk of dying in a fire.[15]

A U.S. Fire Administration report from 2007 indicated that heating devices, smoking, and electrical distribution were the top three causes for fires in rural areas. Working smoke alarms were absent in 73 percent of rural residential properties where fires occurred.[16]

engineering ■ actions that employ physical science to find solutions for reducing the impact or eliminating the risk of fire and that may involve the development of codes, standards, equipment, and systems and construction techniques.

enforcement ■ a systematic approach to adopting community codes and certifying officials with authority to inspect the extent of code compliance and to implement rules and regulations to clarify the application of codes.

Engineering

The second and third of the "Three Es," **engineering** and **enforcement**, are addressed in the codes in use to ensure an acceptable degree of fire safety. Engineering principles and practices are at the core of these codes. Often the evidence for the need of a code and the very specific guidance provided by codes are products of engineering research. Like research, investigation and analysis of structural failure, building collapse, and major fires often result in technical reports with very specific engineering reasons for the disaster and can lead to or even contain specific recommendations for code changes (Table 2-1). An example is the National Institute for Standards and Technology (NIST) investigations of the World Trade Center collapse on September 11, 2001, and the fire at the Cook County, Illinois, Administration Building, October 17, 2003. Engineering firms as well as independent engineering consultants are key players in the development of codes and either directly propose code changes or technically support any group that is promoting a code change.

TABLE 2-1 Major Deficiencies Found in 10 Major U.S. Fires

FIRE INCIDENT	DATE	LIVES LOST	PREDOMINANT CODE ISSUE
Iroquois Theater, Chicago	December 30, 1903	602	• Exit door swing • Combustibility of decorations
Triangle Shirtwaist Factory, New York City	March 25, 1911	145	• Inadequate and locked exits • Accumulations of combustibles
Cleveland (Ohio) Clinic	May 15, 1929	125	• Storage of cellulose nitrate films
Ohio State Penitentiary, Columbus	April 21, 1930	320	• Inadequate egress supervision • Lack of fire protection equipment
Cocoanut Grove, Boston	November 28, 1942	492	• Number of exits • Exit door swing • Combustibility of decorations
LaSalle Hotel, Chicago	June 5, 1946	61	• Corridor protection • Enclosure of vertical openings
Winecoff Hotel, Atlanta	December 7, 1946	127	• Enclosure of vertical openings • Early notification (fire alarm) • Recognition that "fireproof" does not exist
GM Transmission Plant, Livonia, Michigan	August 12, 1953	3	• Concealed combustible spaces • Unprotected steel columns • Lack of fire separations • Lack of sprinklers • Large quantities of heated combustible liquids
Hartford (Connecticut) Hospital	December 8, 1961	16	• Combustible ceiling tile and glue • Delayed fire reporting • Open doors onto corridors
Golden Age Nursing Home, Fitchville, Ohio	November 23, 1963	83	• Lack of fire separations • Lack of fire protection equipment

Data derived from Federal Emergency Management Agency, U.S. Fire Administration, *Report for Beyond Solutions*, 2000, p. 25.

DEVELOPMENT AND ENACTMENT OF FIRE SAFETY CODES

Fire prevention codes are generally administered and enforced by the fire department. Historically, they have been concerned with fire safety issues related to fire protection equipment, to maintenance of buildings and premises, and to hazardous materials, processes, and machinery used in buildings.

Building codes, which are usually administered and enforced by a community's building official, address fire safety requirements with respect to the construction of buildings. Internal agencies vary by community, but in most communities, some person has formal responsibilities to oversee the administration of building codes. Depending on its size and complexity a community may actually have a standalone building department that handles the range of duties necessary for development and construction. Because of possible areas of overlap, a cooperative relationship between the fire department and the building official is essential. It should be noted that neither fire prevention nor building codes address the furnishings of dwellings, which are major contributors to fire fatalities.

Generally, fire safety and building codes, when adopted for the first time in a state or community, do not apply to existing structures or installations except when the enforcement official determines that continuation of the hazard would jeopardize safety of life and property. The inclusion of such a clause does not exempt existing facilities from code application; it merely places an onus on the enforcement official to be in a position to justify retroactive application from a life and property safety standpoint.

Fire prevention requirements in some form or another have been a part of our society for many generations. The inclusion of a document known as a fire prevention or fire safety code in the municipal government framework can be traced in more recent years to the influence of the insurance industry on public fire protection. In the aftermath of the great Baltimore fire of 1904, the National Board of Fire Underwriters developed the Standard Grading Schedule for cities and towns with reference to fire defenses.[17] This schedule had a profound effect on the development and enactment of municipal fire prevention codes.

The National Board of Fire Underwriters (later known as the American Insurance Association) also formulated a model Fire Prevention Code, copies of which were made available to jurisdictions for the asking. Countless localities throughout the United States adopted this code without change as the legally constituted fire prevention code. Although some cities probably would have developed and implemented fire prevention codes on their own, it is doubtful that there would have been anywhere near the current use of fire prevention codes without the availability of this code.

MODEL FIRE PREVENTION CODES

The Uniform Fire Code, National Fire Protection Association (NFPA), and the International Fire Code, International Code Council (ICC) cover subjects typical of those covered in a modern state, county, or municipal fire prevention code. These codes, which have been adopted in most jurisdictions, have chapters covering administration and enforcement, general fire safety requirements, means of egress, fire protection systems and equipment, automatic sprinkler systems, and fire detection and alarm systems. The codes also include occupancy fire safety requirements ranging from assembly occupancies to airports and heliports, and special processes and material handling ranging from hazardous materials and chemicals to safeguards during building construction and demolition operations. The codes also contain chapters of definitions and appendices with more detailed or special information. Like the other model codes, they are updated every few years.

The International Fire Code published by the International Code Council was designed to be used in coordination with the International Building Code. Many communities using the Building Code adopted this fire code as well. Few jurisdictions continue to use an in-house created fire prevention code. For example, New York City had one of the oldest in-house building codes until 2003, when it adopted the family of ICC codes.

MODEL BUILDING CODES

Building codes play an important role in protecting the community from a fire safety standpoint. Code coverage includes structural requirements, seismic protection, means of egress, interior finish, vertical and horizontal opening protection, and many other areas that are directly related to fire protection. As a rule, the building code applies to new construction, whereas a fire code applies to both new and existing construction. The building code official must be concerned with all fire prevention and fire protection measures relating to original construction. Unless building officials and fire officials enjoy a close working relationship, possible conflicts of authority would be detrimental to community service and safety.

Of all functions in municipal government, the building official works most closely with the fire department. Like all municipal departments, the firefighting services need certain tools to perform their duties: fire trucks, water, hoses, ladders, and the like. However, one of the least recognized tools used by the fire department is the city building code. It is imperative for a successful, effective fire prevention program to have adequate building regulations that provide fire departments and building officials the authority to require adequate fire protection measures to reduce fire losses.[18]

The most common model building code in the United States is the **International Building Code (IBC)**, published by the International Code Council. The IBC was developed in the 1990s as a result of a decision by the three regional building codes: BOCA National Building Code (BOCA/NBC) by the Building Officials Code Administrators International (BOCA); Uniform Building Code (UBC) by the International Conference of Building Officials (ICBO); and Standard Building Code (SBC) by the Southern Building Code Congress International (SBCCI). The first IBC was published in 1997. At present, it has been adopted at either a state or a local level in all 50 states, adding up to approximately 23,000 jurisdictions that employ some form of it.

International Building Code (IBC)
■ a model building code published by the International Code Council; the IBC is the most common building code in the United States.

The International Building Code contains a number of chapters that relate to fire protection: use or occupancy, building heights and areas, type of construction, fire-resistant materials and construction, interior finishes, fire protection systems, means of egress, accessibility, roof assemblies and rooftop structures, mechanical systems, and other areas having some bearing on this subject. In addition, the International Building Code references the International Fire Code for many additional fire safety and firefighter safety requirements.[19]

Reference Standards

Model building and fire codes reference a number of national standards to provide design guidance and to avoid redundant model regulations. Many of the standards are published by the National Fire Protection Association, such as the Standard for the Installation of Sprinkler Systems (NFPA 13), and the Standard for Portable Fire Extinguishers (NFPA 10). The Life Safety Code (NFPA 101) is widely used as a stand-alone code by many jurisdictions, including federal agencies. Other national standards referenced by the codes include some developed by Underwriters Laboratories (UL), the American Society for Testing and Materials (ASTM), the International Code Council (ICC), and others.[20]

Mini–Maxi Codes

mini–maxi codes ■ codes that are adopted at the state level and cannot be amended at the local level.

The concept of **mini–maxi codes** has caused a great deal of concern in fire service circles. Under the concept, municipalities and counties that adopt such a code are prohibited from adopting other codes that are less stringent or more stringent than their state code. This approach gains uniformity of application but is seen by many as inhibiting a local jurisdiction from adopting more stringent fire protection requirements, such as stronger sprinkler provisions, which may be needed to address unique conditions.

Prescriptive Codes

To *prescribe* means to lay down as a guide or rule of action. Building and fire safety codes have long been prescriptive in composition, detailing what needs to be done to comply. For example, specific dimensions are included. The code enforcement official defines parameters within which to work, as do the architect, the contractor, and the user.

Performance-Based Design

compliance ■ conformance with either applicable codes or a notice of violation issued or served by an authorized code official.

Building and fire prevention codes in the United States and Canada were developed entirely through a prescriptive method in past years. For example, the codes state that corridors must be at least a certain width to be in **compliance**. Widths and other code features were often established by tests conducted by various agencies and organizations. The Uniform Fire Code offers provisions for achieving compliance with the codes through prescriptive or performance solutions.

Many of the research projects that influenced code requirements were conducted by Underwriters Laboratories, Inc., and the National Bureau of Standards, the forerunner of today's National Institute of Standards and Technology. Tests that determined fire-resistant rating categories were first conducted by the National Bureau of Standards at a large government building slated for demolition in Washington, D.C.[21]

A prescriptive-type code designates specific parameters for construction, which makes it a rather simple task to discern deviations. Recently, the United States joined several other nations—including Australia, Canada, New Zealand, Sweden, Japan, and China—that have found so-called performance-based designs provide greater latitude for the design professional to fashion a structure embracing the specific needs of the client without having to adhere so closely to the predetermined limitations imposed by a prescriptive-type code.[22] A substantial number of new casinos and megahotels recently built in Las Vegas, Nevada, employed performance-based designs because they could not meet the restrictive elements of the prescriptive-type codes.

performance-based design ■ an engineering design that incorporates fire safety solutions by means of alternative materials and techniques to achieve specific goals.

Performance-based design, as defined in a Federal Emergency Management Agency (FEMA) U.S. Fire Administration publication, is an engineering approach to design elements of a building or facility based on performance goals and objectives, engineering analysis, scientific measurements, and qualitative assessment of alternatives.[23] Performance-based designs and codes do not use quantitative requirements. For example, there may not be a specific requirement for a 2-hour fire-resistive enclosure around the stairs, but rather a performance objective such as all persons must be able to exit the structure within 8 minutes of notification of an emergency. Performance-based designs are employed where the proposed use or occupancy of a building or facility cannot be neatly categorized into the occupancy classification and resulting limitations established by the prescriptive building codes. In some cases a combination of prescriptive and performance codes might be utilized.

Performance-based design provides a new set of challenges for the Authority Having Jurisdiction (AHJ). Rather than requiring the building to meet a list of prescriptive requirements, the AHJ must evaluate how the structure and its occupants will perform under fire conditions. This means the AHJ must be familiar with principles of fire behavior, structural performance, human response, and integrated life-safety and fire-protection systems.

It can be a difficult task for the code enforcer to determine whether the design professional (architect/engineer) has developed performance-based criteria that will enable the structure to withstand the ravages of fire. The code enforcer must know a great deal about suppression system capabilities, human reaction capabilities, and anticipated structural integrity retention in the event of fire. In practically all cases a combination of these issues must be considered in making a final decision about the appropriateness of the design proposal. The proposed design must be approved by all of the project's stakeholders, including the code enforcement official. In some cases, the code enforcer may employ the services of an independent third-party consulting service to review and make recommendations on the proposed performance-based design. If any portion of the design configuration fails, there may be only the good judgment of the code enforcer to rely on, because no specific quantitative code violations may have contributed to the incident. Such an incident may well leave the fire and/or building code official "hanging out to dry" with no real avenues of defense for his or her decision, as there is with a prescriptive code.

Generally, the code official has the option of accepting use of a performance-based code or not. It is absolutely essential that the code enforcement official obtain training in how to review performance-based designs or else transfer that responsibility to some other qualified person or agency. Many fire and building code enforcement officials do not feel comfortable administering the application of a performance-based code. They feel that they are on much firmer ground in using a prescriptive-type code.

The International Code Council has developed a Performance Code for Buildings and Facilities. The following codes all include performance-based design options: NFPA 5000, Building Construction and Safety Code; NFPA 1, Uniform Fire Code; and the International Fire Code. In addition, the National Fire Protection Association has produced NFPA Guide 101A, *Guide to Alternative Approaches to Life Safety,* which was originally developed with assistance from the Health Care Financing Administration, a federal agency, with a goal of providing a means of evaluating existing nursing homes and health-care facilities by considering a number of fire safety measures and programs.

ZONING CODES

zoning codes
▪ a land-use planning process that involves the limitation of various types of occupancies to given sections of the community, including, for example, the land's intended use, building size and height, and lot size.

Zoning codes identify the limitations of various types of occupancies to given sections of the community. They also have an effect on fire prevention in a community. Bulk storage of flammable liquids would not, under a zoning code, be found in the middle of the high-value mercantile district, for example. This condition does exist in a number of communities across the country. However, in practically all cases, flammable liquids storage was set up at the location before the advent of zoning or fire prevention regulations.

Zoning provisions also help ensure that adequate clearances are provided across streets. This control affects the potential for a conflagration in the community. Explosives storage and manufacturing are likewise generally very tightly controlled by local zoning requirements.

ELECTRICAL CODES

Electrical codes are primarily fire prevention codes because electrical safety is so closely related to fire safety. The principal purpose of electrical inspections is to ascertain that wiring is safely installed so that people in the structure will not be directly endangered and so that fires will not be started as a result of faulty installation.

There is a greater degree of uniformity of code usage in the electrical field than in any other field of public safety code coverage. The National Electrical Code, developed by the National Fire Protection Association, is incorporated as the electrical code in practically every state. Methods of enforcement vary; however, municipal electrical inspection is probably the most prevalent method employed. In some areas, electrical codes are enforced at the state level, whereas in a few other areas, this responsibility is carried out by a private inspection organization.

Electrical inspections, like building code inspections, are generally conducted on a fee basis, with the fee being sufficient to offset the cost of inspection. Usually, the cost of the inspection service is passed on to the consumer in the electrical contractor's charges for the job. Building and electrical inspections have another feature in common: the necessity for close surveillance at the time of construction to achieve satisfactory code compliance. A delay of a day or two in making such an inspection can result in overlooking a serious condition from a safety standpoint. For example, an enclosure of structural elements completed during that delay might make proper inspection impossible.

In most jurisdictions, the electrical code is enforced at the point of the provision of the electrical service. Most jurisdictions require full code compliance of a building prior to its connection to service from the public utility or power company. This procedure is quite effective and can also be used as an enforcement lever in the inspection of existing electrical installations.

Another factor in electrical code enforcement is requiring electricians to have a high degree of technical competency, including a knowledge of licensure procedures. This requirement is necessary because of the nature of the work: Much of the potentially dangerous wiring is hidden once the structure is completed.

HOUSING CODES

Another type of code that has a bearing on fire prevention is the housing code. In recent years, most larger communities in the United States have adopted housing codes in an effort to ensure the adequacy of housing facilities, especially those in rental properties.

As a rule, housing codes include a number of provisions relating to fire prevention and life safety, such as means of egress and heating appliances. Space heaters are not permitted under model housing codes. The right of entry for inspection under housing code authority is generally broader than that for fire prevention inspectors. Thus the housing inspector has a means of upgrading life safety in occupancies that may be inspected by fire department personnel only on invitation.

Because of the personal nature of housing code enforcement, many jurisdictions have been reluctant to permit the implementation of such a code. Some citizens feel that the obligatory inspection of individual homes under a housing code represents an unwarranted intrusion into privacy.

OTHER FIRE SAFETY-RELATED CODES

Mechanical codes, including heating, ventilation, and air-conditioning codes, as well as plumbing codes, are usually enforced by the same agency in a community. Plumbing codes have a bearing on fire protection, especially when automatic sprinkler protection is included in the plumbing code. The mechanical or plumbing inspector may be responsible for seeing that automatic sprinkler equipment is properly installed, for checking underground connections related to fire protection services, and for checking hydrants, standpipes, and other features related to fire protection water supplies.

Mechanical codes are the codes that cover heating and air conditioning. They have a major effect on fire prevention. These codes usually include coverage of ductwork and other matters related to distribution through heating and air-conditioning systems.

mechanical codes
▪ codes that cover heating and air conditioning.

Historic preservation codes are enacted to encourage the retention of historically significant structures in a community. These codes may permit less stringent fire protection requirements or alternative protection provisions for existing buildings. Similar in nature to historic preservation codes are codes focused on redevelopment, also known as smart growth regulations. These codes ease the way for reuse of buildings, sometimes even having a higher priority than fire and life safety codes. The intent is to enable the redevelopment of existing areas of a jurisdiction that already have infrastructure, where there are opportunities to reduce waste that may occur in demolishing otherwise sound buildings. Because older buildings may have limited egress and fire protection, the jurisdiction's fire prevention bureau must be involved so that solutions can be planned early in the process to satisfy life safety requirements.

FORESTRY CODES

Usually administered at the state level, forestry codes and regulations have a definite effect on fire prevention. These regulations are generally in effect only within a given number of feet of wooded or forested areas, and in many states they are not applicable within incorporated municipalities. Regulations often prohibit open burning during periods of low humidity. They may also require safety precautions in connection with the use of matches and smoking materials.

wildland
■ often a remote land area characterized by natural, sometimes dense vegetation minimally modified by human activity, such as woodlands, forests, brush, and meadows. Wildland can contain dangerous buildups of fuel from vegetation that may or may not be a risk for fire. Trends in migration and development currently bring together increased wildland interface with urban areas throughout the world.

Forestry laws are generally comprehensive and prohibit a wide variety of unsafe acts in wildlands. The term **wildland** is used to describe any natural vegetation environment that has not been modified by human activity. *Wildland* is a readily identifiable fire safety industry term used here to identify a variety of lands, such as woodlands, forest, brush, and meadows. Among acts generally prohibited are leaving campfires unattended; practicing open burning without a permit during certain seasons; operating a vehicle without a muffler in forest lands; and dropping or throwing burning matches, lighted cigarettes, or other burning materials in or near wildland. The operation of railroads in wildland is also closely regulated because of the danger of fires as a result of sparks. Adjacent lands are required to be cleared to reduce this danger.

In many states people who set fires on property they own and permit the fires to escape and become uncontrolled are responsible for payment of all costs for fighting the forest fires they have caused. The fires may have been set originally to destroy debris or rubbish.

Several counties, primarily in the western states, have adopted provisions that require minimum clearances between wildland and structures built for habitation in forested areas. These requirements have proven to be valuable in preventing structural damage in wildland fires.

CODE ADOPTION

A state fire prevention code is in effect in most states. These codes are promulgated by the state fire marshal, the insurance commissioner, the state fire board, or some other agency granted the power to promulgate regulations. Usually, a public hearing is required so that the public, special-interest groups, and fire service personnel have the opportunity to address the fire marshal or fire prevention commission regarding possible problems or advantages they see in connection with implementation of the requirements.

Enforcement

Early transgressors of fire prevention rules may have been persuaded to change their ways by seeing enforcement measures inflicted on their neighbors. However, today some suggest the term *enforcement* may be a bit harsh. *Compliance* may be better because it is the ultimate goal and may be achieved by either education or enforcement.

Fire and building codes represent the "engineering" phase of the "Three Es" equation. But without enforcement, or compliance, as previously noted, engineering aspects might be forgotten. Somehow the architect and builder must know to install automatic sprinklers; occupants of all homes must know that a working smoke alarm is needed.

UNWANTED ALARMS

Since analysis on U.S. fire data began in 1980, the difference between false alarms and reported fires has changed dramatically. Between 1980 and 2012 fires reported to the fire department have gone down from an estimated 3 million to 1.5 million, compared to false alarms that in 1980 were around 750,000 to just over 2 million in 2012. These statistics put 2012 numbers at around two false alarms for every one report of an actual fire.[24] In 2010 U.S. fire departments responded to approximately 2,178,000 false alarms.[25] That is, the fire service was responding to almost 6,000 unwanted calls each day.

Unwanted alarms (sometimes referred to as nuisance alarms) are a significant fire protection concern. They include equipment malfunctions, human error, and unintentional and malicious false alarms. They can be triggered by dust, smoke from cooking, construction activity, steam, and a host of other causes. Nuisance and other unwanted alarms have a negative impact on pubic and firefighter safety and are a financial and workload burden for the responding fire companies. Unwanted alarms can also create a culture of complacency as people lose confidence in the fire alarm system.

Some fire departments have taken action to reduce the impact of the unwanted alarms by penalizing alarm users for excessive false alarms. However, such adverse actions may have unintended consequences. A deferred response to an automatic alarm could lead to increased risk to occupants or extensive fire loss. Penalties for increased frequency of unwanted alarms could lead to building operators shutting down alarm systems even though the alarm system is required by code.

Montgomery County, Maryland, Fire and Rescue Service, Fire Code Enforcement Section, became concerned when workers there realized that fire companies were experiencing nearly twice-daily responses for nuisance alarms to a shopping mall food court. Countywide data showed that during a one-year period, from 2008 to 2009, there were over 4,000 responses to automatic fire alarms, of which over 1,000 (25 percent) were to the same 117 buildings. To address unwanted alarms, Code Enforcement used a Maryland law that provided authority to local jurisdictions to issue civil citations and penalties to a user who set off false alarms. The civil citation could be issued when an alarm was activated as a result of faulty or improperly maintained equipment or false alarms where fire department response exceeded three responses in a 30-day period or eight responses in a 12-month period.[26]

Montgomery County's approach was direct, but customer-oriented. An inspector provided specific data and information about recent alarm responses to the alarm users' address, explaining the state law and all areas of code that related to the alarm system. The intent was to make the alarm users aware that they were over the threshold of Maryland law for excessive false alarms and that any additional alarms would result in official action that would render their alarm system unreliable.

Though the alarm users could have been issued a citation based on state thresholds, they instead were issued written documentation in the form of a notice of violation to provide a report of the educational session and report that there was a problem with their fire alarm. The notice also specified that upon the next false alarm, the system would be deemed unreliable and that the alarm users would be required to establish a fire watch until the system was returned to a reliable state. The approach by Montgomery County had its intended impact with a 50 percent drop in the number of responses to automatic alarms in just one year.

Economic Incentives

Historically, fire prevention and protection efforts were motivated by economic incentives. Community leaders dealing with threats of conflagrations in Europe and, later, in the United States were concerned about the loss of commerce and damage to the community's economy. Present day economic incentives are often part of an integrated risk reduction approach. People are offered a tangible incentive if they choose to participate in the use of a fire protection measure that ultimately benefits the community by reducing fire risk.

The simplest form of an economic incentive is the smoke alarm installation program. Such initiatives are generally administered by the fire department and often include a private partner such as a business or community organization that helps with funding and other resources, such as home improvement businesses, insurance companies, grocery chains, and service organizations. Free smoke alarm installations for those in need help a community increase individual home safety while achieving the goal of increased community safety and preparedness. These types of home-focused fire prevention efforts generally target high-risk areas of the community where people cannot otherwise afford them.

There are many ways to offer economic incentives in support of fire prevention goals. Incentives can be focused on the developers, builders, and tradespeople and building owners. Over the last several decades there has been an increase in offering incentives to entice developers and builders to consider sprinklers even when they are not required by code. Some incentives for developers may include waived fees and other costs, increased distance between fire hydrants and closer building to lot lines. For owners, reduced or waived property taxes may be an incentive for retrofitting a building with sprinklers.

The Mountain Communities Fire Safe Council in Riverside County, California, obtained a Department of Homeland Security (DHS) Fire Prevention and Safety Grant to entice residents to replace wood shingle roofs with Class "A" fire-resistive roofs, if they participated in the community's Fire Safe Council property hazard assessment and agreed to complete some hazard mitigation tasks such as clearing defensible space. The DHS has specifications within the grant that also require some structural improvement, such as ember-resistant soffit and roof vents.

Emergency Response

Emergency response will always be a critical function of community risk reduction. Completely preventing ignition of a fire, while a worthy goal, is not likely to be achieved, because just too many things can go wrong. Immediate emergency response is necessary when other interventions fail to prevent a fire. Additionally, fire suppression response is needed to ensure that any fire ignited is contained as soon as possible to prevent loss of life and property.

Summary

Sound fire prevention practices that produce better fire safety are the product of the original "Three Es": education, engineering, and enforcement, plus the addition of economic incentives and emergency response in the contemporary adaption of prevention practices. Fire prevention is a critical core service within local government, dating to the earliest days of settlement in America. The contemporary all-hazards model of Community Risk Reduction (CRR) needs to be included in any discussion of fire prevention due to its developing role in many communities.

A community with the five methods of intervention—education, engineering, enforcement, economic incentives, and emergency response—is a safer community. Education requires delivering the appropriate safety message to the affected or at-risk group. It is effective but depends on the commitment of local government or local fire service. Engineering involves fixed system and construction standards as a means of preventing and/or controlling fires. It also promotes life safety by providing adequate means of egress. Enforcement ensures that laws, ordinances, codes, and standards are followed. Many buildings would not be built to appropriate standards without adequate enforcement. Economic incentives employ strategies to entice residents and businesses to provide alarms and fire protection even though they may not be required by code. Emergency response serves as the safety net for a community when a fire cannot be prevented and threatens life and property.

Review Questions

1. What U.S. president conducted a national fire prevention conference after World War II?
   a. Richard M. Nixon
   b. Dwight D. Eisenhower
   c. Harry S. Truman
   d. Franklin D. Roosevelt
2. The "Three Es" of fire prevention are:
   a. education, engineering, exercise.
   b. education, experience, enforcement.
   c. exercise, engineering, enforcement.
   d. education, engineering, enforcement.
3. The "Three Es" were first suggested after they were considered for what U.S. community hazard?
   a. Youth sports injuries
   b. Swimming pool drownings
   c. Automobile collision casualties
   d. Home fire casualties
4. Community Risk Reduction identifies two additional "Es." What are they?
   a. Environmental initiatives and emergency road service
   b. Economic rescue and emergency assistance
   c. Economic incentives and emergency response
   d. Engineered products and emergency exits
5. In 2006, how many U.S. cities with populations greater than 1 million had no recognized fire prevention or code enforcement program?
   a. 2
   b. 3
   c. 4
   d. 5
6. In an NFPA report for 2006–2010, what proportion of fire deaths was attributed to smoke alarms that were missing or not working?
   a. Over half
   b. Two-thirds
   c. Exactly half
   d. 10 percent

7. Children under the age of ____ are twice as likely to die from fire than the rest of the population.
   a. 3
   b. 5
   c. 13
   d. 16
8. Fire death rates in rural areas are how much higher than in nonrural areas?
   a. One-quarter
   b. More than twice
   c. 50 percent
   d. Less than 10 percent
9. Which one of the following statements is false?
   a. Building codes address fire safety requirements.
   b. Fire prevention codes cannot be applied retroactively.
   c. Building and fire prevention codes may overlap.
   d. Non–life-safety elements of a building or fire prevention code are generally not applied retroactively.
10. The two model fire prevention codes are:
    a. NFPA Uniform Fire Code and ICC International Fire Code.
    b. BOCA and Uniform Codes.
    c. Southern and BOCA Codes.
    d. NFPA Uniform Fire and Southern Code.
11. Fire prevention codes can be classified as:
    a. prescriptive.
    b. performance-related.
    c. both a and b.
    d. none of the above.
12. Codes other than building and fire prevention that affect fire safety include:
    a. zoning.
    b. electrical.
    c. forestry.
    d. all of the above.
13. Which one of the following is not a major contributor to fire prevention?
    a. Public education
    b. Fire service privatization
    c. Strict adherence to building codes
    d. Abatement of arson by thorough investigation

End Notes

1. Powell Evans, comp., *Official Record of the First American National Fire Prevention Convention* (Philadelphia, Pa.: Merchant and Evans Co., 1914).
2. U.S. Fire Administration, *President's Conference on Fire Prevention: Action Program* (Emmitsburg, MD: USFA, 1947), p. 3. Accessed April 11, 2013, at http://www.usfa.fema.gov/downloads/pdf/47report/actionprogram.pdf
3. Centers for Disease Control and Prevention, *10 Leading Causes of Death by Age Group* (Atlanta, GA: 2010). Accessed April 12, 2013, at http://www.cdc.gov/injury/wisqars/pdf/10LCID_All_Deaths_By_Age_Group_2010-a.pdf
4. Marty Ahrens, *Home Structure Fires* (Quincy, MA: NFPA, April 2013), p. 1. Accessed December 8, 2013, at http://www.nfpa.org/~/media/Files/Research/NFPA%20reports/Occupancies/oshomes.pdf
5. Ibid., p. i.
6. U.S. Fire Administration, *Topical Fire Report Series: Civilian Fire Injuries in Residential Buildings (2009–2011)*, (Emmitsburg, MD: USFA, March 2013), vol. 14, p. 1. Accessed April 12, 2013, at http://www.usfa.fema.gov/downloads/pdf/statistics/v14i1.pdf
7. U.S. Fire Administration, *Topical Fire Report Series: Civilian Fire Fatalities in Residential Buildings (2008–2010)*. (Emmitsburg, MD: USFA, February 2012), vol. 13, p. 1. Accessed April 12, 2013, at http://www.usfa.fema.gov/downloads/pdf/statistics/v13i1.pdf
8. Research Triangle Institute, International City Management Association, National Fire

Protection Association, *Municipal Fire Service Workbook* (Washington, DC: U.S. Government Printing Office, 1977), p. 4.

9. Federal Emergency Management Agency, U.S. Fire Administration, National Fire Protection Association, *Four Years Later–A Second Needs Assessment of the U.S. Fire Service: A Cooperative Study Authorized by U.S. Public Law 108-767, Title XXXVI* (2006), p. 51.
10. James M. Shannon, "First Word," *NFPA Journal* (National Fire Protection Association, March/April 2004), p. 6.
11. Federal Emergency Management Agency, U.S. Fire Administration, *Babies and Toddlers Fire Death Data Report* (Emmitsburg, MD: USFA, 2003).
12. Canadian Association of Fire Chiefs, *Canadian Fire Chief* (Ottawa, Ontario, Summer 2003), p. 29.
13. Michael J. Karter, Jr., *Fire Loss in the United States During 2012 (*Quincy, MA: National Fire Protection Association, September 2013), p. 11. Accessed October 17, 2013, at http://www.nfpa.org/~/media/Files/Research/NFPA%20reports/Overall%20Fire%20Statistics/osfireloss.pdf
14. Fire Analysis and Research, One-Stop Data Shop, *Demographic and Other Characteristics Related to Fire Deaths or Injuries* (Quincy, MA: National Fire Protection Association, March *2010*), p. 7. Accessed November 30, 2013, at http://www.nfpa.org/research/statistical-reports/victim-patterns/demographic-and-other-characteristics-related-to-fire-deaths
15. Ibid., p. 5.
16. U.S. Fire Administration, *Mitigation of the Rural Fire Problem: Strategies Based on Original Research and Adaptation of Existing Best Practices* (Emmitsburg, MD: USFA, December 2007), p. 2. Accessed March 27, 2013, at http://www.usfa.fema.gov/downloads/pdf/publications/MitigationRuralFireProblem.pdf
17. Fire Suppression Rating Schedule (New York: Insurance Services Office, 1980).
18. Personal communication to author Robertson from the International Code Council.
19. Ibid.
20. Ibid.
21. Ibid., p. SM1-8.
22. Ibid., p. SM1-12.
23. Ibid., p. SMI-7.
24. National Fire Prevention Association, *Trends and Patterns of U.S. Fire Losses in 2012.* (Quincy, MA: National Fire Protection Association, November 2013), p. 10. Accessed December 7, 2013, at http://www.nfpa.org/~/media/Files/Research/NFPA%20reports/Overall%20Fire%20Statistics/ostrends.pdf
25. Marty Ahrens, *Unwanted Fire Alarms* (presentation at the National Fire Protection Association 16th Annual Suppression, Detection, and Signaling Research and Applications Symposium, Phoenix, AZ, March 5–8, 2012). Accessed November 30, 2013, at http://www.nfpa.org/~/media/files/research/research%20foundation/foundation%20proceedings/2012%20supdet/1ahrens%20paper.pdf
26. Article—Criminal Law §9–609, Maryland Code Annotated Criminal Law, Title 9, Subtitle 6, Part II, Subsection 9-609 (2012). Accessed April 2, 2013, at http://167.102.242.144/smb/mgaleg.maryland.gov/google_docs$/2013rs/statute_google/gcr/9-609.pdf

CHAPTER

3 Public Fire and Life Safety Education Programs

Tobias Titz/Getty Images

KEY TERMS

OBJECTIVES

After reading this chapter, you should be able to

- Identify the steps in public fire education planning.
- Identify partnerships between civic organization and fire service professionals and how the message of fire prevention is disseminated by these partnerships.
- Identify methods to measure the effectiveness of fire and life safety education in the schools.

A combination of public fire education, fire prevention, and injury prevention, fire and life safety education is receiving more recognition than ever before by the fire services of North America. Although public fire education is now a well-established practice in many communities in the United States and Canada, it is not yet universal, and more effort must be directed to make it so.

America Burning is a report of the National Commission on Fire Prevention and Control that recognizes the importance of fire safety education. It states:

> Among the many measures that can be taken to reduce fire losses, perhaps none is more important than educating people about fire. Americans must be made aware of the magnitude of fire's toll and its threat to them personally. They must know how to minimize the risk of fire in their daily surroundings. They must know how to cope with fire, quickly and effectively, once it has started.[1]

America Burning
■ the report of the National Commission on Fire Prevention and Control that identified America's growing loss of life and property from fire, and the parallel culture of indifference to fire, with comprehensive recommendations on how to begin to turn around the problem.

It is the mission of fire and life safety education units to influence the perceptions of those who are unaware that, even at home, their safety is at risk. A survey conducted by the Society for Fire Protection Engineers (SFPE) in 2011 reported that 70 percent of the respondents felt safer from fire in their homes than in a commercial high-rise building.[2] This perception of safety is not supported by statistics. The U.S. Fire Administration reports that fire departments responded to an estimated 365,000 residential building fires between 2008 and 2010, with an average of 2,560 deaths and 13,000 injuries per year.[3] The life safety difference between residential building fires and nonresidential building fires is dramatic.

Another U.S. Fire Administration topical report, this time concerning fires in nonresidential buildings, listed fire department response to an estimated 302,000 fires between 2004 and 2006, with an average of 135 deaths and 3,050 injuries per year. For every one fire death in a nonresidential building there are 18 fire deaths in a residential building, and for every one injury in a nonresidential building there are four fire injuries in a residential building. The response numbers are similar but the casualty numbers are significantly lower in nonresidential buildings. The difference is attributable to more safety regulation enforcement in nonresidential buildings and almost no additional regulation enforcement in residential buildings once they are built.

To get public fire safety education programs under way, there must be a "spark plug" or champion in the fire department, ideally the fire chief, but at least someone who has attended programs or had experiences that stimulated an interest in fire prevention. This individual can encourage others in the department to recognize the potential of fire prevention. The department's administration must also have a compelling desire to initiate programs. However, it may come about, acknowledgment of the important role of fire safety education among the fire department's responsibilities is a mark of the modern fire service, whether the department is career, volunteer, call, or part paid.

Scope of Fire and Life Safety Education Programs

Fire prevention education is the dissemination of information and the promotion of actions relating to fire hazards and causes in order to inspire the public to take proper precautions. To conduct educational programs, fire department personnel should be trained in public speaking and in staging demonstrations. In fact, several fire departments have incorporated formal training in public speaking into their recruit training programs.

Public fire safety education includes both fire prevention and fire reaction. Burn awareness is also included. Many fire safety education programs have been expanded to include injury prevention as well. The inclusion of other safety subjects must be carefully considered so that the original fire safety message is not lost. Although all of these topics contribute to the welfare of the community, they must be administered in doses that can be easily absorbed by the particular audience. Widespread inclusion of emergency medical services as a fire department function supports the injury prevention concept. The term **fire and life safety education** encompasses this expansion of goals.

fire and life safety education ▪ planned activities focused on promoting, presenting, and making available information intended to change behavior and reduce the risk of fire and other injury-causing events in the community.

Firewise Communities Program ▪ a wildfire planning and mitigation process cosponsored by the USDA Forest Service, the U.S. Department of the Interior, the National Association of State Foresters, and the National Fire Protection Association; it encourages solutions and action at the local level with community, government, and business participation.

Fire Adapted Communities (FAC) ▪ communities that have formed partnerships with homeowners, local agencies, and other organizations to reduce the potential for loss of life and property by providing information and expertise on strategies and actions to mitigate wildfire risks.

Community Wildfire Protection Plan (CWPP) ▪ a local collaborative planning process based on the needs of people living in wildfire-threatened areas that addresses specific issues such as wildfire response, hazard mitigation, community preparedness, and structure protection.

Wildfire mitigation has become a part of many fire departments' goals in public education. Often, state and national forestry organizations share in this responsibility. With the growing prevalence of wildland—urban interface as a major fire problem, fire prevention agencies have added this dimension to their repertoire of activities. Many jurisdictions have adopted programs similar to that of Colorado Springs, Colorado.

Colorado Springs has a Wildland Risk Management Office within its fire department. The fire marshal is a major player in this activity. Much of the program is based on the NFPA's **Firewise Communities Program**, but adapted to local needs. The program involves an evaluation of hazards for homes located in wildfire areas as well as community educational programs on abatement measures that may be initiated by the homeowner. The program limits roofing materials to those that meet fire safety standards, and it also includes an ordinance giving the fire chief the authority to order evacuations in extreme danger situations.[4]

In June 2012, wildfire preparations were tested when a major fire struck approximately 4 miles northwest of Colorado Springs. The Waldo Canyon fire, as it is known, has been called the worst wildfire in Colorado history, killing two residents, scorching over 18,000 acres, requiring 30,000 people to evacuate, and destroying over 300 homes. The **Fire Adapted Communities (FAC)** coalition saw the fire as an opportunity to analyze the performance of the mitigation measures taken in the Colorado area and to compare them with the mitigation strategies that FAC recommends.

FAC is a coalition of partners that includes organizations such as the USDA Forest Service, the Insurance Institute for Business & Home Safety, the International Association of Fire Chiefs, the National Fire Protection Association, The Nature Conservancy, and the National Association of State Foresters. An assessment team from FAC published a comprehensive report finding that the Colorado Springs Firewise Mitigation Plan was in line with the strategies of FAC. Examples of findings that enhance a reduced wildfire risk included a community actively engaged in mitigating wildfire risks, strong evidence of considerable work toward making structures more fire resistive and creating Firewise-recommended defensible space, development planning, and design that includes adoption of codes and ordinances that reduce wildfire risks, as well as a **Community Wildfire Protection Plan (CWPP)** that was being followed. All of the various programs now available to help communities mitigate their wildfire risk depend on public education and community involvement to be successful.

STEPS IN PUBLIC FIRE AND LIFE SAFETY EDUCATION PLANNING

The USFA manual *Public Fire Education Planning: A Five-Step Process,* offers guidance in developing and operating a community risk fire education program, or **public fire education planning**. Any program begins with a **community analysis** to identify fire safety problems and the demographic characteristics of those at risk. The manual also recommends the development of **community partnerships** with groups or organizations willing to join forces and address community risk. The most effective risk reduction efforts are those that involve the community in the planning and the solution. An intervention strategy is then established to begin the detailed work necessary for the development of a successful fire or life safety risk reduction process.

public fire education planning ■ efforts involved in the consideration and implementation of a program to raise awareness or teach new fire safety skills, including analyzing problems, developing partnerships, creating strategies to address priorities, implementing plans, and evaluating progress.

community analysis ■ a process that identifies fire and life safety problems and the demographic characteristics of those at risk in a community.

community partnership ■ a person, group, or organization willing to join forces and address a community risk.

target groups ■ members of a population who are the focus of a specific fire safety message.

After preliminary work in planning and analysis, a community education program's next steps involve implementing the strategy and then evaluating the efforts. Implementing the strategy involves testing the interventions and then putting the plan into action in the community. It is essential that the implementation be well coordinated and sequenced appropriately. The primary goal of the evaluation process is to demonstrate that the risk reduction efforts are reaching the target populations, have the planned impact, and are reducing loss. The evaluation plan measures performance on several levels—outcome, impact, and process objectives.[5]

Target group input at each stage of design and development is essential to the effectiveness of educational program materials. In the design of material, the first goal is to determine message content. Messages should concern specific hazards. They should appeal to positive motives and not be threatening. The messages should show the context of the problem and the desired behavior. The format, whether a wall chart, video, PowerPoint presentation, or folder, should be matched to the message, audience, and resources.

The planner should determine the appropriate delivery time by specifying the target groups and finding out when those audiences will be most receptive to fire and life safety messages. Messages should be scheduled for maximum effect. The program package should be designed and then presented to a sample audience. Once it has been found to be effective, the program can then be implemented: Material can be purchased or produced and distributed, instructors trained and scheduled, and audience participation and cooperation obtained.

Programs should be constantly monitored and should be modified on the basis of the monitoring review. They should be evaluated for effectiveness through review of loss data and educational data. Telephone polls and in-person survey interviews have proved valuable in obtaining such information. Less reliable when measuring the overall effectiveness of a program are voluntary Internet-based surveys. Though often free and easily administered, voluntary Internet-based surveys lack the randomness needed for objective measurement and tend toward self-selection bias.

Each delivery should be evaluated for impact by having the audience complete a survey about the lessons learned. Follow-up evaluation is also effective in determining if any change in behavior has resulted from the education. This type of evaluation may occur in incremental lapses of time after the education was delivered. The results should be favorable if the proper steps have been followed in program development. It is important to know whether the program reduced risk or eliminated hazards in the community.

Home Safety Surveys

The home inspection program is primarily devoted to fire prevention education. This program probably began in the United States in May 1912, in Cincinnati, Ohio, when the fire department started a comprehensive home and business inspection program. By May 1913, the department had experienced a 60 percent reduction in fires. The inspections were made by assigned personnel from each of the 45 companies in the department. There were 80,000 structures in Cincinnati at the time, and the program was aimed at reaching each of them annually.[6] Radio did not exist at the time, and fire apparatus was primarily horse drawn.

VOLUNTARY BASIS

Home inspection programs are primarily fire prevention education endeavors. Although the word *inspection* is used, there is no legal backing for the program in most jurisdictions. Entry into residences is voluntary, and if at any time occupants look on the program as being mandatory and believe they are being forced to make changes, there is a good chance that they will resist. Personnel assigned to home inspection duties should be made fully aware that the program is entirely voluntary. Because of the potential adverse reaction to the term *inspection*, some agencies use the term *home safety survey* to make the process more acceptable to the public. The fire service in the United Kingdom has a robust home safety program that is referred to as a *visit*. Although home inspections were very popular with the U.S. fire service during the 1950s and 1960s, the subsequent addition of emergency medical duties has reduced the amount of time fire departments have to perform home safety agendas.

TRAINING

Fire service personnel must be properly prepared before embarking on home inspection duties. The training program must encompass details of the items to be checked, including potential hazards and smoke alarms, the importance of personal neatness and courtesy, and the public relations aspects of the job. Successful programs have been carried out in volunteer as well as career and part-paid fire departments.

ADVANCE PUBLICITY

To prepare the public to accept home safety visits, the program must receive a considerable amount of advance publicity. That publicity should describe the purposes of the program and mention inspection, with particular emphasis on participation being strictly voluntary.

Planning should encompass schedules for areas to be covered each day, and definite routes should be planned to schedule personnel in a way that makes the most effective use of their time. Current programs involve community members who are trained and participate as partners. These programs are as successful in rural and suburban areas as they are in urban areas. All members should be provided with official identification badges, and they should wear identifying clothing.

HOME SAFETY ASSESSMENTS

Communities are finding that despite the increasing diversity and quantity of emergency calls to fire departments, the benefits of home safety visits outweigh the challenges of conducting them. For several years the United Kingdom Fire Service has been conducting a program known as a Home Fire Risk Assessment. It centers on prearranged residential visits during which on-duty firefighters describe and identify potential risks within a home, offer knowledge on how to avoid risk, and assist the residents in checking smoke alarms and creating a home fire escape plan.

Similar programs are in place throughout many fire departments in the United States. One home safety initiative was delivered as a pilot project in selected cities. This 2010 initiative, known as **Community Risk Reduction (CRR)**, featured home fire safety visits modeled on successful home safety programs in the United Kingdom and Australia. The CRR program focused on high-risk homes identified by the participating fire departments. In Philadelphia, Pennsylvania, 9,000 smoke alarms were installed during the program. Other cities participating in the CRR program included Dallas, Texas; Tucson, Arizona; Portland, Oregon; Madison, Wisconsin; and Vancouver, Washington. The initiative was managed as part of the **Vision 20/20 Project** and was awarded a Federal Emergency Management Agency (FEMA) Fire Prevention and Safety Grant.

Community Risk Reduction (CRR) ▪ local-level commitment to develop internal and external partnerships with the community to implement programs, initiatives, and services that prevent and/or mitigate the risk of human-caused or natural disasters.

Vision 20/20 Project ▪ a grant-funded fire safety project sponsored by the Institution of Fire Engineers that includes five key prevention strategic initiatives: advocacy, marketing, culture, technology, and codes and standards. A sixth overarching strategy provides local-level evidence of positive prevention results.

The Vision 20/20 Project continues to provide resources and facilitate support for CRR by offering training on how to implement a program as well as awareness training for fire operations supervisors. As more fire industry professionals are exposed to training in CRR concepts, we can anticipate it continuing to grow and have a positive impact on home fire safety.

Fire Prevention Education Through Business, Community, and Civic Organizations

One of the best means of promoting fire safety education is through business, community, and civic organizations. These organizations are the essential partners that can become advocates and supporters in fire prevention efforts. Organizations such as the Chamber of Commerce, League of Women Voters, Lions, National Urban League, Kiwanis, and other neighborhood and community associations, for example, are interested in projects that may be of assistance to the community.

Most of such organizations have a wide range of community interests represented among their members. This variety of membership is helpful to the fire preventionist, because the contacts made can result in garnering much community support for the fire department.

ABILITY TO COMMUNICATE WITH ALL PEOPLE

Special efforts must be made to ensure that the fire education message is understood in communities where significant numbers of individuals do not read or speak English. The fire prevention bureau chief must always be prepared to reach any group regardless of language. By including a demographic study as a part of the community's risk hazard analysis, the manager can become aware of the different languages and means of communication needed. This study should include the deaf and hard-of-hearing community and the need to provide appropriate access and ensure use of sign language interpretation for live events and video or televised messages. When developing and distributing, or otherwise marketing, fire safety messages, be aware that advocates for the deaf and hard-of-hearing community do not consider their absence of hearing a disability. It would serve the fire prevention bureau well to seek out community advocates or service providers for the deaf and hard of hearing for consultation on preparing the most effective and appropriate communication.

People who are deaf or hard of hearing, who are blind or vision impaired, or who have mobility issues all face unique challenges during emergencies. Home fire safety programs must include a means to communicate the fire safety message to them. It should also offer a process for installing appropriate alarms so that individuals who are unable to see or hear clearly can become aware of and react to a fire at the earliest possible moment.

Many fire departments have initiated programs that make education more accessible to individuals with disabilities and have installed specialized residential fire alarms. For example, in 2007 a joint program of Fire Protection Publications and Oklahoma ABLE Tech at Oklahoma State University, Stillwater, created *Fire Safety Solutions for Oklahomans with Disabilities.* The program focuses on the two metropolitan areas of Oklahoma City and Tulsa. This comprehensive home fire safety program provides information needed to educate people with disabilities and offers guidelines for installing specialized smoke alarms. The program was established and funded under a FEMA Fire Prevention and Safety Grant and has become a model for communities that want to adopt similar programs.[7]

Fire Safety Clinics and Seminars

The fire safety clinic or workshop has been a valuable aid in fire prevention efforts in some communities. Under this concept, representatives of industry and institutions are encouraged to attend a fire safety program during which all phases of fire prevention and reaction are discussed. Some of these programs include outdoor demonstrations that are most helpful in creating an interest in fire safety.

Seminars of this type are especially valuable if they are aimed at one particular type of audience because it is difficult to develop a meaningful day-long program for a general audience. For example, seminars may be designed primarily for department store personnel, apartment house operators, college dorm resident assistants, or nursing home and hospital personnel.

The Gainesville, Florida, Fire Rescue Department and its Risk Reduction Bureau (RRB) provided a focused public education program for bar employees. Information included public safety, the night inspection program, types of fire, arson prevention, public assemblies, fire codes, actions to take in an emergency, the use of candles and pyrotechnics, and parking lot safety.[8]

In another example, Minneapolis, Minnesota, has successfully employed a task force approach to assembly occupancy safety that includes information-sharing workshops. One such workshop was conducted within 24 hours after the 2003 Chicago nightclub E2 stampede in which 21 people died. The Minneapolis Fine Line Music Café experienced a serious fire that was accidentally ignited by pyrotechnics set off by the band the night after the E2 club stampede. A task force workshop reemphasized critical safety information that helped personnel of the Minneapolis Fine Line Music Café avoid major casualties.

Community Events

Fairs and other organized activities give communities an opportunity to promote fire prevention by setting up displays and booths for public informational purposes. They also give the fire department an opportunity for personal contact with the general public. Participation at a fair may involve staging displays of firefighting equipment and fire safety demonstrations.

Inviting citizens to visit a home that has experienced a serious fire provides a meaningful safety message. Several cities have held open houses at recently burned dwellings to let the public see the ravages of fire. This can be done only with the permission of the homeowner. However, many victims welcome the opportunity to help mold positive attitudes toward fire safety. Fire department personnel may prepare the dwelling by placing signs to indicate point of origin, fire spread characteristics, and other salient features of the fire. Personnel should also be on hand to answer questions. Note that safety of visitors should be a primary concern.

In a slightly different approach, some fire departments use similar fire-event-driven opportunities as teaching moments. In-service fire suppression crews in some communities go door to door with specific facts about a recent fire in the neighborhood and offer education based on that fire. The crews are also stocked with smoke alarms that many departments install for free if a home is inadequately equipped.

Fire and Life Safety Education in the Schools

Opportunities to present fire prevention and reaction messages in schools are limitless. The school system must in all cases be consulted in the planning of any program addressed to school personnel or students.

A common school-fire-prevention effort is an appearance of community fire safety professionals before school assemblies. However, such an appearance has the disadvantage of limiting personal contact and subsequent questions and discussion. A discussion session with smaller classroom groupings and with age-appropriate material can be much more successful. So both types of presentations should be considered.

Technology has increased the number of ways an educator can reach an audience and enhance live demonstrations. In one example of how educators can employ innovative technology for education programs, firefighter Dayna Hilton from Clarksville, Arkansas, produces a fire safety television show that appears on the 24-hour preschool television network Sprout. Her show reaches directly into classrooms live on the Internet-based video technology of Skype. She also extensively streams educational videos on a website, making it accessible to anyone with a computer and an Internet connection.

The overall program for fire education in the school system should address contact at each grade level. Unfortunately, personnel limitations and other factors often restrict the activities to one or two grades.

COORDINATING FOR SUCCESS

Close coordination with school faculty is necessary for the development of successful fire and life safety education programs. Many school systems have a standing policy against including material in the curriculum that has been developed outside the system. A spirit of cooperation with appropriate personnel in the school system may result in modifying such a position to provide at least some fire prevention coverage where no material has been developed in the education system.

Fire and life safety educators are becoming more creative in developing programs and ways to get safety information into preschools and elementary schools, partly because they face the challenge of having schools approve time away from the prescribed curriculum. It seems that fire departments are continually competing with the many demands placed on teachers and schools. Two examples of creative approaches follow, both of them developed by the Lakeland, Florida, Fire Department:

> "Fire Safety Traveling Trunks" provides preschool and kindergarten teachers with all the materials needed to teach fire safety in the classroom. The trunks include items for the various centers that are usually set up in preschool and kindergarten classrooms. The trunks include the NFPA "Learn Not to Burn®" preschool curriculum and such items as child-sized gear for the dress-up center; matching cards, sequencing cards, and counting cards for the math center; an interactive fire safety computer program for the computer center; and audiotapes of fire safety songs for the music center. The trunks also include items to be used during circle time such as a black sheet to practice *crawl low under smoke* and felt flames for practicing *stop, drop, and roll.* The Traveling Trunks were developed to enhance the fire safety education, not to replace the educational programs presented in the schools and centers. Teachers utilize the materials in the trunks that are in their classroom for 2 weeks, and the materials are user friendly.
>
> "Story Time Is Safety Time" is a second program developed for preschool children to fit into "circle time" or "story time" in the classroom. The preschool story

time includes stories, songs, finger plays, and action rhymes, and utilizes a felt board. Story time can also be presented at other locations that host story times, such as the local library, children's museums, and bookstores.[9]

OTHER RELATED PROGRAMS

The Junior Fire Marshal program has been effective in many sections of the United States. Both English and Spanish versions are available. Sponsored by the ITT Hartford Group, the program is designed for children in kindergarten through third grade. It focuses on fire prevention tasks through the use of teacher guides and printed activity sheets. Fire safety in apartments, public places such as schools and theaters, and single-family homes is addressed.

In some areas this program has been sponsored by a local insurance organization, with the fieldwork done by fire department personnel. In most localities the material has been supplied by the school district or fire service, and all field operations have likewise fallen within the jurisdiction of that agency. Membership in the Junior Fire Marshal program is usually obtained by completing a home fire prevention form or by taking some other step that aids the cause of fire prevention. School students enrolled in the program are usually given certificates of fire safety knowledge.

Wildfire Prevention

As previously noted, fire abatement at the wildland–urban interface has become a part of the responsibilities of many fire departments. As the loss of life and property to wildfires increases, so does the cost of managing these fires. Understanding the complex relationship between wildfire behavior and environment can help communities be more prepared to prevent or limit wildfire impact.

Wildfires can be caused by natural events such as lightning and volcanic lava, but they are most often the result of human carelessness and malicious, intentional fire setting. In addition to the benefits of understanding the science of fire, it is beneficial to know its history. Fires in natural-growth fuels can be as necessary as they are feared. In most cases a wildfire, and only a wildfire, can manage or reduce the dangerous buildup of fuels. Periodic fires help control and reduce fuel, ultimately preventing a catastrophic fire; they also increase wildlife and plant habitat and growth opportunity.

Wildfires occur naturally and are an intrinsic part of the life cycle of forest and woodlands. But after trying to tame and develop the wildlands, humans are only just realizing that completely preventing naturally occurring fires has unintended consequences: more fierce fires when they occur and interruption of the natural order of forest life cycles. More research and understanding, as described in the following paragraphs, allow us to plan how to manage fires better.

The small booklet *Fire Ecology of the New Jersey Pine Barrens*, produced by the Whitesbog Preservation Trust, traces thousands of years of evidence of fire in the ancient New Jersey Pine Barrens forest and explains in careful detail the cause and effect of many generations of fires and their positive impact on sustaining the forest and its diverse wildlife. For example, a fire that destroys existing trees also serves as a catalyst for the next generation of trees by melting the sap in pinecones and allowing seeds to be released to start new growth.

Many communities at risk for severe wildfires are beginning to be more proactive in planning for and mitigating the hazards that lead to catastrophic fires.

CALIFORNIA FIRE SAFE COUNCILS

Fire safe councils are grassroots, community-level groups organized to reduce the risk of wildfires in California. The establishment of local fire safe councils is one way to empower efforts at the community/homeowner level to reduce the risk of wildfires. The councils are nonprofit charitable organizations funded through donations, fund-raising initiatives, and grants. For example, many of the estimated 250 fire safe councils have taken advantage of obtaining grants from the U.S. Forest Service, the Bureau of Land Management, and the U.S. Fish and Wildlife Service.[10] The grants offer funding support to conduct community training and public education, to promote fire-resistant materials in structures, and even to provide the manual labor to reduce fuel loads in creating and maintaining defensible space.

The local fire safe councils in California are empowered centrally through the Fire Safe Council. The Fire Safe Council is made up of 50 public and private organizations that include the insurance industry, Pacific Gas and Electric, NFPA, Bureau of Land Management, Orange County Fire Authority, California Fire Marshal's Office, and California Building Industry Association. Mike Esnard of the Mountain Communities Fire Safe Council in Idyllwild, California, believes:

> Waiting for huge fires, spending millions to put them out, and then replacing destroyed property is clearly the least intelligent option. Thinning forests and modifying vegetation in and around communities, as well as hardening structures in those communities, seems the wisest.[11]

Besides focusing on reducing fuel, councils also work to put space between fuels and structures and have made significant efforts to reduce hazards associated with residential construction.

Federal grants can assist fire safe councils with funding and mitigation of hazards. The Mountain Communities Fire Safe Council received funding from the U.S. Department of Homeland Security and FEMA to begin a mitigation project aimed at reducing one of the more notorious hazards in the Wildland Urban Interface (WUI): wood roof shingles. Wood shingles are a readily available fire fuel source because airborne embers can land on them and become wedged between and under them, potentially causing the roof to ignite. Wood shingle roofs can also create an additional source of flying embers, as they quickly burn and dislodge from the roof. Federal grants can be used to replace home roofs with Class A fire-resistant roofs, funded 75 percent by the federal grant and 25 percent by the homeowner. To obtain the grant funds, individual homeowners must first submit to a home inspection that leads to risk reduction actions for the homeowners and an increased awareness of what to look for themselves. Among other federal requirements for the homeowner is installing home exterior vents that will not allow passage of the flying embers. Ultimately, the grants provide for a safer community by employing more fire-resistant structural materials.

The fire safe councils also feature news stories about community members who are actively engaged in making their community safer by reducing the risk of fire. For example, the Mountain Community Fire Safe Council has a unique group of active members, called Woodies, who are available to help others in the community reduce wildfire exposure around their homes. Every week the group assembles and assists community residents who are unable to complete the work themselves. Excess timber from the cleanup is donated to local families to help heat their homes.

Local Government and Community Collaborations

Public interest in crime prevention is generally greater than in fire prevention. Often a high crime rate or fire experience can be a symptom of many other community problems. Communities, their residents, and the infrastructure they depend on can deteriorate if not maintained. The City of Cincinnati, Ohio, has established the Neighborhood Enhancement Program, which is a 90-day collaborative effort with city departments, such as fire, police, housing and building, neighborhood residents, and community organizations. The purpose of this program is to improve the quality of life for neighborhood residents and to engage residents in activities on their own behalf after the first 90 days.

To become involved, communities must prepare a commitment agreement that includes a priority list of projects to complete or problems to resolve, such as removing unsafe vacant buildings; cleaning up streets, sidewalks, and vacant lots; creating parks; and cooling down community "hot" spots. Working together, the city and the community residents work on projects that lead to a higher degree of safety. There is significant benefit in ridding neighborhoods of vacant, unsecure buildings and property piled with combustible waste, in meeting building and safety codes, and in providing activities to occupy youth in a positive way. Accomplishing these types of activities can reduce opportunities for set fires and false alarms and can increase the safety of residents and firefighters. The Neighborhood Enhancement Program has been achieving good results and has received numerous awards. Other cities and communities have adopted a version of the Neighborhood Enhancement Program for use in their neighborhoods.[12]

Publicity Programs

A variety of programs can be devised to capture the attention of the community and to publicize the theme of fire prevention. Following are some ideas that have been used by fire departments in the United States.

CONTESTS

Poster contests have been widely used in the fire prevention field for many years as a way to foster an interest in fire prevention. Contests are generally conducted by the local fire department or fire service association in cooperation with community schools. Some school officials believe that poster contests are inappropriate as a means of achieving fire prevention because the artwork may not be in keeping with educational programs. Any fire department considering a poster contest should first check with school system officials, obtain their approval before proceeding, and check entries for technical accuracy. The posters may be displayed locally.

PROMOTIONAL AIDS AND ACTIVITIES

Place mats bearing a fire prevention message can be used to increase awareness. Diners often read the wording on a place mat while waiting to be served. Table tents with printed messages can also be useful.

Favors such as rulers, pencils, fire helmets, and stickers can help in promoting fire safety. They are usually relatively inexpensive and are appreciated by children as well as adults. They may be passed out at public events as prizes for answering fire prevention questions accurately. Favors should not present a safety hazard to recipients.

The hydrant collar is another fire prevention message-bearer used in some communities. Talking hydrants, which are operated by remote control, have proven to be very effective in imparting fire safety messages to young children.

Placards on bus shelters, traffic safety islands, lampposts, and other city property can be helpful in fire prevention, as well as posters used on fire department vehicles, municipal buses, and municipal garbage trucks. Recently, some fire departments have included a safety message as part of the way they mark their vehicles, such as the community name and the unit's name and purpose. Cities with subway systems have a great opportunity for displaying fire prevention messages through the use of placards. A similar but newer approach to posters on publicly seen vehicles is known as a wrap. It consists of vinyl that has a printed message, often an eye-catching graphic. The vinyl wrap fits most contemporary vehicles, even large delivery vans, transit buses, and subway trains. The benefit is a strikingly visual message that is seen by tens of thousands of people a day as the vehicles move around the community.

Outdoor reader-boards, banners across roads, banner stands in public lobbies, and billboard fire prevention publicity can be helpful (Figure 3-1). Seen weekly, the simple reminder to attend to a smoke alarm could spur action.

A number of cities have opened permanent facilities in which citizens can learn fire prevention measures as well as fire survival procedures. In Milwaukee, Wisconsin, Survive Alive House, a converted recreational field house, provides visitors with an opportunity to practice home fire drills under realistic conditions. Similar to the Milwaukee Survive Alive House is the Fire Department of New York City's Fire Zone in Manhattan. Fire Zone offers education and interactive demonstrations acted out by docent tour guides. There are static displays resembling a fire station and fire apparatus as well as a realistic-looking burned-out residence.

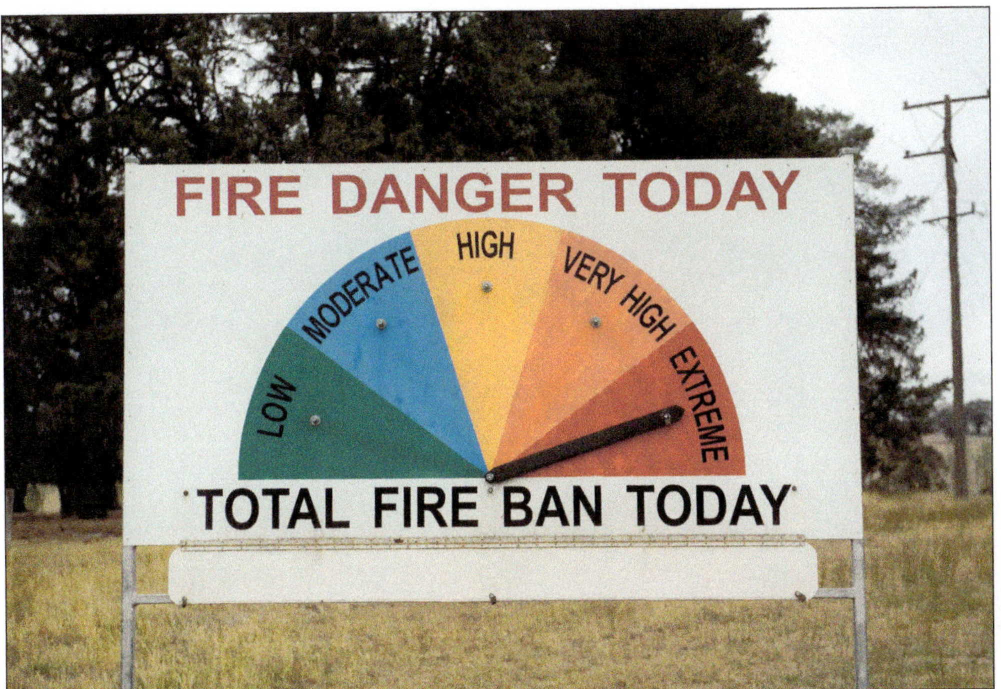

Tobias Titz/Getty Images

FIGURE 3-1 Use of a banner to promote a fire safety message can be effective when placed where many people will see it.

Figure 3-2 shows a firefighter explaining how to choose the safest escape route from a home fire.

One of the most successful citywide fire safety activities is the Great Louisville (Kentucky) Fire Drill, which has been held annually since 1984 on the Sunday of

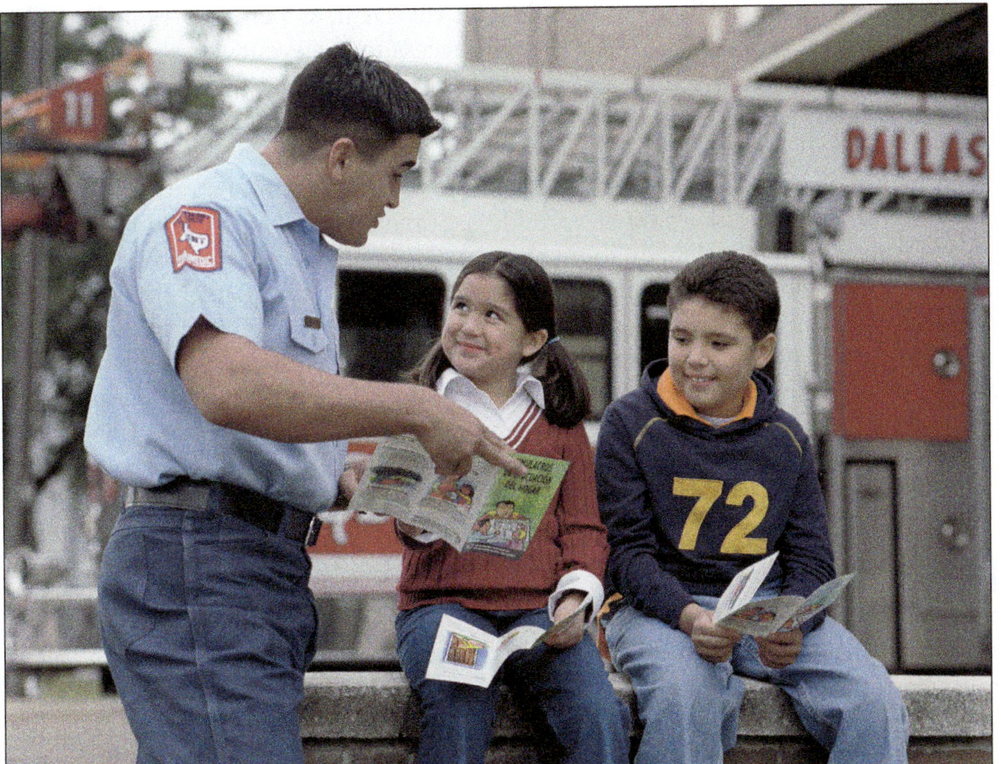

Stewart Cohen/Blend Images/Getty Images

FIGURE 3-2 Fire education experiences for children, especially realistic ones, show them how to safely escape a fire.

FIGURE 3-3 Special regional events like fairs, festivals and parades can provide opportunities to reach target audiences for a key fire safety theme. Events that are strictly dedicated to the safety theme are even better.

Suzanne Tucker/Shutterstock

Fire Prevention Week (Figure 3-3). The drill is designed to educate the public in fire safety by giving participants a chance to practice home fire drills and learn to stop, drop, and roll. They also learn about matches and lighter safety and how to report a fire. Between 12,000 and 15,000 people attend the event each year. Many other exciting activities take place at the Drill, which rotates among the city's parks so that all neighborhoods have access to the event.

Talking robots are used by a number of fire services. They are especially helpful in attracting small children to fire safety messages in public places such as shopping centers and fairs.

In recent years fire departments have put on dramatic demonstrations showing the effectiveness of residential fire sprinklers. Some of the demonstrations are conducted using side-by-side cubes about the size of a small room. The cube is outfitted with typical residential furnishings, which are duplicated in an adjacent cube. One of the cubes is outfitted with residential sprinklers and the other is not. By using a scenario that simulates the ignition of the home furnishings, the audience can clearly see how effective sprinklers can be in a home and how quickly fire spreads without them. A large timer marks the growing seconds as a fire educator announces the events along the critical timeline. It can be a very effective visual demonstration.

Media Publicity

An important part of any fire prevention undertaking is publicity. Media releases must be interesting, direct, and to the point. Material that is uninteresting and vague will not receive newspaper, television, or radio attention. It is necessary for the public information officer, chief, fire marshal, or other individual connected with fire safety publicity to work to develop or locate material that will be accepted and used by the media. A fire chief or fire marshal who understands the benefits of media attention often can use incidents and news events to promote fire safety. Ending a news interview about a fire with an appropriate message, such as "We want to make sure that all homes have a working smoke alarm outside every bedroom area and on every level, and that residents develop and practice a home fire escape plan," can end tragic news about a fire with a positive suggestion for action.

The fire prevention officer should visit the media offices to develop proper rapport with staff. Such a visit may be quite rewarding and can result in long-term cooperation between the media and the fire department. Some fire departments have full-time Public Information Officers (PIO) who are adept at promoting the department's safety message regularly through the department's marketing and information services. Though it may at times be inconvenient, it is often worthwhile to take advantage of late or early morning newscast opportunities to promote fire safety. If the fire chief, fire marshal, or PIO cultivates a good relationship with the media, he or she is more likely to be contacted with future opportunities.

The media should also be encouraged to accurately report fires and, in particular, to include information on contributing factors such as lack of smoke alarms. There is no greater opportunity for disseminating fire safety information than when reporting fires in the community. A regular stream of information that includes news releases about fires can emphasize information the fire department wants to highlight. Modern PIOs embrace technology such as Twitter, Facebook, and YouTube to assist in disseminating information about incidents and other events. News media representatives often take social media feeds and repost them to their network, thereby providing an unanticipated opportunity to reach a much wider audience.

Radio and television stations can contribute a great deal to fire prevention publicity by running public service spot announcements, which are among the most effective means of getting out fire prevention messages. Announcements may be developed by the fire prevention bureau locally or by one of the national fire organizations. Locally prepared announcements are usually the most effective, so some announcements prepared by national organizations are designed to permit wording pertaining to local problems. Messages may address current or seasonal fire problems.

An example of Internet use for fire prevention is the website of the South Jordan, Utah, Fire Department, which announces its babysitting course, high school fire science class, and other fire safety programs. Fire departments are actively using the Internet through social media to promote, educate, and inform the public of the latest news. Facebook, Twitter, Yahoo Groups, and YouTube allow free posting of videos. All of those mentioned and many more are helping fire and life safety organizations communicate in today's world.

Trailers or buses can be equipped to include a considerable amount of fire safety material for demonstration and may be arranged to allow rapid movement of people through the display. Many communities have outfitted such a unit. In some cases the unit is a surplus bus from the municipal transit company. A talking Sparky the Fire Dog® can be added to the fire prevention bus or trailer to attract attention to the unit. A number of the trailers are equipped to demonstrate smoke alarm and residential sprinkler operation as well.

Smoke Alarm Programs

There are many smoke alarm distribution programs. In some instances, smoke alarms are installed in residences by firefighters. This program is very easily incorporated into a home fire safety survey program.

Smoke alarm distribution programs have been—and will continue to be—a mainstay of home safety programs. However, in many homes, smoke alarms are not maintained properly, and in some cases, they are completely ignored. Fire departments want to remind residents about the importance of smoke alarms as well as the need to test them regularly and replace the batteries at least yearly. Optionally, a long-life battery that is advertised to last 10 years is currently on the market. Recent product standards in the smoke alarm industry have suggested that the alarms should also be replaced after 10 years because they could become less effective with age.

Because smoke alarms provide early warning when fire occurs, this type of program can have a positive impact on fire loss data in a community. Any smoke alarm program should be ongoing and should address smoke alarm maintenance, including annual battery replacement or using an alarm with a long-life battery and replacing the entire device after 10 years of service.

Fire Prevention Week

Fire Prevention Week has long been looked on as an opportunity to spotlight fire prevention for all citizens during a concentrated time span. The observance began as Fire Prevention Day through a proclamation by President Woodrow Wilson on October 9, 1920. In 1922, the observance was extended to one week by proclamation of President Warren G. Harding. The date of October 9 was chosen to commemorate the Great Chicago Fire in 1871.

Fire departments throughout the United States and Canada use Fire Prevention Week as the time for aggressive activities aimed at all segments of the population. Radio, television, and newspaper announcements are extensively used at this time.

Volunteer Fire Departments

Many volunteer fire departments provide their citizens with excellent fire safety programs. Because most departments were first organized primarily for fire suppression, the volunteers are really going the "extra mile" to render this service in their communities. Many volunteer fire departments have members with talents and experience in teaching, home repair, and a variety of skills that can be useful in fire prevention activities.

Although it is difficult for most volunteer fire departments to conduct mandatory fire safety inspections, no conflict of interest arises with public education activities.

Volunteer fire services of all kinds—rural, urban, and suburban—usually have individuals who for reasons of health or age can best use their talents in public education work rather than fire suppression. In fact, a number of Maryland volunteer departments have excellent fire and life safety education programs.

Review of Successful Programs

The publication *Proving Public Fire Education Works* from the TriData Corporation provides the results of evaluative research into what makes a good public education program. The publication offers evidence from 77 public education programs in the United States and Canada that made a positive difference in the safety of communities. As described by TriData, certain factors led to successful programs, such as having champions or "spark plugs" to lead the implementation efforts, careful planning to include specific fire safety problems, partners that can help overcome barriers, and marketing to the appropriate audience. Another important aspect of a successful program is the use of evaluation to determine whether goals have been met.[13]

Vision 20/20: Strategic Fire Project, from the **Institution for Fire Engineers (IFE)**, offered communities and organizations an opportunity to submit descriptions of their fire safety programs for presentation before the national symposium, *Models in Fire Prevention*. The symposium was conducted in 2010 in Towson, Maryland, and again in 2012 in northern Virginia.[14] Collectively, several dozen case study models of innovative fire prevention programs from across the country were presented at this symposium. Presentations included documentation on how the programs were evaluated, which supported Vision 20/20's strategic initiative to help fire prevention personnel measure the results of their efforts. The presentations provided a unique opportunity for communities to learn from each other and to share information about fire safety programs that work. The Vision 20/20 Project makes all products involving its grant activity available on its website so others can benefit from this grassroots effort.

Institution for Fire Engineers (IFE)
■ an international organization with the mission to encourage and improve all areas of the science and practice involved in fire safety.

CHAPTER **REVIEW**

Summary

The field of public fire and life safety education offers many opportunities to reduce loss of life and property by fire. The key to success in the field is not always easy to find. Hard work and enthusiastic support are necessary ingredients.

Fire safety education is the dissemination of information relating to fire hazards and fire causes with the hope that the public will get the message and take the proper precautions against fire. It also includes the development of proper fire reaction. Among the avenues used to reach the public is the home inspection program, which apparently had its beginnings in Cincinnati, Ohio, in May 1912. Although the program is called "inspection," homeowner participation is voluntary.

The fire prevention educator can work through civic organizations, such as chambers of commerce, parent–teacher associations, clubs, and churches, by giving talks and demonstrations. Community events—parades, fairs, dances, contests, and other organized activities—offer many opportunities to publicize the message of fire prevention and enlist community cooperation.

Fire safety education in the school classroom offers limitless opportunity to instill lifetime attitudes about fire prevention precautions. Students who are taught to be aware of hazards that cause fires and what to do about them can help the cause of fire safety in the home and in the community.

School programs can take the form of actual classroom instruction or talks and demonstrations before school assemblies. The fire preventionist can also work with young people in group activities outside school, such as in Junior Fire Marshal programs, Sparky the Fire Dog® fire departments, and the Scouts.

Publicity aids for fire prevention can take many forms: contests staged by the fire department; billboard advertising; distribution of signs and posters, book covers, and favors printed with fire prevention messages; innovative ideas such as talking trash cans or fire alarm boxes; or roving fire engines delivering messages through loudspeakers. The possibilities are endless.

One of the most effective means of reaching the public is through media publicity: newspaper, radio, television, the Internet, and social media. The press should be encouraged to report fires accurately and to give specific information on contributing causes. This information can have a tremendous impact on citizens of the community.

Successful fire safety education programs provide evidence of their value to the community. Careful analysis of the problems in the community paired with the right participating partners can lead to greater awareness of the problems and a will to accomplish the hard work needed to reduce the risk of fires.

Review Questions

1. Public fire safety education activities include:
   a. fire prevention.
   b. fire reactions.
   c. burn awareness.
   d. all the above.
2. Wildfire mitigation education is the responsibility of:
   a. local fire departments.
   b. state organizations.
   c. national forestry organizations.
   d. all the above.
3. Public fire education planning involves a ________-step process.
   a. three
   b. four
   c. five
   d. six

4. Which one of the following is NOT considered a home inspection duty?
   a. Exit signs
   b. Smoke alarms
   c. Public relations
   d. Potential fire hazards
5. Fire safety clinics have been valuable for which one of the following audiences?
   a. Nursing homes
   b. Hospitals
   c. Apartment house operators
   d. All the above
6. What group has offered a better chance to promote a fire prevention message than any others?
   a. Worker
   b. Youth
   c. Faith-based
   d. Neighborhood
7. The Firewise and Fire Adapted Communities are programs that suggest wildfire ____ strategies.
   a. tactics
   b. action
   c. mitigation
   d. research
8. Fire Prevention Day is always on _______, which commemorates the Great Chicago Fire of 1871.
   a. September 7
   b. October 9
   c. November 15
   d. December 7
9. Mountain Communities Fire Safe Council is using a FEMA grant to mitigate what notorious hazard?
   a. Fire-prone canyons
   b. Wood shingle roofs
   c. Dead-end streets
   d. Clogged evacuation routes
10. How are home safety surveys usually conducted?
    a. With an administrative search warrant
    b. On a voluntary basis
    c. With the intention of code enforcement
    d. In an effort to prevent juvenile fire setters
11. What training would firefighters NOT be given to conduct home inspection duties?
    a. Aggressive fire code enforcement
    b. Details of items to be checked
    c. Neatness and courtesy
    d. Public relations aspect of the job
12. Which organization sponsors the Junior Fire Marshal program?
    a. National Fire Academy
    b. International Association of Fire Fighters
    c. International Association of Fire Chiefs
    d. ITT Hartford Group
13. Community Risk Reduction programs funded under Federal Fire and Life Safety Grants focus on _____ as identified in program planning.
    a. acute industrial hazards
    b. high-risk pedestrians
    c. juvenile fire setters
    d. high-risk homes

End Notes

1. U.S. National Commission on Fire Prevention and Control Report, *America Burning* (Washington, DC: U.S. Government Printing Office, 1973), p. 105.
2. Society for Fire Protection Engineers, *2011 SFPE Survey Results* (Bethesda, MD: Author, 2011).
3. U.S. Fire Administration, *Topical Fire Report Series: Residential Building Fires (2008–2010)* (Emmitsburg, MD: U.S. Fire Administration, Vol. 13, Issue 2, April 2012), p. 1. Accessed November 30, 2013, at http://www.usfa.fema.gov/downloads/pdf/statistics/v13i2.pdf
4. Cathy Prudhomme, "Wild Wild West," *Fire Prevention—Fire Engineers Journal,* Institution of Fire Engineers (Leicester, UK, 2003).
5. U.S. Fire Administration, *Public Fire Education Planning: A Five-Step Process,* FA-219 (Emmitsburg, MD: U.S. Fire Administration,

June 2008). Accessed November 30, 2013, at http://www.usfa.fema.gov/downloads/pdf/publications/fa-219.pdf

6. Powell Evans, comp., *Official Record of the First American National Fire Prevention Convention* (Philadelphia: Merchant and Evans Co., 1914), p. 150.
7. Fire Protection Publications and Oklahoma ABLE Tech at Oklahoma State University, *Fire Safety Solutions for People with Disabilities* (Stillwater, OK: Fire Protection Publications, 2007).
8. Lise Fisher, "Safety Training for Bar Employees," *Gainesville* (FL) *Sun*, September 18, 2007.
9. Personal communication to author Robertson from Cheryl Edwards, Lakeland, FL, Fire Department, 2003.
10. Personal communication to author Love from Margaret Grayson, Executive Director for The California Fire Safe Council, March 8, 2013.
11. Mike Esnard, *Getting Rid of Our Number One Fire Hazard* (Mountain Communities Fire Safe Council Newsletter, Vol. 13, Summer 2012). Accessed November 18, 2013, at http://www.mcfsc.org/Documents/NewsletterSummer2012.pdf
12. City of Cincinnati (Ohio) Community Development, *Neighborhood Enhancement Program (n.d.)*. Accessed March 22, 2013, at http://www.cincinnati-oh.gov/community-development/neighborhood-development/nep/
13. Philip Schaenman et al., *Proving Public Fire Education Works* (Arlington, VA: TriData Corp., 1990). Accessed March 25, 2014, at http://www.sysplan.com/documents/tridata/prevention/PublicFireEdu.pdf
14. Vision 20/20, *2012 Model Performance in Fire Prevention (*Warrenton, VA: Institution of Fire Engineers (IFE), n.d.). Accessed March 23, 2013, at http://strategicfire.org/page.cfm/go/2012-Model-Performance

CHAPTER

4 Enforcing Fire Safety Compliance

Billy Morris Director of Fire Science Cisco College

KEY TERMS

American National Standards Institute (ANSI), *p. 68*

ASTM International, *p. 68*

control of occupancy, *p. 75*

incapable of self-preservation, *p. 79*

plan-review program, *p. 65*

structural control, *p. 72*

OBJECTIVES

After reading this chapter, you should be able to:

- Describe the legal authority for code enforcement.
- Describe the importance of plan review and its relation to code enforcement.
- Identify how control of sales and use assists in fire code enforcement, and give some examples.
- Recognize how structural control is used as a means of code enforcement, and give specific examples of structural control.
- Identify how control of occupancy is used as a means of code enforcement, and give an example of controlling occupancy.
- Describe compliance and abatement procedures and why assurance of compliance is preferred over court action.
- List some fire safety considerations in specific special types of occupancies.

Foremost among the responsibilities of the fire prevention bureau is the enforcement of the municipal fire code. Some of the approaches that have proved to be effective in developing fire safety codes and regulations are examined in this chapter. Inspection

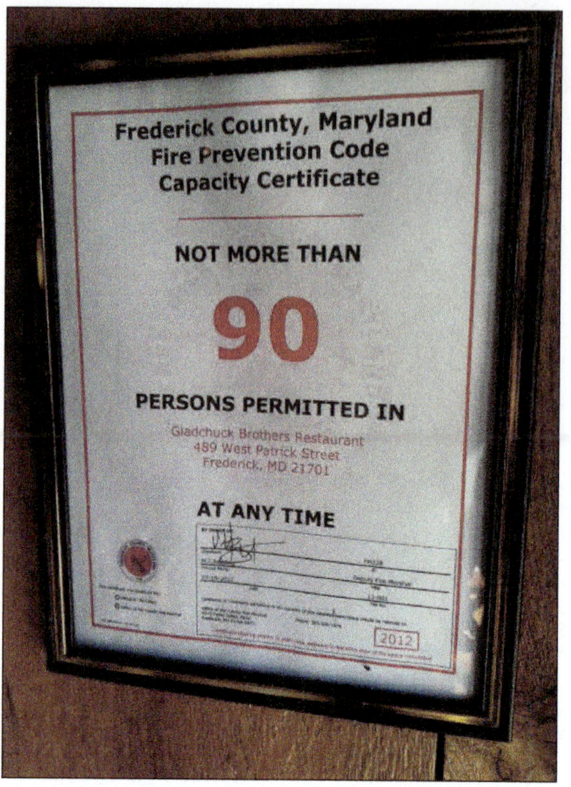

Mike Love

FIGURE 4-1 Occupant capacity certificates are required to be visibly displayed in assembly occupancies as a constant reminder of the maximum number of people allowed in areas of a building.

is an intrinsic part of most of the approaches, but because the subject covers so much ground and deserves special attention, it is treated separately. Engineering principles have a major role in code development (Figure 4-1).

Publicity for Fire Codes

Though some local government regulations require public notification of hearings and pending regulations, publicity regarding fire safety and building codes and the enforcement of such codes is not used as often as it could be to educate the public. In fact, press publicity of code enactment is too often limited to occasions when a requirement is imposed to abate a specific hazard that has come to light as the result of a tragedy.

As one example of effective use, the fire prevention bureau should seek press coverage in connection with hearings for enactment of fire prevention codes. Additional publicity options include the use of mailing lists, fax blasts, and social media to notify customers and stakeholders of where they can find information, such as a government blog or website. It never hurts to reach out with a presentation at the Chamber of Commerce meetings or offer some well-publicized workshops to present an overview of new regulations and provide an opportunity to "meet and greet" and get to know some of the stakeholders. Although such an announcement of new regulations may bring out individuals who are opponents of the legislation, publicity gives citizens a clearer idea of fire code coverage. It is always a better outcome in the code management world to have a positive relationship with stakeholders, who often are just looking to understand what is expected. Media publicity on code enforcement procedures is an excellent way to reach the public.

Signs placed in public buildings indicating that certain activities are violations of the fire safety code (such as those indicating "No Smoking" or occupancy capacities) remind citizens of fire safety. Although potential dangers from careless smoking are obvious, many may not understand maximum occupancy capacities.

In the years after the tragic fire at The Station nightclub in West Warwick, Rhode Island, in 2003 only a few states made sweeping fire safety rule changes for such occupancies, although several other states and cities increased controls over indoor pyrotechnics. The data again emphasize that code changes aimed at strengthening requirements in a given occupancy seldom are made beyond the boundaries of the state in which the tragic fire occurred.[1]

Plan-Review Program for Fire Code Enforcement

A number of municipal fire departments have a **plan-review program** for the examination of plans and specifications for buildings to be constructed within their jurisdiction. Qualified individuals are assigned to check plans and specifications for conformance with fire and building code requirements. Usually, the reviews are coordinated with those carried out by the local building department. Plans are received by the building department and are then routed through the fire department as a matter of procedure.

plan-review program ■ prior to new construction or building modifications, an evaluation by certified officials that ensures compliance of architectural or engineering plans with building and fire codes.

A plan-review program can best be established through a municipal directive or state law mandating that all plans for construction be reviewed by the fire department's fire prevention bureau. Florida, for example, has such a law.

The scope of plan reviews carried out under those arrangements is not always as broad as it might be. In a number of jurisdictions, the review is confined to checking the locations of fire hydrants and standpipe connections and the accessibility of apparatus. Major items of concern in fire protection, such as exit ways, interior finishes, basic construction, and sprinkler layouts often are not checked by the fire department.

Personnel need professional training to carry out those responsibilities. They cannot be expected to participate in detailed reviews of plans and specifications and in conferences with architects, engineers, and builders without having an adequate background. Fire service personnel who have been assigned to plan-review duties without proper training and background have caused a loss of prestige professionally for the departments they represent.

UTILIZING FIRE PROTECTION ENGINEERS FOR PLAN REVIEWS

Ideally, a fire department will have a fire protection engineer available to review plans and specifications and to consult with architects, engineers, and builders. The desirability of assigning to plan-review duties professional personnel who can speak with authority in the engineering field cannot be overemphasized. Smaller municipalities may find the addition of a fire protection engineer to the payroll cost-prohibitive. More than 100 fire protection engineers are assigned to fire departments in the United States, usually in a civilian, nonuniformed force and in a staff rather than a line position.

In addition to reviewing plans and specifications, fire protection engineers are frequently involved in long-range departmental planning related to station locations, response districts, and the like. They also may have considerable input in the development of fire prevention codes and regulations. In view of their assignment in staff positions, they usually are not responsible for directing inspection programs and other line functions, but they are called on to assist with highly technical code issues such as those posed by performance-based codes. Codes have increased the demand for fire protection engineers.

In smaller municipalities, where the possibility of employing a fire protection engineer may be remote, the service might more logically be provided at the state level. The state fire marshal's office may have a fire protection engineer on staff who can be made available to assist with plan reviews in jurisdictions that find it difficult to have an engineer on staff. This arrangement has an advantage from the standpoint of workload and definitely makes the most efficient use of the available talent. Another advantage to having a state-level fire protection engineer conduct the plan review is that the review will consider both state and local fire safety codes plus sections of the building code that pertain to fire protection. Some small jurisdictions contract with nearby cities or with consulting firms to provide this service on an as-needed basis.

Plan-review programs should, if at all possible, provide for review of preliminary plans. Some jurisdictions do not permit personnel to take the time to review plans in the preliminary stage. However, review at that point is usually beneficial.

The entire direction of a project from a fire protection standpoint can often be fairly easily changed at the time of a preliminary review, while a change at a later stage is usually considerably more difficult and many times more costly.

Preliminary consultations are usually voluntary on the part of the prospective permit applicant, but the benefits of review and comments can identify potential omissions that may lead to delays in a project. Public relations are also improved by an early review of plans.

Fire departments that do not conduct reviews of preliminary plans undoubtedly end up spending more time reviewing final plans than would have been necessary had a preliminary review been conducted. Problems that might easily have been corrected if detected in a preliminary review can develop into major problems at the time of a final review.

Some departments have developed very specific checklists of expectations for minimum requirements for their permits. These checklists may be available for review on the fire prevention bureau's website as well as offered as part of the preliminary consultation. Often the checklist of requirements is integrated in the permit application, so it serves multiple purposes. Any information in the form of meetings and checklists reduces ambiguity and raises the potential for a successful construction or fire protection system installation.

Most architects and engineers appreciate having a plan-review service available because it serves as an additional level of design review. Plans should be submitted in such form as to permit detailed review of the architectural and engineering documents.

Although consultations with architects during a plan review are necessary and desirable, the reviewer should have the opportunity to study the plans in detail alone, without interruption or distraction. Notes on the review should be kept on file, and narrative comments should be made to the architect. In some communities, the shortage of personnel has made it necessary merely to mark comments on plans and return them to the architect. This arrangement is undesirable because it permits no retrieval of notes and recommendations once the plans have been returned.

ARCHIVING OF CRITICAL DOCUMENTS AND PLANS

Building construction and fire protection plans submitted as part of an approved project or permit are a critical piece of history that must be saved by the building or fire code official. After many years this can become a challenging management concern due to space limitation and the necessity of a system to protect and retrieve the records when needed. If there is ever a question regarding what was officially approved for a building, the plans may be the only legal history available to answer that question. For example, consider this scenario:

> A complaint is received about inadequate exits in an assembly occupancy where plays have been held for 25 years. The inspector consults the plans on record for the occupancy and notes all the original required exits. The inspector then observes that, in fact, two of the exits have been filled in with a masonry wall. Those exits are still needed and, for the occupancy to operate, the owner is required to restore the exits.

Sometimes building owners make changes to a building that compromise structural safety or fire and life safety, and it may be necessary to review the original plans and other records, such as approved code modifications and inspection records. So despite the challenge created by an ever-increasing mass of historical records, an efficient means of managing the challenge is necessary. There are a number of alternatives to storing paper files today.

Many communities facing limited storage space for records have resorted to technology solutions. An intermediate step for small to medium communities is to have images of documents transferred to microfilm or microfiche. The process requires an in-office camera designed for microfilm to photograph the documents or the availability of a business that provides that service. Usually, the fire prevention bureau has one or more film readers, so staff can retrieve documents. The readers have the ability to send a file to a separate printer so a hard copy can be produced. Medium to large communities that have proportionally larger document storage needs may use other technological solutions, such as dedicated document management systems. The systems have become more common as electronic storage has become less expensive. Today the use of cloud data storage businesses to secure electronic documents is becoming more common.

The affordability of electronic storage and its nearly limitless capacity make it an efficient method to safely store critical records. Most of the current electronic-document-management systems run on an office personal computer and are compatible with the fire prevention bureau's suite of business computer software and applications. Documents can be searched with keywords or numbers such as street name, address, and building name.

A well-planned information technology (IT) system that includes scheduling or calendar programs, word processing, databases and spreadsheets, email, voice mail, fax, and electronic imaging is an example of the minimum needs in an efficient fire prevention bureau. Pressure for high performance in turning around building plans for review is intense. The business process of the fire prevention bureau must be nearly parallel in capability and preferably integrated with the building official's office. It has become an industry standard to ensure a turnaround time guarantee to builders seeking approval to build. Many building officials guarantee a 24-hour turnaround of plans and inspection appointments, so an independent fire prevention bureau must offer the same or come under scrutiny as the reason for building and development delays.

TRAINING OF FIRE SERVICE PERSONNEL FOR PLAN-REVIEW FUNCTIONS

A community that cannot hire a fire protection engineer or obtain such services from the state or county can still have an effective plan-review program by preparing other personnel for such duties. In most fire departments, at least one person, possibly a firefighter, can be trained in reading construction drawings and in fire protection requirements. This person provides at least some representation of the fire service in the plan-review process.

Many avenues are available to fire departments in acquiring training for staff, including through the National Fire Academy, ICC, and NFPA, for example. Any one of these agencies offers general training as well as more specific training that leads to higher proficiency in managing a community's fire code needs. Volunteer departments can solicit help from their communities, possibly recruiting someone with engineering credentials. A retired engineer, for example, may be interested in contributing time to training and work activities as a volunteer. Unless the fire department has a person already on board or can hire or vote into membership a volunteer who has the skills and certification needed, the department should expect a member to spend a couple of years to become proficient at plan review.

ASTM International
■ a voluntary consensus standards organization founded in 1898 that publishes standard fire tests of building construction and materials and recommends practices for preparing fire test standards. Formerly known as American Society for Testing of Materials.

American National Standards Institute (ANSI)
■ a clearinghouse for voluntary safety, engineering, and industrial standards nationally.

CODE REFERENCE FOR PLAN REVIEWS

The fire protection engineer or other person carrying out plan-review responsibilities should reference an authoritative document to provide justification for decisions. Not every circumstance will be found in any particular code. Thus, referenced authoritative documents may include the national consensus codes, state or locally adopted codes, published engineering decisions such as those from **ASTM International** or the **American National Standards Institute (ANSI)**, or manufacturer's specification manuals. As previously mentioned, many building code requirements are applicable. If fire safety–related portions of building codes are to be enforced by a fire department review person, prior arrangement must be made with the building official to prevent conflict and misunderstanding.

CORRELATING PLAN REVIEW WITH INSPECTION

Plan reviews are only as effective as the inspection follow-up to ensure full compliance with requirements noted by the reviewer. Requirements set down by a plan reviewer may not always be carried out by either the architect or the builder (Figure 4-2).

Unfortunately, omissions cannot always be readily detected in the field because the deficiency may be enclosed during construction and may not be revealed until a fire occurs. Inspections should be scheduled as necessary to coincide with the completion of various stages of construction so that the inspector can be at the site to detect any deviations from specifications. For example, providing a builder or fire protection system installer with a written list of the fire prevention bureau's requirements ensures a clear understanding of what the customer should expect. (In this case the term *customer* refers to anyone needing plan approval or inspections from the fire prevention office or building department, such as the builder, architect, developer, or a sub-contractor.) Some fire prevention bureaus have several preplanned requirement forms that differ according to the type of construction and/or fire protection system. Included are blanks to be filled in for the dates of inspections, critical areas of concern such as penetration in fire-rated barriers, and notes to the builder from the plan reviewer indicating needed corrections.

COOPERATION IN PLAN REVIEWS

Cooperation among several agencies is absolutely essential for the purposes of the plan review. Agencies include the building department, the fire department, state-level agencies, and any other governmental agencies involved in the plan review. In a number of jurisdictions, all agencies having responsibility for plan review and

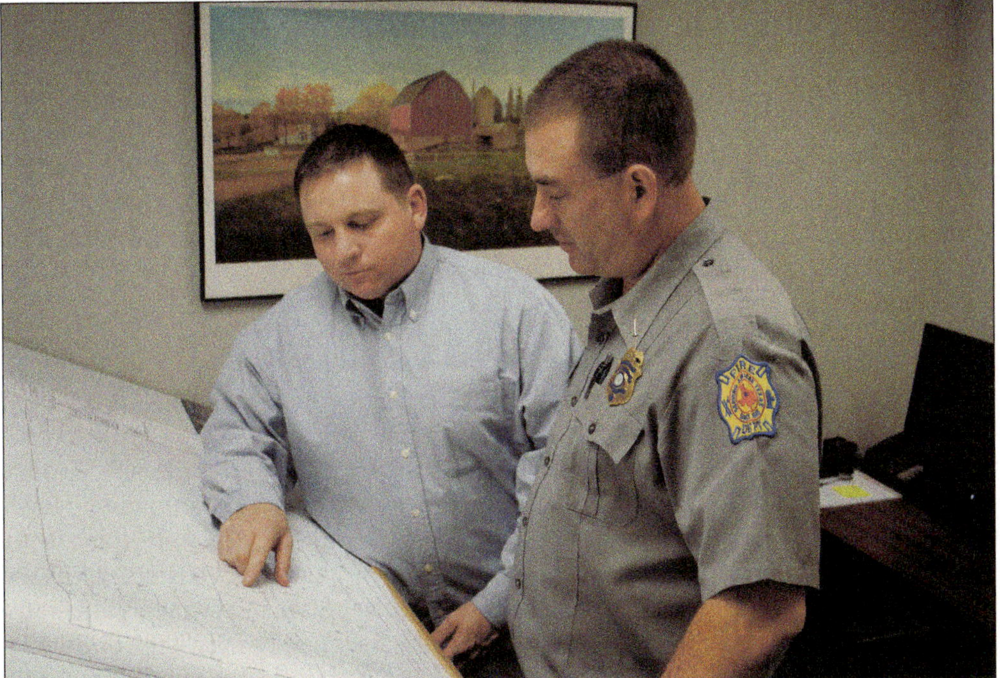

FIGURE 4-2 Plan-review meetings with builders or fire protection systems installers can offer an opportunity to clarify the expectations of the fire prevention code. Plan reviewers often provide specific notes and checklists to reinforce requirements.

Billy Morris Director of Fire Science Cisco College

code enforcement meet periodically to review plans and to discuss problems with architects, contractors, and others involved in a project. Some jurisdictions also include zoning, planning, traffic control, environmental services, and law enforcement representation.

Many fire departments participate in "one-stop shops," where local review agencies function under one roof to facilitate the expedient processing of regulatory provisions. As described previously, there is great pressure on local governments to provide the least amount of delay in processing building and development approvals. One-stop shops and highly integrated building and fire code offices reduce some of the steps and time spent by builders to gain approval for plans.

Control of Sales and Use as a Means of Fire Code Enforcement

Controlling the sale and use of products or services is another means to enforce fire codes. Most fire codes contain provisions that give the fire prevention bureau responsibility for controlling sales of explosives, fireworks, fire extinguishers, fire alarm systems, and other related devices, as well as general control of the handling of sales of gasoline and other hazardous materials. Maine and several other states require that all electrical devices sold in the state bear the seal of a nationally recognized electrical testing laboratory.

CONTROL OF GASOLINE AND OTHER HAZARDOUS COMMODITIES

Control of the sale and use of gasoline is found in practically all fire prevention codes. Included are required procedures relating to the handling, storage, and dispensing of gasoline. Specific requirements include the type of container into which gasoline may be pumped, how gasoline can be dispensed, the prohibition of smoking, and requirements that engines be turned off during dispensing. Control of sales also includes the sale, handling, and storage of other dangerous and hazardous commodities.

To assist firefighters, some jurisdictions require the clear identification of hazardous materials storage locations. Generally, the fire prevention bureau is responsible for enforcement of such mandates.

CONTROL OF FIRE ALARM SYSTEMS

Control of the sales of fire alarm systems, including smoke alarms, is a public safety function. A fire alarm system is sold to the public for fire safety purposes and must be properly designed to function in the event of an emergency. From a public safety standpoint, the average person may have no way of knowing if a fire alarm system is functioning properly until there is an emergency. In the case of practically every other type of household equipment, including toasters, TVs, and DVD players, the buyer can try the product at home to see if it will perform satisfactorily. Some communities require that all fire alarm systems in the community be approved by the fire department.

CONTROL OF EXPLOSIVES

Controlling the use, storage, and sale of explosives has long been covered in fire prevention codes in hundreds of municipalities, as well as in fire codes of counties and states. Explosives storage is considered in zoning regulations as well. The control of the sales of explosives may include requirements for identifying people who purchase explosives and recording the names and addresses of purchasers. Control in this field is also a function of the federal government through the explosives regulations enforced by the Bureau of Alcohol, Tobacco, Firearms, and Explosives. Many communities authorize and assign responsibility to the fire prevention bureau for investigation of explosions in their fire prevention code. Both the International Fire Code and the Uniform Fire Code outline this authority.

Security Measures

Explosives have been a useful tool in the development of the United States but have also been a very destructive and violent weapon. Notable bombings with explosives resulting in the deaths of innocent people can be found as far back as the 1880s. Explosives are still a tool of choice for highly destructive and violent intent as exemplified by the bombings at the Murrah Federal Building in Oklahoma City and the World Trade Center bombings in the 1990s.

Regulations may limit the sale of explosives to bona fide purchasers and may thus preclude acquisition of explosives by those intent on using them for terrorist and other destructive purposes. Theft then becomes a major problem, and secure storage of explosives must be ensured. Regulations for storage magazines usually specify theft-resistant and fire-resistant construction and security locking devices. In some instances, security fencing may also be required. As terrorist activities intensify, the need for stringent security measures increases as well.

Control of Fireworks

Another field of responsibility usually assigned to the fire service—and that often involves police agencies as well—is fireworks control. This control responsibility probably has as many varieties of approach as any subject in the fire prevention field. There are many differences in requirements relating to fireworks from state to state and even within counties in a given state. Fireworks are a danger primarily to personal safety, though they may also present a hazard with respect to fire safety. As in the field of explosives, training is needed to carry out an effective fireworks control program.

Control Through Limitation of Sales

Control of fireworks is usually effected through limitations on sales and ownership permits. However, a major problem exists when one jurisdiction permitting sales adjoins a "closed" jurisdiction. The public can simply cross the state line or municipal boundary to purchase fireworks and then return home. Public sale of certain types of fireworks is permitted in unincorporated areas of a number of states, but major cities and towns in those states do not permit possession of fireworks without a permit. Enforcement of local regulations is difficult where items may be purchased nearby. Public education programs have been used in an effort to control the problem.

Legal Ban on Fireworks

Fire service organizations have been in the forefront of attempts to require a legal ban on the public sale or use of fireworks. In fact, fire service organizations are probably primarily responsible for restrictive legislation in the majority of states. An organization that has had a major interest in fireworks legislation is Prevent Blindness America. The society is especially interested in eye injuries, but its effective work has not been limited to that interest. The organization has successfully worked in a number of states to promote prohibitive legislation. It also cooperates with the National Fire Protection Association to enhance public awareness of the dangers associated with use of fireworks.

Public Fireworks Displays

Although public fireworks displays with permits are allowed in practically all jurisdictions, it is incumbent on the fire marshal's office to monitor those events. Provisions should be made for closing down the display in the event of problems. In some jurisdictions, individuals who set off displays are required to be certified by the fire prevention agency.

Enforcement of Regulations

Enforcing explosives and fireworks regulations brings fire prevention bureau personnel into contact with members of the public with whom they usually have little contact. The explosives control field, for example, involves a considerable amount of contact with the construction industry. A close relationship must be maintained with these individuals to ensure safe operations.

Likewise, fireworks control, if effective, may create a number of adversarial situations. The fire prevention inspector may be placed in the position of having to seize fireworks and file juvenile petitions against middle school or high school students. Probably most other contacts by the fire prevention official with people in that age group are of a positive nature. The confiscation of fireworks that have been legally purchased in another jurisdiction is not well received and definitely places the enforcement official in an adversarial situation. As a result, fire safety personnel may be reluctant to follow through with full enforcement of the regulations.

Structural Control as a Means of Fire and Building Code Enforcement

structural control
▪ a means of enforcement effected through the plan-review function coupled with inspection before occupancy to be sure that all fire safety requirements have been fulfilled.

Structural control of buildings—another means of enforcement—is effected through the previously mentioned plan-review function coupled with inspection of structures, before and upon occupancy, to ensure that all fire safety requirements have been fulfilled. This inspection may be conducted jointly with building officials. Structural control includes inspection of materials, all ancillary systems and equipment, and any other factors relating to the building.

FIXED MANUFACTURING EQUIPMENT

Fixed equipment used in processing and manufacturing is included in structural inspections for fire code enforcement. However, the equipment may have design features that can lead to fires. Such hazards must be identified in the inspection process. Occasionally, the degree of fire danger may increase as the equipment is used.

INTERIOR FINISHES

Control of a structure includes the interior finishes. This important aspect of fire protection often is overlooked during fire safety inspections. Proper evaluation requires analysis of materials used in a structure to determine that the materials will not readily burn or, if they are ignited, that rapid spread of fire will not ensue.

FIXED FIRE PROTECTION EQUIPMENT

Inspection of fixed fire protection equipment is another part of the job of control of structures. Tests and inspections at the time of occupancy are quite important. As in other phases of fire protection inspection, personnel assigned to this responsibility should be familiar with the testing procedures necessary to ensure proper operation. Fire prevention codes extensively cover requirements for the maintenance and testing of fixed fire protection equipment. This level of work effort, on the part of both building owner and fire prevention personnel, is intended to prevent the failure of protection when it is most needed, during a fire.

REINSPECTION

Control of structures must be maintained by continuing inspection of buildings to note changes that take place from time to time. The addition of a partition or a change in the swing of a door, as insignificant as it may seem, could make a major difference at the time of a fire. All such changes can be noted in a structural control inspection program. Reinspection of buildings should also include periodic inspection and supervision of the testing of fire protection equipment.

TEMPORARY CONDITIONS

Inspection for the control of structures must recognize temporary conditions that can impede egress in the event of a fire, such as exits or corridors blocked by furniture, cabinets, or other materials. It must also determine that exits are not being locked when the building is occupied. This kind of inspection is similar to inspections carried out for control of occupancy, a discussion of which follows.

ELECTRICAL EQUIPMENT

Another phase of the inspection procedure related to the control of structures is checking electrical equipment and devices. Although fire safety inspectors do not usually carry out a full range of electrical inspection duties, some elements of electrical safety are noted as a matter of routine. Frayed wiring, obvious overloads, unidentified circuits, extension cords, arcing or overheating equipment, and inadequate fusing are conditions within the scope of the fire inspector's responsibilities. Again, it is vital to have a good working relationship with the building department, which usually offers electrical inspection services. Fire prevention bureaus can establish referral procedures that involve the right inspectors from other disciplines to check out concerns about electrical hazards and other systems.

Control of Ignition Sources

Without an ignition source, fuel is just another material that is not burning. Consider the availability of smoking materials that can start an unwanted fire. The U.S. Centers for Disease Control and Prevention (CDC) reported that in 2010 an estimated 45.3 million adults in the United States smoked cigarettes.[2] A 2013 NFPA report, *The Smoking-Material Fire Problem,* noted that smoking materials started 90,000 fires in the United States in 2010 and a similar number in 2011.[3]

Smoking material is a very prevalent ignition source, but there are ways to control it. When evaluating a facility, fire inspectors and others involved in identifying and recommending ways to reduce risk often use a process called "fault tree analysis" to find risks such as ignition sources.

Fault tree analysis can help solve a problem by predicting relationships and actions that can lead to some failure. A simple description of a fault tree analysis starts with understanding a system (storage of waste materials from a furniture store, for example) and predicting any failures (an unwanted fire in the store) that would impede success of the business. This process then requires adopting positive actions (new safety procedures) or eliminating any destructive factors (inappropriately discarded smoking material) that could play a role in an unwanted outcome (fire). Lastly, it involves deciding what actions to take to avert the failure.

A business such as a furniture store can generate large amounts of bulky waste. Sometimes the materials need to be stored until they can be disposed of. A fire on the loading dock of the furniture store, for example, could endanger customers and employees and cause loss of business, which would be considered a failure that should be eliminated. Through use of the fault tree analysis a list of actions can be developed to eliminate the risk, including, for example, strictly prohibiting smoking in the area; posting signs that warn of the fire hazard and that remind people that no smoking is allowed; creating checklists for store managers that include inspecting the area for evidence of smoking, such as cigarette butts or ashes; offering periodic training and awareness for employees that emphasize the fire hazard associated with the combustible waste and need for vigilant care with smoking materials; and providing a lounge or designated smoking area away from the hazard.

Use of problem-solving systems such as the fault tree analysis can become a learned behavior to the point where the fire inspector uses them without expending a great deal of energy. This behavior is learned through training and experience. More detailed information can be located in *NFPA 550: Guide to the Fire Safety Concepts Tree.*

FIRE SAFE CIGARETTES

In addition to electrical and other energy-related sources of ignition, cigarettes and other types of smoking material have long been responsible for starting fires. Too often such fires result in fatalities. In fact, cigarettes are faulted with causing, over the years, the deaths of thousands of people and dollar losses from fire in the billions. The loss of life and property was compelling enough to motivate the U.S. Congress to pass the Cigarette Safety Act of 1984 (P.L. 98-567). The legislation directed research and funding to reduce the cigarettes' ignition of upholstered furniture. The next step was taken in 1990, when Congress passed the Fire Safe Cigarette Act of 1990 (P.L. 101-352), which ultimately created a standard test method to determine the ability of cigarettes to cause ignition.

NIST conducted the research that led to the development of ASTM Standard E2187-02b and later ASTM E2187-04, Standard Test Method for Measuring the Ignition Strength of Cigarettes. In 2000 New York State passed the first legislation in the world to require limits on cigarettes with reduced ignition propensity. ASTM E2187-04 is featured as the referenced test to measure how manufactured cigarette are judged to be in compliance. Now Canada and the United States have implemented "reduced-ignition-propensity-cigarette" legislation. Future research and analysis will seek to reveal the impact that reduced-ignition-propensity cigarettes have had on fire loss.

Control of Occupancy as a Means of Fire Code Enforcement

Control of occupancy is an enforcement method that determines the adequacy of egress and provides fire safety personnel on scene to ensure an orderly evacuation in the event of fire. It includes determining and posting occupancy capacities within structures, coupled with the periodic presence of fire prevention bureau personnel to ensure that the posted capacity is not being exceeded, and that all exits are usable. Occupancy also includes contents storage concerns.

control of occupancy ■ an enforcement method that controls occupancy load through determination of the adequacy of egress and through provision of fire safety personnel on scene to ensure an orderly evacuation in the event of fire.

Fire departments and fire prevention bureaus must be aware of the types of events occurring in their communities. Attention to media that advertises events provides leads to events that could exceed the threshold of safety. Venues with occupancies classified as assembly are frequently sought for a variety of shows, conferences, and special events. What an event planner brings into the occupancy should be scrutinized, as high-flame-spread contents and activities such as unregulated cooking can raise the risk to visitors. Collaboration and expectations that are well communicated to the assembly managers can help keep risk to a minimum.

Requiring permits for events is one way to keep the fire prevention bureau in the information loop. It also helps to make fire prevention bureau staff aware of how they can detect potentially crowded events, such as concert by a popular performer.

Fire prevention and fire department staff should also be aware of a growing industry of illegal occupancies for assembly, such as events in buildings that are not equipped to safely support events. Examples of illegal event venues include large single-family houses and spaces in warehouse districts. Some are able to escape the notice of enforcement officials. Fire department personnel may become aware of these events when they run calls nearby and should especially question an unusually large crowd in a warehouse district late at night.

An additional example of control of occupancy is the assignment of trained crowd managers to public events to see that aisles are not blocked and that overcrowding does not occur. The Life Safety Code has very specific training and responsibility requirements for crowd managers and assembly occupancy staff. Crowd manager training includes knowledge and understanding of techniques to manage crowds, such as metering, a strategy sometimes used to control the rate of arrivals and the degree of crowding. Training includes how to keep aisles and egress paths clear, how to complete fire safety checklists, and how to maintain the number of occupants within the established occupant load limits. It also involves learning to operate fire protection equipment on the premises and training employees of the places of assembly in the use of fire extinguishers.

Legal Aspects of Fire Code Enforcement

In compliance and abatement procedures for fire safety code enforcement, the ultimate goal is the improvement of the facility from a fire safety standpoint rather than imposition of fines or other punitive measures. There is little satisfaction in having fines levied or jail sentences imposed if in fact the facility remains unsafe. Court actions are time-consuming and should be considered a last resort rather than a primary means of enforcement in fire prevention.

Some fire prevention personnel have the authority to issue official notification of violations. In some cases fines may be involved. This procedure enables the fire department to immediately correct hazardous conditions and is somewhat analogous to a police officer's issuing a traffic citation or ticket to an errant motorist. The logic is that a fire code violation is endangering others.

In addition to serving legal orders, fire departments sometimes obtain injunctions or other sanctions to bring about prompt correction of a serious fire hazard. Fire department personnel who have enforcement responsibilities should meet with the attorney for their jurisdiction to learn about specific avenues available for enforcement. It is far better to obtain this information in advance of an incident than to wait until life is in jeopardy in an occupancy. In addressing target risks, such as safety compliance for bars and nightclubs, some communities have used an effective task force approach that partners key enforcement agencies such as the fire prevention bureau, the police, jurisdictional attorneys, and the health department. Enforcement teams can monitor occupancies with random visits during the peak times of operation. Providing meetings, workshops, and regular communications for the occupancy managers and staff can effectively clarify expectations, increasing voluntary compliance and potentially reducing the need for active enforcement efforts.

Decisions handed down by several state supreme courts have a direct bearing on fire code enforcement. These are in addition to U.S. Supreme Court decisions involving proper identification of inspection personnel and permission to inspect. The U.S. Constitution does not address fire code enforcement; states have that power.

In a number of such cases, courts have held that building and fire code inspection and enforcement are discretionary functions of other branches of government and entail no statutory duty of due care for the benefit of individual citizens or specific groups.[4] In one case, a state supreme court remarked that a failure in code enforcement represented no more legal liability on the city's part than there would be if a judge failed to make a decision or made a wrong one. Likewise, the city is generally not held liable for economic losses that may occur as a result of enforcement failures.[5]

In one case, the city was held not subject to recovery of damages in the deaths of and injuries to several motel guests who were trapped during a fire. Their escape was thwarted by improper stairway enclosures that were in violation of the code. Negligence was alleged on the part of the city inspectors in issuing the permit for renovation that included the improper enclosures.[6]

Courts have also taken the position that general negligence principles govern municipal liability when a building official is negligent in the performance of prescribed duties, regardless of the "public" nature of those duties. These decisions cover the performance of fire code officials as well and mean that fire prevention personnel cannot rely on any generally perceived immunity to protect them or their employers from successful litigation.

An appeal heard by the Kentucky Supreme Court, *Grogan v. Kentucky*, held that the City of South Gate and the Commonwealth of Kentucky were not liable for failure to enforce fire safety codes. The court held that government was free to enact laws without exposing the taxpayer to liability for failure to enforce those laws. The suit related to the Beverly Hills Supper Club fire that took 165 lives on May 28, 1977.[7]

On the other hand, courts have held that local government can be held liable for its actions when it issues a certificate of occupancy for a building with knowledge that the structure contains blatant safety and fire code violations. In one case, a court held that the town had a duty to refuse to issue a certificate of occupancy under such conditions and that such a rejection did not require the exercise of judgment or weighing of competing factors.[8]

Hiring an incompetent inspector can likewise cause a municipality to be liable for damages. A village was subject to liability for its board's alleged negligence in having hired such an inspector. The inspector issued a building permit, construction began, and the inspector then revoked the permit for unspecified reasons. The court held that a municipality is just as liable as a private employer for harm caused by its negligence in hiring an incompetent servant.[9]

Another decision held that the owner of a motel was responsible for building a structure in accordance with codes. A fire victim's representative had sued the motel for construction defects contributing to the fatality.[10]

Condemnation of Unsafe Structures

In addition to activities usually thought of as fire prevention measures, statutes in many states and ordinances in a number of cities provide fire and building enforcement officials with the power to condemn unsafe buildings. This power is granted to permit the fire or building official to order the demolition of structures found to be hazardous if the owner fails to take corrective action. Generally, the structures must be of combustible construction for the provisions of the statute or ordinance to be imposed. It is presumed that a combustible building could endanger other properties if it caught fire.

Most condemnation statutes and ordinances require that the building's location would endanger other properties if the building were ignited. In some jurisdictions, exposure to a telephone line or other utility line is sufficient to invoke the provisions of the statute.

These ordinances and statutes usually contain a requirement that the structure be unoccupied. There is also a requirement that the structure be open for trespass and therefore so situated that individuals can enter the building without difficulty. In some cases, arrangements may be made for boarding up the building, to secure it from the possibility of entry by trespassers, except by force.

The reasoning behind condemnation provisions is that the properties contribute to the fire risk within a community because they are open for trespass and are of combustible construction that endangers other properties. Such structures are open invitations to arsonists and are fire hazards. Many fires are set by juveniles in such locations, and in recent years vagrants setting fires to keep warm or cook.

Most condemnation ordinances contain requirements for the demolition or repair of the structures in question, followed by the initiation of a legal suit to recover costs or other expenditures by the government entity, possibly including the recovery of the court costs incurred. In these situations, owners had a responsibility to keep the structure safe but failed to do so, leading to the action.

Other property owners in the neighborhood usually welcome the condemnation of structures as a fire prevention measure. Most buildings that are candidates for condemnation are unsightly and therefore lower property values in the neighborhood. Drug users and other undesirable occupants frequent some. When the property is owned by an estate, it may be difficult to locate all the heirs in order to serve them the required notice. No work should begin on the demolition of the building until it has been definitely determined that the legal owner has been contacted.

The Uniform Fire Code and International Fire Code both have provisions giving fire officials authority to reduce the hazard posed by a vacant structure. Many steps can be taken to reduce risk, but planning and communication with a building's owners or agents should be a priority. Discovery of how the building will be used in the future is critical. For instance, proposed redevelopment often take time for planning and approvals.

There are many ways to make vacant buildings safer by boarding them up and providing security as well as removing combustible contents and debris or vegetation around the exterior. Fire prevention bureaus need to keep tabs on activity around vacant buildings, such as nuisance fires, and be proactive in reducing occurrences so complacency does not allow fire frequency to increase. Again, early contact with owners and agents with written notification of fire safety code expectations can reduce the fire risk posed by vacant buildings.

A team or task force approach can be helpful in managing the risk of unsafe structures. Such a group should involve building officials because they have responsibilities that include evaluating structures and have some of the earliest knowledge of any land-use plans. A system could be set up to monitor vacant structures and the task force could meet or otherwise communicate and share status information about potential unsafe structures. Other potential members of a task force could include, for example, representatives from the tax assessment office, police department, and public works. Public works may ultimately be responsible for the boarding up and security of a structure and the disposal of materials if the building is deemed too unsafe to remain in place.

Fire Safety Considerations in Special Occupancies

Fire prevention and life safety measures in health-care facilities and high-rise structures have all too often come about only when public concern has been aroused in the aftermath of tragic fires. These areas are now receiving more attention at all levels of government.

LONG-TERM-CARE AND NURSING HOMES

Nursing home safety is an example of the evolution of fire prevention and protection requirements pertaining to a specific type of occupancy. Because the potential victims are generally **incapable of self-preservation** and are physically or mentally incapable of fending for themselves in an emergency, the relative safety of nursing homes is based almost entirely on the effectiveness of code requirements.

incapable of self-preservation
■ a condition associated, for example, with reduced cognitive ability, inability to ambulate independently, illness, and medical treatment that makes a person unable to take appropriate actions when faced with life-threatening hazards.

The term *nursing home* has become a generic name for any facility in which aged and disabled individuals are housed. There are federally recognized terms for these facilities based on the level of care rendered. State terminology varies as to title for some classifications. A common denominator is that, without proper safeguards, such facilities have a potential for serious fires involving fatalities.

The nursing home field is a developing one because of the increase in average life expectancy. Programs such as Medicaid and Medicare that provide financial assistance to older adults have likewise encouraged the development of nursing homes and other facilities for the extended care of the aged. The Medicaid program is primarily designed to assist those who have low incomes and resources, whereas the Medicare program provides assistance to those who have participated in the Social Security program through employment and made contributions during their working years. All these factors have combined to create a great demand for nursing home care for older adults.

As the industry expanded, some communities gave little thought to the fire safety of the structures. In many cases, old, large dwellings that were no longer suitable for family occupancy were converted into nursing homes. In some jurisdictions, code officials equipped with adequate codes required the provision of automatic sprinkler protection and other safeguards. In some cases, strict requirements, such as sprinkler protection, enclosure of vertical openings, and other safeguards, caused nursing home operators to situate their facilities outside the city limits in an area where regulations were not so stringent.

A number of fires that resulted in multiple fatalities have occurred in substandard structures in practically every section of the United States. In some states, fires brought about modifications in fire prevention regulations, including requirements for automatic sprinkler protection and other fire safety features for all other nursing homes. Several jurisdictions found, only after a major nursing home fire, that there was no agency in the community or state that had responsibility for fire safety enforcement in such facilities.

The interest of the federal government in fire safety increased because of the previously mentioned programs for federal financial assistance to many older adult citizens. The U.S. Congress, as well as the U.S. Department of Health, Education, and Welfare (now the Department of Health and Human Services), recognized in the late 1960s that many of those citizens were not being housed in safe environments. A number of states were not ensuring a maximum degree of safety. In addition, there was disagreement about the best methods for providing adequate safety.

Fires resulting in multiple fatalities continued, and pressure to do something increased. Eventually, in 1971, federal regulations went into effect. They require automatic sprinkler protection and other fire protection safeguards in many nursing homes. These regulations have had a great impact in jurisdictions that had not previously required such protection. As could be expected, many of the same appeals were heard again, and it was necessary to discontinue funding in a few cases to enforce compliance.

Currently, the federal program is in full effect for skilled- and intermediate-care facilities. Enforcement of the regulations is federally funded and carried out by state personnel, usually through the state fire marshal's office or state health department.

BOARDING HOMES FOR OLDER ADULTS

Adequate safeguards for older adult residents of board-and-care homes, or boarding homes, as they are generally called, are far from universally provided. Many older adults are housed in old buildings of frame or ordinary construction, and very few are provided with automatic sprinkler protection or any other meaningful protective devices. A U.S. House of Representatives report on health-care facilities made recommendations with respect to this problem, but progress in adequate safeguards will probably be limited until public interest is sufficiently aroused. The interest shown by Congress may serve to hasten action, but it is unfortunate that additional lives will probably be lost before adequate safeguards are required in all jurisdictions. The problem is aggravated because control in some states for facilities of this type is somewhat limited.

Congress authorized a special study of health-care facilities as a project of the Committee on Government Operations of the House of Representatives, which was carried out in 1972. The committee's report strongly recommended increased life safety for all health-care facilities, including extended-care facilities and so-called boarding homes that house primarily older adults.[11] The committee recommended the continued use of federal financing as a lever to improve fire safety in health-care facilities.

During a recent 7-year period, the National Fire Protection Association documented 67 deaths in approximately 3,300 nonconfined board-and-care facility fires.[12] The data property code for residential board-and-care facilities in the

National Fire Incident Reporting System (NFIRS) has not been in place long enough to allow an effective analysis of trends, but with the evidence of over 3,000 fires and 67 deaths, it appears to be an area of fire safety concern. Residential board-and-care facilities include long-term care facilities, halfway houses, and assisted-care housing facilities. The code excludes nursing facilities that provide 24-hour nursing care for four or more persons. It is anticipated that the new data collection focus will allow the identification of hazards associated with these facilities and may suggest improvements in procedures for reducing risk.

RESIDENTIAL SPRINKLERS

The development of mandatory requirements for residential sprinklers has taken a long and tortuous route. Although the value of automatic sprinklers has been recognized for more than 100 years in industrial, commercial, and institutional occupancies, little credence has been given to the application of the same principles to residential occupancies. The one exception was the use of automatic sprinklers in hotels and large apartment houses.

In the late 1970s the city of San Clemente, California, pioneered requiring automatic sprinklers in residences located beyond usual response limitations for fire apparatus. The city declared that residences built more than 5 miles from the nearest fire station had to be equipped with residential automatic sprinklers. The organizations developing the standards and the sprinkler manufacturers had to adjust to this new demand. The sprinkler industry began producing suitable equipment on a limited basis while a standard was developed and published.

The old concept of "a man's home is his castle" has made this type of regulation difficult to implement. In addition, many homebuilders' groups, including the National Association of Home Builders (NAHB), have strongly opposed any measure that would bring about an increase in the construction cost of new dwellings. They feel that even a minor increase in cost will reduce the number of people who are able to buy new homes.

A number of jurisdictions have adopted regulations similar to those in San Clemente. In fact, San Clemente and its home state of California have subsequently changed regulations to require sprinklers in all new residences. Currently, all model building and fire safety codes in the United States require residential sprinklers in new homes.

In 2013 with the addition of two more cities, the State of Illinois reached 90 communities to pass residential fire sprinkler requirements in one and two-family homes.[13] Such a requirement had been imposed in many jurisdictions in Maryland, including the two high-population centers of Montgomery and Prince George's counties and the capital city, Annapolis. Maryland has since adopted the 2008 International Residential Code, which makes residential sprinklers a requirement in all new residential construction.

A 2008 change to the International Residential Code requires residential sprinkler installation upon adoption of that code. Cobb County, Georgia, a leader in residential sprinkler installation, uses construction or economic incentives rather than a specific ordinance as the method for securing sprinkler installations. The building official, with fire department support, has granted a number of incentives in multifamily residential construction requirements that result in such buildings being constructed less expensively with sprinklers than without. Walls between multifamily residential units are required to resist fire from spreading from one unit to another, and so are lab tested for up to a 2-hour endurance. A 2-hour fire rated wall is more expensive than a 1-hour rated wall because it is made of more robust materials and takes longer to install. Historical evidence supports that sprinklers would suppress a fire before it can spread and gives the fire prevention official confidence that 1-hour rated walls could be used in place of 2-hour fire rated walls, resulting in a cost saving. Practically all contractors building multifamily residences in Cobb County are using that alternative means of construction, thereby effecting a savings in construction costs. Of course, the life expectancy of the building and its occupants from a fire safety standpoint is much greater, and the apartment units are more marketable.

A report by the Fire Protection Research Foundation noted that incentives could be a key to wider implementation of residential fire sprinklers. The report documents examples of incentives and their value. The group and the magnitude of incentive value received varied among developer, builder, and owner. Incentives were categorized as financial incentives, such as reduced impact fees or property taxes; on-site design flexibility, such as reduced fire ratings on building assemblies; and off-site design flexibility, such as increased fire hydrant spacing and longer approach to dead-end streets. Values of the incentives were reported to have been in a range from around $100 to over $1,500 on average per building lot.[14]

Canadian cities are also requiring sprinkler systems in new construction, including residences. Westmount, Quebec, imposed this requirement in 1989, as did Quebec City. Vancouver, British Columbia, later became the first major North American city to require sprinklers in all new buildings.

An example of code development was the successful effort by the Florida Fire Chiefs Association and other organizations in that state to secure legislation that retroactively requires automatic sprinklers in certain existing hotels, motels, and time-share occupancies. With the support of the hotel and motel associations, League of Cities, and other interested groups, the fire chiefs promoted this legislation. It came about without a multifatality fire in Florida as an impetus. However, the sponsoring agencies recognized the value of the tourism industry to the state and the devastating effect that such a fire could have on business. Previous attempts by the sprinkler industry to get this type of legislation enacted had failed, partly because the attempts were perceived as self-serving. A concentrated effort, plus a reasonable time period for compliance, brought about enactment in 1983.

Another consideration in the development of codes is evaluating new conditions, innovations, occupancies, and processes. As an example, the advent of the "hotel–hospital" brings a challenge from a code application standpoint. These facilities are designed for individuals who no longer require close hospital care but have recovered to the extent that they can exist in a homier, less expensive setting near the hospital. The latter feature is considered important because of the possibility of a relapse. The code developer and enforcer is concerned about appropriate fire safety measures for this new type of occupancy. The question before them now: Is this entity considered "institutional," or is it considered "residential"?

Summary

Adequate fire and building codes are not sufficient to provide fire safety. Plan review by competent people is essential. Enforcement is also needed. Fire code enforcement can be accomplished through control of sales and use, especially items such as explosives, gasoline, and fireworks. Structural control is a means of fire and building code enforcement. It includes fixed manufacturing equipment, interior finishes, and fixed fire protection equipment. Communities have the legal right to impose fire safety and code requirements. Often, it takes a disastrous fire to impact fire code enforcement. On rare occasions, some communities are proactive such as those that have adopted comprehensive sprinkler ordinances to protect their residents.

Review Questions

1. The Station nightclub fire in West Warwick, Rhode Island, was caused by:
   a. careless smoking.
   b. indoor pyrotechnics.
   c. grease fire.
   d. electrical short.
2. The ideal person to conduct plan reviews for a fire department is a(n):
   a. fire protection engineer.
   b. senior fire inspector.
   c. architect.
   d. battalion chief.
3. Which department in most cities needs to cooperate closely with the fire department on plan-review issues?
   a. Police
   b. Zoning
   c. Building
   d. Finance
4. Which is not typically an item controlled by sales for fire safety?
   a. Explosives
   b. Fireworks
   c. Fire alarms
   d. Cigarettes and cigars
5. Fireworks that some states permit to be sold commercially are known as _____ fireworks.
   a. open
   b. closed
   c. conservative
   d. consumer
6. Inspections of structures before and upon occupancy are considered part of:
   a. structural control.
   b. educational control.
   c. plan review.
   d. fire safety maintenance.
7. Court actions for compliance should be considered:
   a. always necessary.
   b. the responsibility of the fire chief.
   c. a partnership with the police.
   d. a last resort.
8. Nursing home occupants are uniquely at risk because:
   a. they are not capable of self-preservation.
   b. the owners seldom cooperate.
   c. the staff is not trainable.
   d. most fire department response is inadequate.
9. The fire protection engineer within a fire department is usually:
   a. a civilian.
   b. nonuniformed.
   c. holding a staff rather than line position.
   d. all the above.

10. Technology to reduce the ignition hazards of cigarettes was enabled by:
    a. new chemicals.
    b. the tobacco industry.
    c. acts of Congress.
    d. less flammable upholstered furniture.

End Notes

1. "One Year After Deadly Club Fire, Few Changes Seen," *Tallahassee* (FL) *Democrat,* February 15, 2004.
2. Centers for Disease Control and Prevention, *Vital Signs: Current Cigarette Smoking Among Adults Aged ≥ 18 Years—United States, 2005–2010,* Morbidity and Mortality Weekly Report 2011. Accessed December 3, 2013, at http://www.cdc.gov/mmwr/preview/mmwrhtml/mm6035a5.htm?s_cid=%20mm6035a5.htm_w
3. John R. Hall, Jr., *The Smoking-Material Fire Problem* (Quincy, MA: National Fire Protection Association, July 2013), p. i.
4. *Trianon Park Condominium Association v. City of Hialeah*, 468 So. 2d 91 (1985, FL).
5. *Island Shores Estates Condo. v. City of Concord*, 615 A.2d 629 (1992, NH).
6. *Hoffert v. Owatonna Inn Towne Motel, Inc.*, 199. N.W. 2d. 158 (1973, MN).
7. *Grogan v. Kentucky*, 577 S.W.2d 4 (1979, KY).
8. *Garrett v. Holiday Inns, Inc.*, 58 N.Y.2d 253 (1983, NY).
9. *Lockwood v. Village of Buchanan*, 18 Misc. 2d 862 (1959, NY).
10. *Northern Lights Motel v. Sweeney*, 561 P.2d 1176 (1977, AK).
11. *Saving Lives in Nursing Home Fires: Sixteenth Report by the Committee on Government Operations*, Issue 92, Part 1321 of 92d Congress, 2d session, 1972.
12. Ben Evarts, *Structure Fires in Residential Board and Care Facilities* (Quincy, MA: NFPA Fire Analysis and Research, May 2012).
13. "Ninety Communities in Illinois Have Home Fire Sprinkler Requirements with Additions of Hazel Crest and Niles," PR Newswire, 2013. Accessed April 29, 2013, at http://www.prnewswire.com/news-releases-test/ninety-communities-in-illinois-have-home-fire-sprinkler-requirements-with-additions-of-hazel-crest-and-niles-203662261.html
14. National Fire Research Foundation (Newport Partners, LLC), *Incentives for the Use of Residential Fire Sprinkler Systems in U.S. Communities—Final Report,* 2010, pp. ii–iv. Accessed December 8, 2013, at http://www.nfpa.org/~/media/Files/Research/Research%20Foundation/Research%20Foundation%20reports/Suppression/rfincentivesresidentialfiresprinklers.pdf

CHAPTER

5 Fire Safety Inspection Procedures

CandyBox Images/Fotolia

KEY TERMS

common fire hazards, *p. 96*
exit interview, *p. 94*
exposures, *p. 90*
International Fire Service Training Association (IFSTA), *p. 86*
National Fire Protection Association (NFPA), *p. 86*
See v. City of Seattle, and Camera v. City and County of San Francisco, *p. 89*
special fire hazards, *p. 96*

OBJECTIVES

After reading this chapter, you should be able to:

- Identify the components of a thorough inspection.
- Identify the frequency of occupancy inspections.
- Understand the right of entry and which pertinent case law applies to entry.
- Recognize who should accompany the inspector on an occupancy inspection.
- List how the inspector should conduct the inspection of the occupancy.
- Discuss the importance of the exit interview.
- Identify the legal and moral responsibilities of the inspector.
- Recognize the importance of the reinspection.
- List some hazards specific to types of occupancies.
- List three common types of possible conflicting agency inspections.

Inspection is a key function in the enforcement of fire laws and regulations. Duties of the fire prevention inspector call for knowledge and competency that can be acquired only through proper training and education in code requirements and

National Fire Protection Association (NFPA) ■ an organization that aims to reduce the risk of fire and other hazards by offering consensus codes and standards, educational and training materials, and fire safety advocacy.

International Fire Service Training Association (IFSTA) ■ an organization that identifies needed areas of training materials and supports those needs through the development and validation of the training materials for the fire service and other related services.

inspection procedures. Other requirements are good judgment, keen observation, and skill in dealing with people. Fire prevention inspections may be conducted by all levels of government including federal, state, and local, as well as businesses and private organizations, including insurance companies, universities, and hospitals. As a representative of the fire department, the fire prevention inspector has the opportunity to build good public relations and to educate the public about the need to observe the rules for preventing fires.

This chapter outlines the basic principles that apply to the actual inspection procedures, whether conducted for control of structures, for control of occupancy, or for a combination of purposes. For more detailed information regarding inspection procedures, publications of the **National Fire Protection Association (NFPA)** and the **International Fire Service Training Association (IFSTA)** are excellent sources.

Preparation for Inspection

Preparation should include a positive attitude on the part of the inspector. The inspector must know why the inspection is being conducted and what to look for during the inspection. This is an important phase of preparation, one that, unfortunately, is often overlooked in training and in planning for inspection programs. Another requirement is the acquisition of the necessary equipment, such as flashlights, cameras, notebooks, data loggers, and suitable clothing. In some departments the inspector is also required to carry manuals, codebooks, and other publications that may have a bearing on the work. Laptop, tablet, smartphone, and palm computers are useful for this purpose. When connected to a wireless network, access is possible to Internet resources, records archives, and the fire prevention bureau's resource files. It is always a good idea to have appropriate materials in the car for reference when needed.

KNOWLEDGE OF CODES

One of the prime requisites for inspector competence is a thorough knowledge of the local fire safety code, as well as other relevant municipal or state codes. The inspector may not necessarily be fully conversant with each point in the fire prevention code but should be familiar enough with the provisions and requirements to make general observations and recommendations during the inspection tour, with subsequent detailed study where necessary. At the site, the inspector might refer to code material carried in the vehicle. In some instances, special problems may require more detailed analysis or consultation with supervisors after the inspector has returned to the office.

TIME REQUIREMENTS

Under no conditions should an inspection be undertaken when the inspector is pressed for time and is thus unable to make a thorough inspection. An official inspection, however haphazard, may lull the owner or property manager into a false sense of security and subject the inspector and fire department to legal action. It is important to plan and schedule inspections with a realistic expectation of what can be accomplished.

The need for thoroughness cannot be overemphasized. All areas of the structure must be covered adequately, even when entry to certain spaces is difficult.

INSPECTION FREQUENCY

Inspection frequencies vary considerably from one jurisdiction to another. Availability of personnel is a major factor in determining the amount of time between inspections. Another factor is the perceived degree of hazard in the occupancy. For example, institutions are usually inspected more frequently than storage warehouses. The potential for finding deficiencies may be a factor in scheduling inspections even within a given class of occupancy. Those facilities in which violations are usually found may be scheduled for more frequent visits than those where fire safety problems are seldom found. It is helpful to be able to collect and analyze deficiency data for inspected occupancies to understand the types of problems that continually come up. Knowing this kind of information helps in the planning of the frequency of inspections.

After the Hamlet food processing plant fire on September 3, 1991,[1] North Carolina's building code council added a mandatory fire safety inspection schedule to the state's Fire Prevention Code. The state recognized that the fire, a very preventable one, might not have taken 25 lives if inspections had been required. The Hamlet facility had never been inspected in its 11 years of operation under Imperial Foods. In North Carolina, local governments are now responsible for conducting inspections of occupancies.

North Carolina also requires annual inspections of hazardous, institutional, high-rise, assembly (except churches and synagogues), child care, and common areas of residential occupancies. Assembly occupancies with occupancy potential of fewer than 100 must be inspected every 3 years. One- and two-family dwellings are exempted. Industrial and educational occupancies, except public schools, must be inspected at least once every 2 years. Public schools are required to have at least two fire inspections per year. Business, mercantile, and storage occupancies and churches and synagogues require inspections at least every 3 years. Local jurisdictions may require more frequent inspections.[2]

Another measure instituted subsequent to the Hamlet fire was the North Carolina State Department of Labor's adoption of the NFPA's Life Safety Code as part of its regulations. These provisions cover most places of employment.

INSPECTION PRELIMINARIES

Before leaving the office or fire station to inspect a facility, the inspector should review reports of any past inspections at that facility, which are of material value in conducting an inspection and must be considered in an evaluation of the facility. A property owner is certainly justified in being alienated by a fire inspector who fails to recognize past efforts. Most property owners are proud of their efforts to correct deficiencies, and their work should be recognized. By reviewing past reports, the inspector can also get a better idea of the construction and general layout of the building, and the time taken to review the file will be rewarded by a more expedient inspection.

If another inspector made the previous inspection, it may be necessary to confer with that individual to ensure a clear understanding of all items listed in the inspection report. If the inspector going out on the job finds some previously reported items confusing, the property owner is also most likely to have difficulty understanding what is meant.

MANNER OF DRESS

While the standard for many years has been a fire department uniform, there is now more interest in plainclothes dress for fire safety inspectors. An example is the increasing use of polo-like shirts and cargo slacks. Both can be purchased from suppliers of uniforms and work clothes, and they offer functionality, comfort, and safety. Many fire departments have begun to use them as their day-to-day fire operations uniform. It is helpful to have the department's insignia and name imprinted or embroidered on the shirt as an extra means of identifying an inspector. Some municipalities have no uniform requirements, or the fire department is a part of an agency that does not traditionally use uniforms. There is no evidence that inspectors who do not wear uniforms are any more or less effective. The main concern is that an inspector present professionally and that the clothes do not detract from the purpose of the inspection.

TAKING NOTES

The inspector should be prepared to carry a notebook, clipboard, recorder, or computer to make notes of all findings during the inspection. After a full day of inspecting many types of facilities, the resulting notes should be orderly and easily understandable. It is becoming common to use computer applications to record inspections on forms that include drop-down menus and checklists for the required information. This approach can reduce the amount of note taking, but there should be a text box available for remarks that are not covered elsewhere. In spite of technology, there will always be a need to write down some observations and questions for later research. These notes might cover items that need correction and their specific locations. For example, such an item might be an OSY valve in the basement to the left of the stairs that needs to be locked in the open position.

Well-prepared inspectors use professional notepads, checklists, or computer software to make notes during an inspection. The evolving use of smart phones, handheld palm-size computers, and tablet computers offers hundreds of innovative applications for note taking, photo and video integration with text, and even the conversion of handwriting to typed text. These new and quickly developing technologies offer efficiency by combining multiple functions in one electronic device.

Identification and Permission to Inspect

The approach to and contact with the owner or operator of the property are important factors in establishing a favorable climate as well as a legal basis on which to conduct an inspection. Fire safety inspections of businesses, under the Fourth Amendment of the U.S. Constitution, may be conducted only if a representative of the business or the premises consents or if there is an official administrative warrant to inspect. Fire prevention bureau or fire department inspectors may enter public areas to contact a building representative within the public space. On entering the structure, the inspector should immediately present identification to personnel in the building. If at all possible, the manager, superintendent, or other person in authority should be notified. Such identification ensures compliance with the guidelines of ***See v. City of Seattle,* and *Camera v. City and County of San Francisco***, the U.S. Supreme Court decisions pertaining to the right of entry, presentation of adequate identification, and request for permission to make an inspection. The decisions handed down in those cases make it imperative that fire inspectors comply with all constitutional requirements.

See v. City of Seattle, and Camera v. City and County of San Francisco ▪ significant court cases holding that inspectors must not enter nonpublic areas of a business unless permission is granted; these decisions expanded the boundaries of the Fourth Amendment of the U.S. Constitution beyond just dwellings.

The cases referenced substantiate the property owner's right to refuse the inspector entrance to nonpublic areas except in situations involving imminent danger. Imminent danger or what is referred to as "exigent circumstances" in the law covers those conditions where it would be unreasonable for the fire department to have to obtain a warrant to enter a building, for example, when smoke is issuing from the structure and people may be inside. However, the inspector must obtain a warrant to enter nonpublic areas when refused permission to enter.

On occasion, a delay is caused by the absence of those in authority at the property when the inspector arrives. The inspector should wait for the person in authority and then present identification and request permission to conduct the fire prevention inspection. Under no circumstances should the fire inspector make an inspection without proper identification and permission. When permission is granted to make the inspection, the inspector customarily requests that a representative of management go along on the inspection tour. In most cases, management readily agrees.

Occasionally, a fire department's response to an alarm creates an opportunity for the observation of fire code violations. Response to a false alarm gave the Toledo, Ohio, Fire Department an insight when it found sprinkler system plumbing to be rusty and in need of replacement. The subsequent violation notice was appealed on the grounds of an allegation that the department's entry for the inspection was in violation of provisions of the U.S. and Ohio constitutions, which require permission or search warrants. The appeal was denied because the entry was not for the purpose of finding evidence for a criminal prosecution.[3]

The Inspection Tour

For many years, it has been customary to start an inspection at the top of the structure, including a walk out on the roof to check roof structures with respect to fire safety. The inspector should check the elevator machinery room and should note whether the roof is used for occupancy, which is becoming popular with roof-top dining, bar areas, and other uses where people may assemble. If the roof is in use, a check should be made to ascertain that proper egress exists. The inspector and the building representative should observe the condition and construction of the roof, as well as make note of any fire protection problems at that level. This is also a good opportunity to view adjacent structures to evaluate any exposure hazards to the structure being inspected. (At some point, the inspector should also look around the outside of the building.)

The inspection party should then move to the top floor of the building and make a thorough inspection of each section of that floor. Progress through the building should be systematic, and the same principles should be applied down through the entire structure, including the basement levels.

The inspector should be aware of any diversionary tactics used to keep him or her out of certain areas. The building representative may try to steer clear of hazardous areas or those that are not up to standard. The motivation for attempts to reroute the inspector is not necessarily avoidance of hazardous conditions but may be a desire to keep secret certain proprietary processes, machinery, or research work. Industry often has a great deal at stake in the development of new processes or products. An unthinking fire inspector can cause a great deal of hardship to an industry by passing on such proprietary information. The building representative serving as the escort may need to obtain special permission for the fire inspector to go into certain spaces in which such processes are being carried out. This may mean a delay. However, it is important that the inspector wait for permission.

In making the tour through the building, the inspector should make it a point to think about everything noted. One of the major weaknesses in fire inspections is that many inspectors look at hazards, unusual conditions, or exit deficiencies but somehow do not register them mentally as problems. Studies of fires with major loss of life are replete with examples of this phenomenon. This is not necessarily a result of inadequate training. Understanding the significance of what one observes is a skill that takes practice to develop.

Conversation with the building representative serving as the escort can cause the inspector to overlook important items. If the person accompanying the inspector is carrying on an extended conversation about sports activities or television programs, for example, the inspector can easily walk by a hazard without taking notice.

The building must be considered both for possible development of a fire within it and also from the standpoint of potential spread of fire to other structures. Careful consideration must also be given to anticipating occupant reaction in the event of a fire.

Although it is impossible to determine every factor that must be considered in examining a building, the salient points of inspection may be classified as follows: exposure hazards, potential fire hazards outside and inside the building, potential fire causes, potential avenues for spread of fire, water supply, life safety features including means of egress, and alarm and extinguishing systems. Other outside considerations may include for example, exit doors free of obstructions, fire department connections for the sprinkler or standpipe in good working order as well as availability of fire hydrants, fire department access, and clearly marked address as required by the authority having jurisdiction.

exposures
■ areas of a structure where its proximity to other structures or combustibles could endanger the building in question if involved in a fire.

POSSIBLE EXPOSURES

The inspector should take full note of all building additions or accessory spaces, as well as hazardous locations and **exposures** (the proximity of other structures or combustibles that might, if involved in a fire, endanger the building in question).

Time spent this way can be most helpful to the inspector in assessing the overall fire safety of the facility.

The inspector should also take into account the construction of adjoining buildings, the relationship of windows and other openings between the two structures, the existence of protection such as fire sprinkler systems within the exposure, and the height and area of the structure being inspected. Corrective action in the form of outside fixed protection, wire-glass windows, or other safety features might be recommended in the fire prevention report as a result of this review of conditions outside the building.

POTENTIAL FIRE HAZARDS

The fire safety inspector should check for the existence of combustibles that, if ignited, could cause the spread of fire. No blanket statement such as "Improve general housekeeping" can be recommended in this connection because of variables that must be considered. Hazards that present a danger of fire include flammable and combustible liquids and gases, oxidizing agents, explosives, and acids. Other fire hazards include materials subject to spontaneous heating, combustibles, and explosive dust, as well as ordinary combustible materials in which rapid fire spread might be anticipated. The last category includes foam rubber, plastic packing material, cotton batting, paper, and other finely divided material. Management of waste disposal is a critical factor that should be considered. Recommending a more frequent pickup or disposal schedule could reduce accumulation of a hazardous amount of waste. The inspector must realize that all such hazards may exist as a necessary and integral part of the operation of given occupancies.

The fire inspector may expect to reduce the number of hazards in some locations. In others, however, the solution may involve providing fire sprinkler protection, separating fire hazard areas to reduce fire spread, or other steps designed to reduce the possibility that a serious fire could develop. The inspector must be resigned to the fact that it is impossible for most occupancies to operate without having at least some potential hazards on the premises.

The fire inspector must be able to evaluate problems created by combustibles in light of the potential fire spread. Certainly, a furniture factory, for example, cannot be told to eliminate all combustible materials. The extent of storage of combustibles in any location within the factory is directly related to protection available, construction, and possibilities for ignition of materials. The trained inspector must use judgment in deciding on appropriate action in each inspection. Combustibles are essential to the operation of many industries and mercantile establishments.

POTENTIAL CAUSES OF FIRE

During an inspection, the inspector should observe sources of ignition that could cause a fire and take steps to minimize this danger. Again, it must be realized that it is impossible for the fire safety inspector to effect safeguards that will eliminate all possible fire causes.

Sources of ignition that are frequently responsible for fires include electrical arcs and sparks; lightning; static electricity; friction through grinding, polishing, cutting, and drilling operations; various types of chemical reactions; and open flames. Many other sources exist, including improper heating devices, smoking, and use of torches. Sources of ignition are present in most occupancies.

Certainly, an inspector should exert every effort to reduce to an absolute minimum any potential causes of fire in a structure. Conditions can change very rapidly. In fact, it is entirely possible for abated cause conditions to return almost as soon as the inspector has left the building. For example, a fire inspector may admonish an individual for smoking in a hazardous area only to have the person light up again as soon as the inspector is out of sight. Fire prevention inspection and control should therefore emphasize the need to restrict ignition sources to actual operating requirements, to provide safeguards in installation and handling, and to clear surrounding areas of combustible materials in which fire could spread.

POTENTIAL SPREAD OF FIRE

During the inspection tour, the inspector should consider the ways in which fire might travel within the building. Structural factors that might contribute to the spread of fire are the location and condition of elevator shafts, stair towers, light wells, laundry and trash chutes, pipe chases, conduit openings, vertical and horizontal ducts, and dumbwaiters. Other structural factors include combustible ceiling tiles, combustible interior finishes, windows that might permit vertical spread of fire outside the building, exterior combustible walls, shafts, and any other unprotected openings. Access to adequate means of egress is an important consideration.

Major causes of horizontal fire spread include the absence of fire division walls and partitions in basic construction and lack of or improper fire doors in existing fire walls. Combined with those structural factors, the presence of combustibles and hazardous materials, poor stock storage methods, and lax housekeeping practices contribute to the spread of fire. The inspector with a fire suppression background is in a much better position to accurately evaluate fire spread factors than is one who is not familiar with firefighting procedures and fire spread characteristics.

A proper fire safety inspection of the conditions just mentioned requires a thorough knowledge of building construction as well as an understanding of the travel characteristics of fire. Unfortunately, many fire inspectors do not consider possible means of fire spread within a structure. On their trip through the premises, their primary purpose is noting observable conditions that might cause a fire, such as the presence of combustibles, careless smoking, locked exits, and careless handling of flammable liquids. The mark of a trained, competent fire inspector is a comprehension of fire spread possibilities inherent in the design and construction of the building.

WATER SUPPLIES AND EXTINGUISHING SYSTEMS

In touring the building, the fire inspector should check on water supplies for fire suppression purposes and observe standpipe connections, sprinkler systems, and any other fixed extinguishing systems. Both model fire prevention codes in the United States clearly require that fire protection systems be maintained and be in operating condition at all times (Figure 5-1).

Determining that these systems are functioning properly should be a part of every inspection. Although many inspectors do not consider this an inspection function, it should be mandatory in every fire safety survey. The fire inspector should not open the valve or otherwise test the device but should request that the building

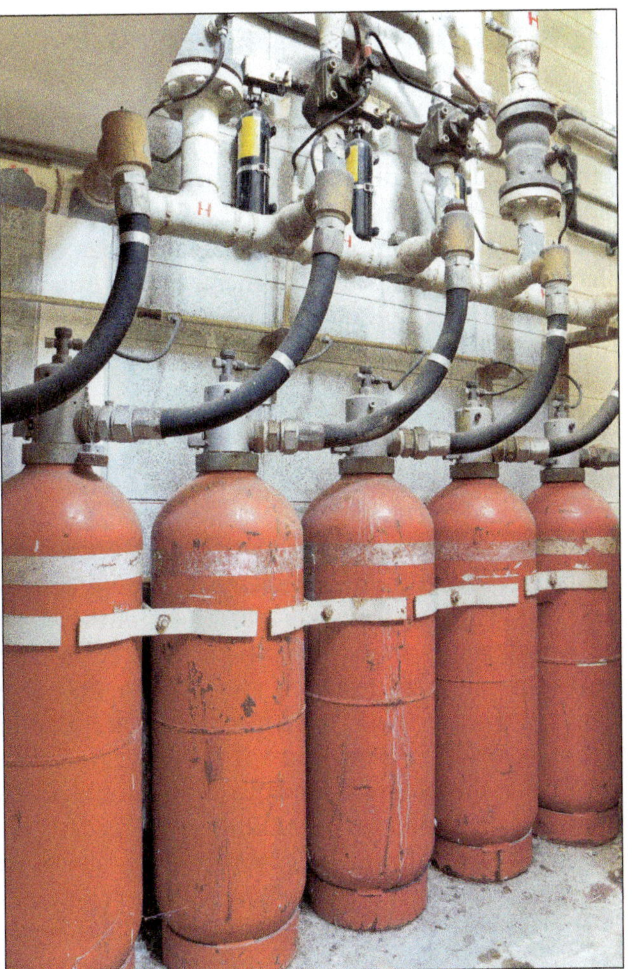

Noppasin/Shutterstock

FIGURE 5-1 Fire inspectors are responsible for and must be knowledgeable about the wide range of fire protection systems they will see on inspections.

operator or owner's representative do so. The facility operator should assume the liability in actually operating a valve. The operator has a responsibility to satisfy the fire department that all fire protection devices on the property are operable.

Correcting Violations During Inspection

It does not make sense for an inspector to record items, such as blocked fire doors, without getting immediate corrective action. As many violations as possible should be corrected during the inspector's visit.

Another example is permitting a sprinkler system valve to remain in a shut position rather than having it opened. However, the system valve may sometimes be shut for some valid reason. This should be checked out, verified, and provision made to return the system to operating condition as soon as possible.

In the prepared report, the fire safety inspector should mention items that were corrected during the inspection as a reminder to management to avoid repetition of violations. Inspectors may specifically require building management to certify that a fire protection system is in compliance and operational in accordance with the accepted standards for the protection system in question.

Discussing Inspection Findings

In some cases it is desirable for the inspector to discuss specific problems as they are discovered on the tour. In other cases it is better for the inspector to record findings and then discuss the locations at the completion of the inspection with those responsible for correction of any violations. There is no reason to go into a detailed explanation of violations with a person who is not responsible for compliance, such as an usher in a theater. However, if the safety director of an industrial plant is accompanying the inspector, the violations should be pointed out and discussed during the tour.

The Exit Interview

exit interview
▪ the closing portion of a fire safety inspection, during which a review is conducted with a person of authority or a responsible management representative to ensure that corrections are understood and will be made.

If someone other than the owner or property manager has accompanied the inspector on the inspection tour, arrangements should be made for an **exit interview**. This review is conducted with a person of authority or a responsible management representative to discuss the findings of the inspection and to get assurance that violations will be corrected. If the person in authority is not available for this interview, the inspector will have to come back on another day. In any event, the exit interview is an important part of the inspection and should not be omitted.

During the interview, the inspector should advise the management representative of all the problems noted during his or her visit. If it is necessary to return to fire headquarters for further technical information, arrangements should be made to discuss these matters in person or by telephone.

It is unwise to state requirements orally that are not later substantiated in the prepared report. The property owner could make changes based on orally imposed requirements only to find that the inspector, after returning to the fire station, has decided they were not necessary.

If, on return to headquarters, the inspector finds justification for making any substantial changes in recommendations after management has reviewed them in the exit interview, the changes should immediately be communicated to the person interviewed, by either telephone or personal visit.

Requirements imposed by the fire inspector should be so clear and concise that they adequately portray to an untrained person the true picture of what is needed for compliance. In explaining the conditions during the exit interview, the inspector should give the owner or property operator the opportunity to offer suggestions on corrective actions. Compliance with requirements is often more readily attained if the owner or property manager comes up with the solutions to the problems.

Technically, it is sufficient for the inspector merely to state that a certain thing must be done because the code requires it under a given section. However, an oral statement may not give the property management much satisfaction or understanding of the real need to comply with the fire code for life safety and protection of the property. By explaining as clearly as possible the reasons behind provisions of the code and how they apply, the inspector will make code enforcement much more palatable to management. It may not be necessary to go into great detail on all points, but at least a summary explanation should be given.

The inspector must refrain from conveying a threatening attitude about the necessity of fire code compliance. The inspector who settles for "I don't know why but the code says it has to be done" leaves the impression of not being well trained or of not being fully convinced of the importance of the inspection.

The property owner may be required to spend a great deal of money to comply with fire inspection requirements, so the fire inspector has a professional obligation to be sure that the stated requirements will correct the violation without unnecessary expenditures. One foolish or impractical recommendation can offset the justifiable aims of the inspection. Management that is not pleased with the inspection will take full advantage of any improper or impractical requirement in attempting to downgrade the entire report. The inspector should clearly identify items that are merely recommendations as opposed to those that are mandatory.

How the exit interview is conducted has important public relations aspects for the cause of fire prevention. A knowledgeable and friendly presentation of the findings can go a long way toward making management receptive to compliance with the code.

Report of Inspection

A report of all inspections must be transmitted to the property owner or occupant. Not long ago the standard for inspection reports was a preprinted form or letter. Now a vast array of computer inspection applications that are components of fire department data-reporting systems or stand-alone fire inspection applications are becoming more prevalent and could change the way inspections are reported. These systems are used with public safety communications systems or commercial business systems on handheld computers, computer tablets, and laptops that are connected directly through wireless communications. While these systems are catching on and offer benefits, it is still pretty standard to document inspections on paper.

Once an inspection has been completed, a written or computer-generated report is left on the premises with the property owner or occupant. Report preparation depends on the resources available in the fire department. Preprinted forms that require the inspection information to be written in or computer forms filled in during the inspection and then printed out at the end of the inspection are two time-saving efficiencies.

Often the inspector has good reason to return to the fire department office to refer to documents for technical information, to give more thought to the requirements specified, or to confer with supervisory personnel on the more complex problems. One disadvantage, of course, is the greater time lag between the completion of the inspection and the time the property owner receives instructions for compliance with inspection requirements. A fire always may occur during this intervening period, although this probability is somewhat remote. Therefore it seems worthwhile to take the time to prepare a report, with copies of all correspondence to be kept for reference for subsequent inspections or possible legal action.

It is extremely unwise to make a statement or to include in a report words to the effect that the building is free of hazards. Even the most competent fire inspector may not be aware of many of the conditions that could bring about a serious fire, because they are concealed behind walls, in enclosures, and in other locations where observation is impossible. One way to cover such possible omissions is to include in the report some general statements about or reminders of areas of ongoing concern, such as maintenance of fire protection systems and housekeeping. In any case, documentation is critical, and if a hazard is not written up, or an inspection form has not been completed, the inspection may just as well have not occurred.

Some larger fire departments have fire prevention bureau personnel who are specialized in certain phases of fire prevention work, for example, complex industrial or institutional occupancies or flammable liquids and gases, electrical conditions, and fixed fire-extinguishing systems. The fire inspector on general assignment may consult with a specialized inspector in preparing the report, or a supervisor may review the inspection report.

Reinspection and Procedures to Enforce Compliance

Normal inspection procedures followed by most municipal fire departments include a reinspection to ascertain compliance with the fire inspection report. An approximate date for the reinspection should be indicated in the report.

The enforcement of compliance with the fire code requires the submission of a report of violations and recommendations for corrective action. The report usually allows a specific number of days for completion of the work. The dates for anticipated completion should take into account the complexity of the job, the availability of materials, and other related factors. For example, the correction of housekeeping violations is normally required in a very short time, whereas the installation of a stair tower is given an extended length of time.

A reinspection is conducted at the end of the specified time to determine whether the work is under way or has been completed. In some cities, the reinspection is considered the beginning inspection for any legal action taken to enforce compliance. If corrections have not been made, the property owner is given a formal legal order to complete the work within a given number of days or face a possible fine or jail sentence. Normal procedure includes issuing a citation or obtaining a warrant for the arrest of the individual who has not completed the corrective requirements within the time specified in the legal order.

It is wise to have a uniform time schedule for correction of typical violations, for example: an out-of-service fire alarm or sprinkler system, 72 hours with fire watch provided in the interim; a damaged or missing smoke alarm, 24 hours; or an imminent threat to life, immediate correction.

common fire hazards ▪ hazards found in practically all occupancies, such as those related to normal use of electricity and heating equipment, as well as to the existence of everyday combustibles.

special fire hazards ▪ hazards a fire inspector associates with conditions unique to an occupancy, such as combustible liquids or gases in an auto repair shop.

Classification of Hazards

In the category of hazards, a breakdown separates those referred to as **common fire hazards** from those referred to as **special fire hazards**. The common hazards are those found in practically every occupancy, such as those related to normal use of electricity and heating equipment. Smoking is included in this category. Special hazards are those the fire inspector associates with conditions unique to a given occupancy.

Hazards in Various Types of Occupancies

Reviews of occupancies subject to inspection may note particular hazards and causes characteristic of each type. Familiarity with these hazards and causes helps the fire safety inspector know what to look for in each type of occupancy.

INSTITUTIONS

Institutions are facilities that house or treat people who would be unable to leave in an emergency without assistance. Examples include hospitals, nursing homes, and prisons. Long considered the highest risk occupancy from a life safety standpoint, the institution may present a number of unusual situations. A hospital, for example, contains flammable liquids and gases as well as moderate quantities of combustibles in storage areas. From a life safety standpoint, the institution presents a major evacuation problem in that a majority of the occupants are not usually able to respond to a fire alarm unaided. Personnel within the facility, aided at a later time by fire department personnel, will need to evacuate such a structure or move patients to a place of safety.

Common hazards found in such facilities include mattresses, bed clothing, and other combustibles normally found in patient rooms. The clothing worn by patients may become a hazard if it is not made of a fire-retardant material.

Other potential fire causes are magnified under institutional conditions. The use of a small amount of flammable gas, necessary in certain medical procedures, becomes a major hazard around electrical appliances. The possible results of an explosion and fire in an institution may be much greater than in an industrial occupancy. In addition, the increased awareness of and concern about the spread of bloodborne infections and other communicable pathogens have significantly increased the installation of hand sanitizer dispensers. The alcohol-based sanitizer can be flammable, should be evaluated for safe placement, and should never be located in evacuation corridors.

PLACES OF PUBLIC ASSEMBLY

Another classification generally associated with life safety problems is a place of public assembly, an occupancy generally associated with large concentrations of people—and the possibility of a stampeding crowd. These circumstances make fire code inspection of this type of occupancy especially important. Conditions in a large public auditorium are quite different from those in an amusement park or a restaurant.

A small fire in the stage area of an auditorium that would have little chance of spreading could cause sudden and rapid evacuation and large loss of life under conditions of improper egress facilities, improper lighting, or overcrowding. The fire inspector conducting a routine inspection may visit a place of public assembly during a time of very limited occupancy. It may be difficult to visualize the conditions that exist when the place is filled with people. Public assembly occupancies include art galleries, grandstands, tents, libraries, exhibition buildings, funeral homes, churches, skating rinks, transportation terminals, dance halls, and theaters.

Large civic centers and arenas capable of seating thousands of people for special events can present significant problems. Hazards and causes vary considerably in this broad occupancy classification. However, the one overriding concern is to ensure a means of immediate egress in the event of fire or other emergency.

In one major incident, a small amount of liquefied petroleum gas, estimated at approximately 3 pounds, was responsible for taking 74 lives in the Indianapolis Coliseum explosion in 1963.[4] In this case, the gas cylinder was located in a storage room adjacent to a refreshment stand inside the structure. Gas apparently escaped and was ignited by a heater for popcorn that was located nearby. The explosion in the confined storage area ripped out heavy concrete and wreaked havoc in the building. The storage of liquefied petroleum gas might have been acceptable from a fire safety standpoint in some other occupancy, but not in a public assembly occupancy. This structure was protected by a well-equipped fire department, and structural deficiencies did not contribute to the disaster.

In public assembly occupancies, fire safety inspectors have usually been successful in getting corrective action to eliminate major fire problems. Constantly recurring conditions that increase the likelihood of problems are overcrowding and locked exit doors. The tendency of some groups to attempt to enter facilities of this type without proper authorization may bring about severe overcrowding and may result in locking of exits to preclude illegal entry.

In restaurants, major fire hazards result from deep-fat frying and grease buildup in ventilation ducts. Protection systems are available for hoods and ducts. However, many older restaurants have not installed such protection. Carelessness in cleanup procedures can result in a hazardous accumulation of grease, which provides immediate fuel if ignited. Hood and duct fire protection systems and the maintenance and cleaning such systems need may require more frequent inspections or self-inspection and submission of a certification of compliance.

The fire in West Warwick, Rhode Island, that took 100 lives in The Station entertainment venue focused attention on public assembly occupancies. That February 20, 2003, fire resulted in major code changes in Rhode Island.[5] As after other tragic fires, the resulting code changes often occur primarily in the state in which the incident happened. The state of Rhode Island and the town of West Warwick each agreed to pay $10 million to the survivors of the fire.[6]

MERCANTILE OCCUPANCIES

Mercantile occupancies usually contain many fire hazards and causes. The trend toward large undivided areas for product merchandising has increased the possibility of major fire losses in retail stores. There is also the possibility of delay in notifying the fire department because a fire may not be readily detected in a large area.

Common hazards increase in number in such merchandising outlets, and the counters may contribute a great deal of fuel to any fire. Fortunately, a majority of large retail stores, generally those in excess of 15,000 square feet, are required to install fire sprinkler protection at the time of construction.

In considering the overall potential loss from a fire in a mercantile establishment, the fire inspector should consider water damage. Because merchandise that has been damaged by water usually cannot be sold, the inspector might point out measures that could be taken to reduce the possibility of such damage. This kind of advice may not be required under the code. However, it can be quite helpful in reducing loss in the event of a fire. It may be feasible for the store operator to place merchandise in the storeroom on skids or pallets up off the floor to reduce water damage potential.

Flammable liquids and gases may also be found in mercantile occupancies. Common causes of fires, such as defective heating equipment and smoking, may be factors in such occupancies. Retail stores may have special hazards associated with the occupancy. An example is a hardware store, where there is a good possibility that black powder may be present. Another example is a store where charcoal briquettes are sold. This material is subject to spontaneous heating. Hazards of contents in mercantile establishments may not be readily recognized. The presence of furniture containing foam rubber was evidently a major factor in the rapid fire spread in the 2008 Sofa Super Store fire in Charleston, South Carolina, that took the lives of nine firefighters.[7]

STORAGE OCCUPANCIES

Closely associated with the mercantile occupancy is the warehouse. If material that is stored poses a special hazard, such as large quantities of aerosols or combustible fibers, additional fire problems are created. Normally, warehouses and other storage occupancies are subject to in-and-out human occupancy. Therefore there is not as much chance that a fire will be started by human carelessness. However, this advantage may be offset by the possibility of delayed discovery because of the limited time of human occupancy.

Problems of water damage become even more acute in the warehouse, and serious consideration should be given to provisions that will preclude water damage if sprinklers or hose lines are operated. This consideration may not prevent all water damage, but at least damage in areas away from the immediate fire area will be abated.

Outdoor storage facilities such as lumber yards and coal yards serve the same purpose as warehouses. Such facilities may be subject to fires that result in large losses and may even be more readily ignited because of their accessibility.

OFFICE BUILDINGS

Office buildings constitute yet another type of occupancy. Common hazards are usually fairly easy to find in such structures. Common causes are also quite prevalent. A major problem in this class of occupancy is the tendency to overlook combustibles. Personnel occupying an office, as well as even some fire inspectors, may ignore the fuel potential in the paper accumulations in office buildings.

The office occupancy also has a higher degree of life safety hazard than is usually recognized. Although the density of population is not as great as in a public assembly occupancy, in many large office buildings the density is fairly close to that found in a school or hotel.

Consideration must also be given to the inclusion of other types of occupancies in a building that is primarily an office building. A restaurant and its inherent fire problems, for example, may present a life safety and fire hazard in a structure that would otherwise be considerably safer.

RESIDENTIAL OCCUPANCIES

The residential occupancy is where the largest number of fire fatalities occur. Generally, municipal fire department inspectors are not responsible for legal inspection of one- and two-family private dwellings or of individual apartments, although many cities provide such inspections in a voluntary program. The fire department inspector is responsible for the inspection of public areas of larger residential occupancies, such as apartment houses, hotels, motels, dormitories, and rooming houses. The inspector and the owner/operator are responsible for making sure that people occupying such structures are housed safely. Multifamily structures occupied as condominiums pose a greater enforcement problem.

Hazards normally found in publicly occupied facilities are usually common ones. Likewise, fire causes generally fall into the same classification. The structures, by nature of their operation, contain combustible materials in the form of room furnishings. Causes are present in the form of smoking materials, heating appliances, lights, television sets, and other electrical appliances. It is impractical to eliminate those hazards and causes, although certainly, electrical devices can be held to a safe limit.

INDUSTRIAL PLANTS

Fire hazards and causes are so varied in industrial occupancies that a uniform checklist is difficult to develop. However, a few of the hazards common in industrial occupancies are friction of moving parts, sparks from static and metal-against-metal contact, overheating of equipment, chemical reactions, and highly flammable materials. Explosions are a periodic problem. Explosions can be caused by fires in proximity to exposed tanks of flammable liquid and compressed gases; fire exposure to chemicals such as ammonium nitrate and other oxidizers; dusts such as saw dust, sugar, and corn milling that can become airborne; and the manufacture of highly explosive materials such as fireworks. Fires can start by friction, sparks, and overheating created by mechanical equipment in the industrial process, such as conveyor belts, shredding equipment, and gears. Fires also can start as a result of chemical reactions and spontaneous heating. Industrial activities such as welding, metal cutting, grinding, and the melting of metals for casting can start fires. There is also the hazard of industrial raw materials, the products themselves, and even their waste materials, all of which may be abundantly available fuel sources for a fire.

CandyBox Images/Fotolia

FIGURE 5-2 The potential for dust explosions in an industrial facility can be easily overlooked during a fire safety inspection.

Industrial occupancies offer all the hazards of any other occupancy plus the risk added by unique and sometimes dangerous processes (Figure 5-2). Throughout history very dramatic examples of industrial fires have left a critical impact on communities through pain, suffering, loss of life, and sometimes loss of a community's main employer. While there is no guarantee that tragedies such as the fires in the Triangle Shirtwaist Factory, Imperial Foods, and the West Fertilizer Company can be prevented, efforts must be made to reduce their likelihood and severity through organized fire prevention activities.

Legal and Moral Responsibilities of the Inspector

The inspector has a legal responsibility to check on all provisions contained in the code the agency is enforcing. In addition, however, the inspector has a moral responsibility to point out other hazards or problems that are not specifically covered in the applicable code. An example is an item not required for compliance because of a grandfather clause in the municipal charter or fire prevention code ordinance. The inspector who personally believe that such a condition is dangerous is powerless to require correction but nevertheless has at least a moral obligation to advise the owner or occupant of the unsafe condition.

After a major U.S. fire in 2003 in West Warwick, Rhode Island, in which almost 100 people were killed, an examination of the most recent inspection report by the local fire department indicated no violations in the facility. The inspection dated to 3 months before this tragic fire. At that time, the inspector failed to cite the glued-on foam insulation to which a major loss of life was attributed. Failing to observe those conditions would not have made the inspector liable, because Rhode Island law grants immunity to fire marshals for omissions or actions performed in good faith during the conduct of their duties.[8] Grand jury investigations of this fire made public the fact that the building had received a clean fire safety report only a few months before the fire, yet the conditions that contributed to loss of life had existed at the time of the inspection.[9]

An occupant of a facility that has been inspected may feel that the building is completely safe from fire when the inspector leaves the premises or when an inspection report states that all deficiencies have been corrected. Generally, the inspector is looked on as capable of providing complete and accurate information on fire spread possibilities, possible points of origin of a fire, and other related information, including potential interim compliance measures such as a roving fire watch.

FIGURE 5-3 Inspection of complex fire protection systems requires specialized training.

Gwoeii/Shutterstock

The inspector has a moral obligation to correctly observe and report all conditions relating to fire safety. This may include looking into accessible voids for concealed problems such as combustible ceiling tiles that, if ignited, could create serious restrictions to evacuation due to smoke generation, high heat, and premature collapse of ceilings.

Although representatives of fire insurance carriers inspect many larger industries and institutions, the fire service should not rely on those agencies to shoulder the responsibility of fire prevention. Inspections should be conducted and reported as if no other individuals have any responsibilities for inspecting the place. The fire department inspector is probably the only one entering the premises who has a legal responsibility for conducting an inspection and taking steps to ensure compliance with code requirements.

Most fire safety inspections, other than those by the fire department, are advisory. An insurance representative can bring about compliance only by speaking for the insurance company. The insured can seek insurance coverage from some other carrier if the current company's inspection requirements are deemed too severe. The insured may also elect to pay higher premiums to keep coverage in effect without making the changes recommended by the insurance inspector or to merely accept the risk by canceling insurance coverage (Figure 5-3).

Conflicts with Other Agency Inspections

Both in the exit interview and in the preparation of a report, it must be remembered that the property owner or operator may be receiving fire-related recommendations from sources other than the fire prevention bureau. The result may be conflicting recommendations or requirements for the same deficiencies. The fire insurance representative might recommend one solution to a problem, and the fire department might recommend another. The insurance inspector's solution may be as good as or even better than the fire department inspector's. In such a case, representatives of the two agencies should meet and try to work as cooperative partners when inspections overlap. But the fire department inspector should be aware of the possible superiority of the solutions developed by inspectors from insurance companies or by management itself.

The inspector should also be aware of the impact of the critical and complicated work the business must do to meet compliance requirements and, in setting the reinspection dates, should allow enough time to complete the work. The inspector may set interim reinspection dates for projects large enough to require the business to find contractors. Reasonable fire prevention bureaus may allow a couple of years or more for some projects, such as the retrofitting of fire sprinkler systems, which may require the business to seek funding help, as well as extra time for design and architectural work. If the work is critical to life safety, alternative measures may need to be devised to come as close as possible to meeting compliance before the final work is completed.

The fire prevention bureau inspector certainly has the final authority, although some other individuals inspecting facilities also may have legal authority in the fire prevention field. The management that receives two conflicting legally constituted fire safety inspection reports must convene a conference of all interested parties in order to resolve the conflict.

Another potential source of irritation to property owners and managers is vague, nonspecific requirements (for example, fire extinguisher height "so as to be safely removed," as opposed to a specific mounting height).

Sometimes an inspector does not receive a warm welcome on arrival to make an inspection or meets with a lack of cooperation during the exit interview. These circumstances may not stem from a resentment of fire prevention inspectors. The average industrial plant, mercantile establishment, or institution is subjected to a multitude of inspections in addition to fire safety inspections from public agencies, fire department prefire planning surveys, and sanitation inspections: inspections for occupational safety, fire and casualty insurance, and types of inspections. As many as 15 or 20 different inspection agencies, public and private, may call on a large industrial plant during the course of a year.

OCCUPANCY SELF-INSPECTION

In some occupancies, inspectors may be employed full time by the plant, the institution, or the mercantile establishment, or inspections may be carried out by individuals from the corporation's headquarters office, in which case the inspector travels from plant to plant to conduct fire safety inspections. These inspections are very similar to those by the public fire service inspector, because this employee is responsible for ensuring complete compliance with the company's fire safety regulations. The company's regulations may parallel the jurisdiction's fire safety regulations, and possibly exceed them. Such employees usually have a thorough knowledge of the business or industry involved, and in fact, their specific knowledge of company processes often exceeds that of the public fire service inspector, who must learn about all types of occupancies.

CONSULTANT FIRE SAFETY INSPECTION

Some mercantile chains and larger industries hire an outside consultant to carry out fire prevention duties. The consulting firms assign personnel who are familiar with the given type of industry and have great familiarity with the processes or procedures of the organizations being inspected. If properly supported by the employing company, this type of program can be most effective.

The consultant concept is used by a number of major concerns that are self-insurers. To be successful, a self-insurance program must include stern enforcement of fire protection requirements, either by company employees or through an outside engineering firm.

INSURANCE CARRIER INSPECTION

The average mercantile establishment, institution, or industry is also faced with the possibility of an inspection by a representative of the insurance carrier, whose purpose may be to determine the insurance rates to apply to the property. Inspections may also be conducted to determine possible limits of policy coverage and to tailor the policy to the specific situation encountered.

An insurance inspection may also be carried out for the purpose of rendering engineering services. This inspection may be by a representative of an individual insurance company or by a representative of an insurance group, such as the HSB Professional Loss Control group or the FM Global Group. The primary purposes of such an engineering inspection are to ensure that the fire protection equipment in the plant is in proper working condition and to identify any major hazards. Changes in procedure, layouts, construction, and protection are also noted.

Inspections for engineering purposes include testing fire protection equipment and available water supplies, which may require the tripping of dry pipe valves and flow tests on hydrants, tests of stationary fire pumps, and tests to ensure proper operation of all fire protection equipment in the structure. Although these inspections are most helpful, they interrupt the routine of the plant, and management must provide personnel to open valves, reset dry valves, and so forth, as a part of the tests.

OTHER GOVERNMENT AGENCY INSPECTIONS

In addition to the previously mentioned private inspections, the industrial plant or mercantile establishment is inspected from time to time by other governmental agencies. The health department, for example, will probably inspect the facility for overall sanitation and for food-handling procedures. The health department will also inspect institutions to determine compliance with standards relating to nursing care, patient health, administering drugs, and the like.

The labor department or department of labor and industry in practically every state has a responsibility for employee safety. A U.S. Department of Labor inspector may conduct these inspections. This responsibility may cover the provisions of the Occupational Safety and Health Act of 1970. The safety inspector usually takes immediate action on conditions encountered that affect the safety and well-being of the employees, including basic fire prevention.

The local industrial plant may have prefire planning inspections by personnel from firefighting companies separate from fire prevention inspection activities. Regardless of coordination and timing, personnel from the plant should take the time to accompany the inspectors in carrying out these duties.

The federal government may send inspectors into certain plants through the provisions of the Port Security and Waterways Act of 1972. This act provides for U.S. Coast Guard inspection, regulation, and enforcement in facilities adjacent to navigable waterways. The Coast Guard is empowered to enforce compliance, which includes fire safety, in such facilities.

A federal inspector may likewise inspect a building for fire protection purposes if the plant has contracts for providing services or materials to federal agencies or has processes subject to federal inspection, such as food preparation. The inspections are designed to ensure the continued safe production of the products under review and may include coverage of fire safety items.

Although all the preceding inspections are necessary, they consume a considerable amount of management's time as well as that of maintenance personnel. Plant management may not welcome the fire inspector because of the multitude of inspections. Time taken by plant personnel to accompany the inspector and by management to discuss requirements is time away from production.

CHAPTER REVIEW

Summary

A key function in the enforcement of fire laws and regulations under the jurisdiction of the fire department is the fire safety inspection. Besides having a thorough knowledge of code requirements and competency in inspection procedures, the fire safety inspector should have good judgment, keen observation, and skill in dealing with people.

Before going out on an inspection, the inspector should be prepared with the proper tools: a flashlight, a camera, and a notebook, data logger, recorder, or computer for taking notes on the inspection tour. Code books and handbooks should be available in the car for reference. The inspector should also look over any reports of previous inspections of the facility for background information to enable proper follow-up on earlier recommendations. Under no condition should an inspection be undertaken when the inspector is pressed for time and unable to make a thorough inspection.

On arrival at the site, the inspector must present identification to a person in authority and secure permission to make the inspection. These requirements are a legal obligation as well as a courtesy. Court cases have challenged the legality of inspections made without proper identification and permission. It is also customary for a representative of management to accompany the inspector on the inspection tour.

The tour usually starts at the top of the structure with an inspection of the roof and then proceeds to the top floor and on down through the entire structure, including the basement levels. A systematic inspection should be made of all spaces on every floor. If access to certain spaces is not immediately possible, the inspector must wait until permission for entry can be arranged.

The objective of a fire prevention inspection is to discover hazards and make recommendations or impose requirements to abate or eliminate them to the greatest extent possible. Conditions to look for are exposure hazards, potential fire hazards, exit deficiencies, or any unusual conditions. Besides noting the obvious potential fire causes or storage of flammable liquids or gases, the competent inspector should look for latent hazards, such as ignition sources from operating equipment, and should be able to recognize fire spread possibilities inherent in the design and construction of the building. Under no condition should the fire inspector leave the premises without being sure that all fixed fire protection equipment is in proper working order as required by the applicable fire prevention codes. And aside from the mandatory responsibility to ensure compliance with legal requirements of the code, there is a moral obligation to point out to property owners any potential hazards not specifically named in the code.

As many easily correctable violations as possible should be taken care of during the inspector's visit. Changes necessary for compliance with the fire code should be discussed with management at the end of the inspection tour. Recommendations should be clear and concise and should be carefully weighed to make sure they will correct the situation and not put the owner to needless expense.

A report should be made of all fire prevention inspections to give the owner notice of violations and to have a record on file for follow-up of recommended corrections. This report would also be used for reference in possible legal action to enforce compliance. An owner is usually given a specific time in which to complete the requirements. If reinspection finds the work not completed, a legal order is issued to comply or face a possible fine or jail sentence.

Among general considerations, the fire safety inspector should be aware of hazards peculiar to various types of occupancies, such as places of public assembly, institutions, mercantile establishments, warehouses, office buildings, hotels, and residential occupancies. For example, such hazards include blocked exits or passageways in public assembly places, grease in restaurants, and the hazards of fire in institutions with bed patients or older adult residents.

The inspector should also be aware of possible conflicts with inspections carried out by insurance carriers and other local, state, and federal agencies and with industry self-inspections. Because the various inspections to which a property is subject may put an owner in a quandary, the fire inspector should be prepared to cooperate in working toward an accommodation of all interests without compromising fire safety.

Review Questions

1. Detailed information about inspection procedures can be found from the NFPA and:
   a. IFSTA.
   b. IAFC.
   c. IAFF.
   d. IAAI.
2. Inspection frequency is based on all the following *except*:
   a. availability of personnel.
   b. degree of hazard in the occupancy.
   c. potential for finding deficiencies.
   d. fire suppression forces.
3. Before conducting an inspection, the inspector should:
   a. review past reports.
   b. confer with the previous inspector.
   c. review the code book based on occupancy.
   d. all of the above
4. An inspector should enter a nonpublic area to conduct an inspection only after receiving:
   a. permission from the proper authority.
   b. certification as an inspector.
   c. orders by a superior.
   d. a favorable weather forecast.
5. Among combustibles, inspectors need to consider all of the following *except*:
   a. acids.
   b. metals.
   c. explosives.
   d. oxidizing agents.
6. All of the following are sources of ignition to be considered during inspections *except*:
   a. fuel leaks.
   b. electrical arcs.
   c. open flames.
   d. static electricity.
7. Absence of fire division walls and presence of improper fire doors contribute to:
   a. vertical fire spread.
   b. horizontal fire spread.
   c. lateral fire spread.
   d. explosions.
8. Special hazards are those the fire inspector associates with conditions:
   a. found in practically every occupancy.
   b. rarely found in any occupancy.
   c. unique to a given occupancy.
   d. that deserve recognition when discovered.
9. In regard to reinspections, the time allowed for the anticipated completion of fire code violations is:
   a. always short.
   b. dependent on the complexity of the violations.
   c. always the same.
   d. none of the above.
10. A fire prevention inspector's results may conflict with:
    a. a fire consultant inspection.
    b. an insurance carrier's inspection.
    c. other government agency inspection.
    d. all the above

End Notes

1. U.S. Fire Administration, *Technical Report Series: Chicken Processing Plant Fires: Hamlet, North Carolina, and North Little Rock, Arkansas* (Emmitsburg, MD: USFA-TR-057/June/September 1991), p. 7. Accessed November 3, 2013, at http://www.usfa.fema.gov/downloads/pdf/publications/tr-057.pdf
2. 2012 North Carolina Fire Code, Section 106, pp. 10–11. Accessed April 23, 2013, at http://ecodes.biz/ecodes_support/free_resources/2012NorthCarolina/Fire/PDFs/Chapter%201%20-%20Scope%20and%20Administration.pdf
3. *City of Toledo v. SETO, Inc.*, 81 Ohio Misc. 2d 1 (1996, OH).
4. The Indianapolis Public Library Digital Collections, *Indiana State Fairgrounds Coliseum Explosion: October 31, 1963*. Accessed October 30, 2013, at http://digitallibrary.imcpl.org/cdm/coliseumexplosion/collection/ffm
5. Seth Cline, "A Decade Later, Nightclub Fire Lives On," *U.S. News and World Report*, Press Past, February 20, 2013. Accessed April 24, 2013, at http://www.usnews.com/news/blogs/press-past/2013/02/20/a-decade-later-nightclub-fire-lives-on
6. Eric Tucker, "Rhode Island, Town Reach $20 Million Deal in Club Fire Suits," *USA Today*, August 19, 2008. Accessed April 24, 2013, at http://usatoday30.usatoday.com/news/nation/2008-08-18-4281531984_x.htm
7. City of Charleston Post Incident Assessment and Review Team, *Firefighter Fatality Investigative Report: Sofa Super Store, 1807 Savannah Highway Charleston, South Carolina, June 18, 2007* (Phase II Report, May 15, 2008), p. 36. Accessed October 30, 2013, at http://www.iaff.org/hs/LODD_Manual/LODD%20Reports/Charleston,%20SC%20-%20Sofa%20Super%20Store%20Report%20Final.pdf
8. Michael Rezendes, "Questions Still Linger over Fire Inspector's Role," *Boston Globe*, February 17, 2013. Accessed October 30, 2013, at http://www.bostonglobe.com/metro/2013/02/17/questions-linger-for-town-fire-inspector-larocque-station-nightclub-blaze/N7OhlfONs8ZuuuH2v09mGP/story.html
9. Kate Zernike, "Nightclub Inspections Found Many Problems, but No Foam," *New York Times*, March 4, 2003. Accessed October 30, 2013, at http://www.nytimes.com/2003/03/04/us/nightclub-inspections-found-many-problems-but-no-foam.html

CHAPTER

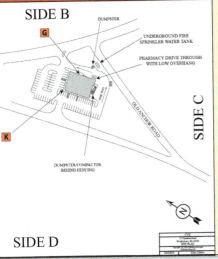

6 Preparing Fire Service Personnel for Fire Prevention Duties

Greg Jakubowski

KEY TERMS

fire protection master plan, *p. 112*

National Board of Fire Service Professional Qualifications (NBFSPQ), *p. 119*

NFPA 1031 Professional Qualifications for Fire Inspector, *p. 120*

NFPA 1033 Standard for Professional Qualifications for Fire Investigator, *p. 120*

NFPA 1035 Professional Qualifications for Public Fire and Life Safety Educator, *p. 120*

prefire planning surveys, *p. 111*

OBJECTIVES

After reading this chapter, you should be able to:

- Explain the lack of emphasis on fire prevention in the United States.
- List the four reasonably expected components of a fire protection master plan.
- Explain the benefits of fire prevention training for all fire service personnel.
- Discuss the importance of prefire planning.
- Describe the importance of having properly trained fire inspectors.
- Identify the educational needs of fire inspectors.
- Summarize the relationship between the fire service and the National Professional Qualifications System.
- Identify pertinent professional standards applied to fire inspectors.
- List the components of a comprehensive review of a community's fire problem.

Many colonial era communities initially emphasized fire prevention through the appointment of a fire warden. However, the primary motivation soon became that of fire suppression. The majority of postcolonial era communities started fire services as the result of experiencing a major fire.

Many people believe that the fire department's obligations have been met if the department responds to all fires when called and brings them under control. Fortunately, the importance and effectiveness of fire prevention have been acknowledged by the fire service and the public, and prevention is once again becoming a major fire service function.

Lack of Emphasis on Fire Prevention

In the past, training programs in the fire service placed limited emphasis on fire prevention. For the most part, they concentrated on fire suppression, whether the trainee was to join a career, volunteer, or combination-type fire department. Fortunately, the fire service currently is trending toward including fire prevention training as part of many basic training programs for both career and volunteer firefighters.

The U.S. Fire Administration offers the web-based Self-Study Course for Community Safety Educators that departments can assign to new firefighters while they are completing other probationary training and orientation programs. The course provides an introduction to fire prevention and knowledge that helps increase the student's effectiveness in conducting fire and life safety education in the community. The Fire Chiefs Committee of Metropolitan Washington DC's Council of Governments adopted this training to introduce new firefighters to fire safety education with a plan to have rookie firefighters complete the self-study training as part of their probationary training package. This training was seen as a cost-effective way to expose new fire service members to the essentials of fire and life safety education.

Major fire conferences designed primarily for fire service audiences have not included fire prevention to any great extent. And with some recent exceptions, national periodicals in the fire service field have also given less coverage to fire prevention than to other fire service activities.

There are probably other reasons than those already mentioned why fire prevention has not been well recognized in the overall fire service picture. Motivation is certainly limited because of the less glamorous nature of the tasks performed. Certainly, the exciting activity of the fire scene is far more challenging and rewarding to many people than performing a building's fire prevention inspection or staging a demonstration before a group of 6-year-olds.

The fruits of one's labor are also much more readily recognizable in fire suppression efforts than in fire prevention work. The individual engaged in fire suppression is rewarded with direct evidence of something accomplished. The control and extinguishing of a major fire are highly visible, whereas the rewards of fire prevention work are intangible and not so easily measured.

Individuals assigned to fire suppression and emergency medical duties seldom find themselves in confrontation with the public, which generally welcomes their assistance. So, yes, there are occasions of conflict, but generally, firefighters and

EMTs are called because they are needed. In contrast, the fire preventionist is often in confrontation with others, especially during code enforcement and arson suppression. Citizens do not usually appreciate advice to extinguish cigarettes, provide automatic sprinkler systems, remove obstructions from exits, and complete other fire safety requirements. They are, in fact, often alienated by such efforts. This alienation is probably another reason for the reluctance of some fire service personnel to accept fire prevention as a major fire department activity.

Fire Prevention Training for All Fire Service Personnel

In preparing fire service personnel for fire prevention duties, it is important to consider the needs of the individual firefighter. Motivational factors also should be considered.

One important incentive is the promotion of positive and proactive image of the fire service because of the specialized knowledge and skills required to do the job. Firefighters serving in fire prevention assignments are often working at levels similar to those of department staff officers. Generally, they perform at a higher standard of decision making, problem solving, writing, and self-sufficiency on the job than is expected in the operations environment of firefighting. Both branches of the career have challenges and benefits. Technical training and proficiency are needed to understand the hazards of industrial processes and the uses of dangerous chemicals, fire protection strategies, and the evaluation of construction methods. Fire prevention training opens up new fields of specialization for fire service personnel and widens their range of activities in the service.

The firefighter also gains knowledge helpful in fire suppression duties. In conducting inspections and **prefire planning surveys**, the firefighter has an opportunity to learn about conditions in the community. Prefire planning surveys are reviews carried out primarily to familiarize fire department personnel with building locations and floor plans, as well as sources of potential danger or other unusual features. This knowledge can be of great help to an incident commander as well as the firefighters who have to enter a building during a fire emergency.

prefire planning surveys
■ on-site visits to buildings and occupancies for the purpose of gathering and documenting information that will ultimately be used in prefire plans and building information files.

A vigorous fire safety program that places fire service personnel in frequent contact with the people of the community can do much to enhance the image of the fire department. It may help erase some preconceived notions about the work of the fire service. Too often, the picture of the firefighter in the public mind is of an individual leaning back in a chair in front of the fire station waiting for a fire. If the public can be brought around to thinking of the firefighter as an active public service employee who is on the street preventing fires when not engaged in firefighting or emergency medical activities, certainly the overall good is served.

Public contacts with fire service personnel who are discharging fire prevention duties, especially if the contacts are favorable, have other benefits. As citizens become more aware of the hazards of fire and gain confidence in the fire service, they are more inclined to call the fire department without delay in the event of an incipient fire, no matter how small, instead of trying to cope with it themselves, risking serious injury or greater property loss.

Many communities have found that after they stepped up their fire safety program, the number of fire calls declined, presumably because fire prevention advice was being followed. There is little likelihood that this decline reduced the loss per fire, but any reduction in the number of fires means an overall gain to the community.

Public respect for the fire service increases because of the community involvement aspects of fire prevention work and the opportunities they provide to build good public relations. This increased public respect can improve the support of the fire department in its efforts to obtain funding for needed apparatus, personnel, and new stations as well as other upgrades in the services rendered. Provision of emergency medical services also has this effect.

Master Plans for Fire Protection

fire protection master plan ■ a comprehensive analysis and descriptive report about a community's fire problem, which includes subsequent preparation and implementation of a plan that addresses the fire protection needs of the jurisdiction, with an economic analysis of the costs associated with the recommendations.

Successful preparation of fire service personnel for fire prevention duties requires a consideration of the entire fire protection picture in the community. Individuals undertaking this work will have a much better understanding of the importance of their roles if they are aware of the concepts embodied in a **fire protection master plan**. A comprehensive evaluation of the entire fire problem in a community leads to the preparation and implementation of the fire protection master plan, which addresses the needs of the jurisdiction.

America Burning, a report of the National Commission on Fire Prevention and Control, contains a recommendation that states:

> The Commission recommends that every local fire jurisdiction prepare a master plan designed to meet the community's present and future needs in fire protection, to serve as a basis for program budgeting, and to identify and implement the optimum cost–benefit solutions in fire protection.[1]

Legislation enacted pursuant to this report also strongly encouraged the promotion of the planning process and defined a fire protection master plan as follows:

> . . . one which will result in the planning and implementation in the area involved of a general program of action for fire prevention and control. Such a master plan is reasonably expected to include: (1) a survey of the resources and personnel of existing fire services and an analysis of the effectiveness of the fire and building codes in such area; (2) an analysis of short- and long-term fire prevention and control needs in such area; (3) a plan to meet the fire prevention and control needs in such area; and (4) an estimate of cost and realistic plans for financing the implementation of the plan and operation on a continuing basis and a summary of problems that are anticipated in implementing such plan.[2]

Fire prevention, including fire and building code enforcement, zoning provisions, fire safety education, installation of fixed fire protection in structures, and a myriad of other fire protection–related areas, is considered in the master plan. If the plan is to be successful, its development must involve community leaders as well as fire service personnel. Ultimately, the master fire protection plan serves as a road map to what is needed in a community in regard to fire protection and fire safety, while the concept of Community Risk Reduction (CRR) and the "Five Es" provide the actions and activity needed to accomplish the plan.

A master plan results in an organized means of defining and implementing a level of fire protection in a community that meets the needs identified by the citizens, coupled with legally mandated requirements. A comprehensive database, including demographics, property evaluations, fire experience, and several other elements, is needed to develop and implement the plan.

Acceptable life and property risk levels should be established in the plan. Direct and indirect costs of fire protection need to be considered. For example, many businesses damaged or destroyed by fire do not rebuild in the same community, so part of the tax base is lost. The cost of treating burn victims is another example.

Although the cooperation and assistance of a wide variety of agencies and organizations are necessary to prepare and execute a master plan, the fire department is always the primary instigator and the agency that "rides herd" to be sure the job is completed.

Those fire service personnel engaged primarily in fire prevention work seldom are in the limelight for their efforts. An appreciation of the community's master plan will enhance their personal pride in the major role they play in the safety of their fellow citizens.

Howard D. Tipton, former administrator of the National Fire Prevention and Control Administration and the recognized initiator of master planning for the fire service, provided this summary of the concept:

> The plan provides an organized means to document the current fire protection elements and services provided to the community and the future service required by the community in order to achieve the goals and objectives. Definition of the basic elements of fire protection (e.g., water supply, zoning, suppression, prevention, life safety, etc.) and establishment of desired types and levels of services provide a basis for community awareness of the extent of the services offered. Further, the community is made aware of the risks involved and the degree of private sector participation required to complement the public sector services for meeting the goals and objectives.[3]

Recruit Training Programs

An example of an excellent training program in the field of fire prevention is that of the Dallas, Texas, Fire Department. As part of the recruit training school, each new member is required to undergo an additional 20 hours of training in fire prevention and related subjects. The fire prevention subjects are designed to give the recruit an appreciation of the worthwhile purposes of fire prevention duties. During the fire prevention phase of the training program topics typically covered include, for example, objectives of fire department inspections, fire hazards and causes, inspection techniques and procedures, prefire planning and inspection, fire cause determination, and fire protection systems.

The New York City Fire Department likewise has a program for indoctrinating new personnel in fire prevention methods. It includes fire prevention training at the fire academy and in the station as part of its overall program. The recruit is teamed with a company officer or senior firefighter during the Building Inspection Safety Program and is given the opportunity to apply the fire prevention principles he or she has learned. This hands-on training with experienced personnel allows any weaknesses to be revealed and remedied through additional fire prevention drills and discussion.[4]

Louisville, Kentucky, includes 40 hours of class on fire prevention and public education in its 720-hour recruit training program. Subjects include methods of conducting home and nonresidential building inspections, technical fire codes, home fire safety lessons, and public fire education. As a final examination, recruits are required to conduct fire safety classes for high-risk groups in the city.[5]

Many cities assign fire suppression companies to conduct code enforcement inspections. Training must be provided to participating fire company personnel. Some states require instructors, as well as inspectors, to be certified.

Prefire Planning

There is a very close relationship between prefire planning and structural control inspections. Some fire departments have attempted to combine these visits; however, this is not advisable because of the differing purposes of the activities.

In the purest form a prefire plan is established to assist in the management of emergencies at complex structures or ones considered target hazards. This plan may include, for example, specific apparatus assignments; tactical objectives; property plot drawings, floor plan drawings, and identification of key building systems; descriptions of fire loading, and fire flow predictions; and predesignated water supply instructions. Prefire plans are most useful for those occupancies and buildings where day-to-day operations are not enough to manage an incident.

Prefire planning surveys are carried out primarily to gather and document critical information that will go into the plan but can also serve to familiarize fire department personnel with building locations and floor plans, as well as sources of potential danger or other unusual features. Some fire departments collect information on all commercial, industrial, institutional, and multifamily structures and record the information in a format that allows retrieval during emergencies. The benefit of conducting building site visits and collecting information includes knowledge of the locations of sprinkler control valves, standpipe connections, and other information related to the fixed fire protection equipment in the building. The location of hazardous commodities that may endanger firefighting personnel is also noted (Figure 6-1).

Prefire planning surveys and fire prevention inspections are devised for two different purposes, and combining the two into one inspection may place the property owner in a rather difficult position. The owner is interested in providing fire suppression personnel with information helpful in fighting a fire but is probably not as interested in having fire personnel enter for the purpose of bringing about corrective action. If the same firefighters are assigned to carry out both responsibilities at the same time, there is a possibility that the two undertakings will conflict with each other.

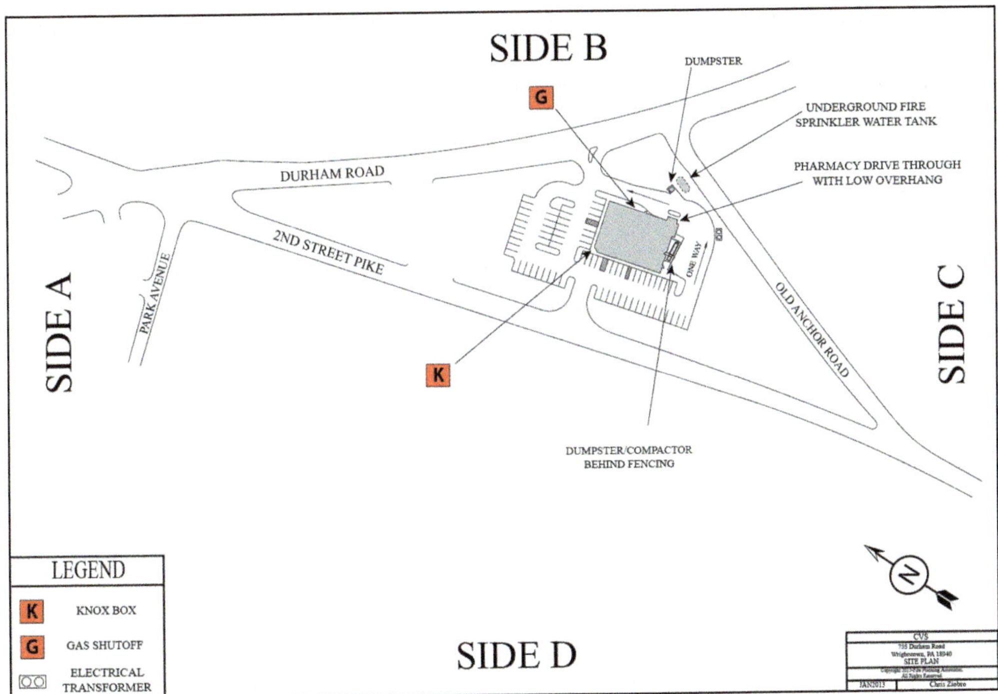

FIGURE 6-1 Information gathering can result in clear and informative site plans that help operational firefighting units to effectively control a fire.

It could be helpful to prepare well-thought-out informational brochures from the fire department explaining the different inspection and information-gathering surveys conducted during visits to buildings in the community. Company officers and lead firefighters that are well trained and informed about the purpose of inspections, and the dispersal of program brochures ahead of inspections, can go a long way to increasing community relations while accomplishing the gathering of information.

Fire Prevention Training Assignments

Another approach to fire prevention training is to assign all personnel to fire prevention functions at some time during their career in the fire department. In several departments, for example, personnel promoted to the rank of lieutenant must, during their tenure at this rank, serve in the fire prevention bureau for a specified period of time. This procedure has been found to be quite useful in helping individuals moving up through the rank structure to gain an appreciation for fire prevention. In addition, this practice does a great deal to preclude the possibility that an individual will reach the position of chief officer without having served in fire prevention.

Greater emphasis is being given to the use of all active fire service personnel for fire prevention duties. A time may come when fire prevention will be a recognized duty assignment for anyone entering the fire service, as it is for wildland firefighters. Such an approach will go a long way toward reducing fire losses.

Many states have training programs to prepare persons to conduct fire inspections, fire investigations, and fire safety education activities. Awarded after completion of training, some states offer certification or credentials that document an individual's achievement of required skills, knowledge, and abilities.

The Committee on Fire-Fighting Services of President Harry S. Truman's 1947 Conference on Fire Prevention gave considerable attention to fire prevention. An opening statement illustrates:

> Traditionally, fire departments have been considered as fire-fighting organizations; but today fire departments are recognized as agencies seeking to prevent fire as well as extinguish them. A fire prevented is better than one extinguished. Fire prevention work by fire departments is prompted by the lively concern firemen have for avoiding unnecessary loss of life.[6]

The statement continues by listing inspections, educational campaigns, enforcement of public safety laws, and fire investigations as the elements of this effort. Plan reviews and door-to-door voluntary home inspections are mentioned as desirable fire prevention activities. The committee recommended that a community's fire prevention bureau function directly under its fire chief and that fire companies periodically inspect all commercial and residential properties in their district, recognizing that many jurisdictions were already doing this. Committee recommendations tracked those issues closely.

The committee's first recommendation on training called for continuing, systematic, up-to-date training programs for firefighters that would prepare them to discharge their responsibilities in firefighting and fire prevention. The report also recommended more standardization of fire service training from state to state. More than 65 years later, standardization remains a problem, with different states requiring from 20 hours to 200 hours to earn fire inspector certification to do the same job.

Fire prevention activities by volunteer fire departments were also addressed by the committee:

> In addition to their regular duties, the department personnel should become a self-constituted educational force to disseminate the fundamentals of fire prevention, as gained from their personal experience in fighting fires and from publications on the subject which should always be at their disposal.[7]

Recommendations made by the committee included having volunteers make regular inspections to correct potential fire hazards, providing driveways to facilitate access to farm ponds for water supplies, and using newspaper publicity for fire prevention.

Training in Fire Prevention Inspection

In addition to indoctrinating and training general fire service personnel in fire safety duties, there is a need to train full-time personnel in the fire prevention bureau in inspection procedures. As an example of early training in fire prevention duties, the following instructions were given to fire inspectors in the Philadelphia Fire Department in 1913:

Major fundamental principles in the training of fire inspectors:

1. The work of fire prevention may impress you as a fad or a joke at the outstart, but your matured view will be that it is both interesting and serious and in comparison with fire extinguishing, it stands as a Bureau of Police to the Detective Bureau.
2. Your sound judgment of corrections to remove fire dangers still has the backing of your superior officers against all protests, and to secure the desired result, you are to permit no hindrance in your work through fear or favor, friends or enemies.
3. Be courteous, explain in detail your mission, and do not quarrel with the occupants of any building. Remember there is a distinct advantage over the man who has lost his temper to the man in an argument with self-control. You are representing the dignity of a department of your city and must in no case initiate a quarrel; if you cannot, report the facts to the officers in charge of the fire prevention work and have them open this avenue for you.
4. Make friends with the engineer, janitor, superintendent, or any other person in charge of the building. These men have considerable information on the property condition and influence in having promptly corrected conditions that you desire improved.
5. This new responsibility calls for intelligence, sound judgment, broad views, and being amenable to reason. Narrow-mindedness and bombast have absolutely no place in this work and every new and practical idea bearing on the end ultimately to be achieved should be promptly grasped and utilized.
6. Gently and firmly decline the friendly cigar or other small article usually offered in good faith, but occasionally used by occupants or owners of property to put the inspector in a bad light. In particular, guard against the least suspicion or suggestion of graft in this form; by adhering to strictly business dealings you are in a stronger position should your work be questioned or reviewed.[8]

Those principles are as applicable today as they were in 1913. They should be considered and adhered to by everyone assigned to fire inspection duties. In addition to avoiding the possible temptation to accept graft, as noted in item 6 above, the fire inspector must adhere to a strict code of ethics, not only declining the "friendly cigar" but avoiding the use of city time to visit theaters or taverns as a guest while on duty. Fire inspectors have been seen and photographed leaving department stores with bags of merchandise while on duty, thereby reducing public confidence in their judgment.

The more an inspector knows, the more valuable are the inspections. A major problem in effective fire prevention inspection is failing to see the significance of what is being inspected. This failure may be the result of lack of training in fire safety inspection, or it may simply be the individual's inability to register problems seen during the inspection. An inspector may report that a complete inspection has been made by going around and checking only on fire extinguishers and on enforcement of smoking regulations, when in reality an effective inspection has not been conducted. The building occupant may well feel the building is safe because the inspector found no violations.

Educational needs include training in inspection procedures, familiarity with fire and life safety codes, training in plan reviews, an appreciation of building code fire protection provisions, an understanding of health codes as related to fire safety, and an understanding of court decisions affecting the field. The fire inspector should also have a general knowledge of the procedures used to determine the causes of fires. Information obtained from legitimate websites may also be helpful.

In smaller fire departments, where it may not be practical to establish formal programs, training often occurs entirely on the job. In this case, it is essential that the trainee be assigned to work with an inspector who is fully qualified to perform inspection duties. Such training might also take place in another city that has a larger bureau, or at a state fire academy. The National Fire Academy also offers extensive training in this field.

Many larger departments have formal training programs for fire safety inspectors. The case study method often employed in these programs consists of studying and analyzing histories of actual field problems in fire prevention and fire code enforcement.

All publications pertaining to changes in the field, including books and magazines relating to new processes, new fire protection equipment, and the like, should be routed through fire prevention inspectors. Weekly conferences and training sessions are also helpful.

Preparing fire service personnel for basic fire prevention duties does not require the extensive training needed to assume technical fire inspection duties. Some training programs, including home inspection and basic fire prevention education, are effective with relatively simple training methods. The less complex

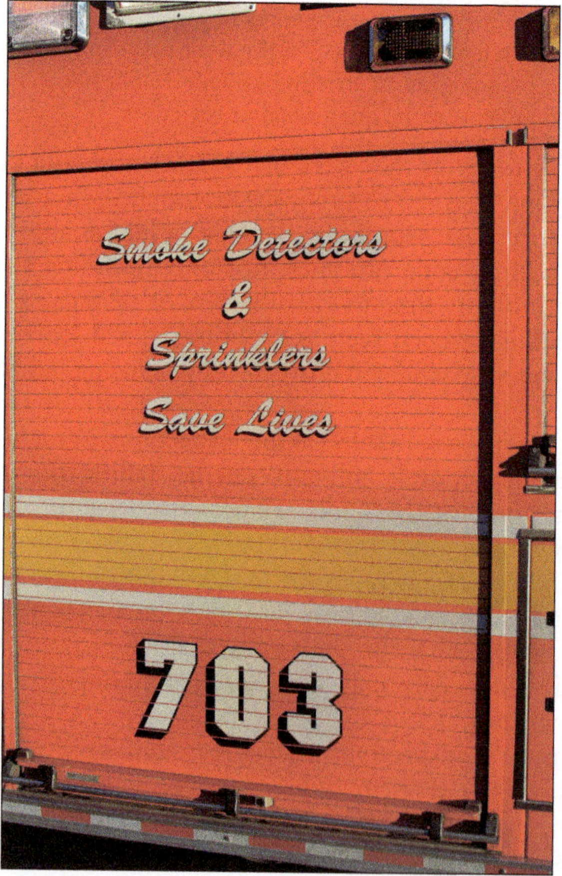

FIGURE 6-2 Fire apparatus bearing a fire safety message as part of its graphics scheme can effectively reach a large number of residents and promote a fire prevention program.

inspections often can be used as stepping stones for new inspectors, giving them an opportunity to become familiar with the process of inspection on a small scale, and then advancing as experience is gained (Figure 6-2).

National Professional Qualifications Board

The Joint Council of National Fire Service Organizations, a now-defunct group composed of the major national fire service organizations, agreed in 1972 to establish standards for professional competency in the fire service. Committees were formed to develop and prepare recommended minimum standards of professional competence for four groups: firefighters, fire inspectors and investigators, fire service instructors, and fire service officers. The Joint Council then established the National Professional Qualifications Board to oversee the project and validate the standards developed by the four committees. The board was also responsible for establishing and maintaining a nationally coordinated professional development program for the fire service.

In 1990 the National Fire Protection Association assumed responsibility for oversight of the Professional Qualifications project. Committees were established and currently operate under the National Fire Protection Association's procedures for establishing standards. Proposed standards and revisions are submitted to the association for final adoption. The standards may be used by states and communities as part of their certification requirements. The system is designed to encompass all duties of the uniformed fire service members and certain civilian positions.

The Joint Council established in its plan a certification process in which each state would make decisions on individual certifications within their jurisdiction. The **National Board of Fire Service Professional Qualifications (NBFSPQ)** accredits state certification boards and other agencies that certify persons to NFPA standards. The International Fire Service Accreditation Congress also accredits state programs. Certification at each level, starting with Fire Fighter I, is earned when individuals successfully complete certain job performance objectives. The individual's progress is based on an ability to master the requirements and demonstrate proficiency to an examiner and/or on a written exam. Of course, promotions are limited by the number of job vacancies in the fire department.

National Board of Fire Service Professional Qualifications (NBFSPQ) ■ an organization, commonly referred to as the Pro Board, established to oversee a means of acknowledging firefighter and other related fields of achievement, primarily through the accreditation of organizations that certify members of public fire departments.

The professional qualifications system does not attempt to assign ranks to levels of qualifications. It expects each local jurisdiction to establish its own rank structure.

The professional qualifications system expects the training programs to prepare fire service personnel to achieve the job performance objectives of the standards. The system also envisions a controlled independent testing procedure for candidates.

The original standard, NFPA 1031 Professional Qualifications for Fire Inspector, was designed to ensure the competency of personnel.[9] The standard is not designed to be used for personnel other than those in the fire service. It does include "direct entry" by individuals who are not firefighters, however.

NFPA 1031 Professional Qualifications for Fire Inspector ▪ a standard that identifies professional performance qualifications for fire inspectors and plan examiners.

NFPA 1033 Standard for Professional Qualifications for Fire Investigator ▪ a standard that identifies professional performance qualifications for fire investigators.

NFPA 1035 Professional Qualifications for Public Fire and Life Safety Educator ▪ a standard that identifies professional performance qualifications for public fire and life safety educators, public information officers, and juvenile fire setting interventionists.

Because of a growing tendency to employ individuals specifically for assignments in fire inspection, investigation, and public fire safety education, in 1987 the original standard was divided in three: **NFPA 1031 Professional Qualifications for Fire Inspector, NFPA 1033 Standard for Professional Qualifications for Fire Investigator**, and **NFPA 1035 Professional Qualifications for Public Fire and Life Safety Educator.** (NFPA 1037 Professional Qualifications for Fire Marshal was added soon after NFPA published it in 2007.) This arrangement provides qualification requirements for individuals entering directly into the specialist areas.

The standards do *not* require a separate position for each job classification or level of progression. Though the hiring of a fire inspector, a fire investigator, and a public fire educator is found only in the larger fire departments, hundreds of smaller fire departments assign only one person to all three jobs; this person must meet the requirements for all three positions. In other communities, two of the three duty assignments may be the responsibility of one person, while another person performs the third. No matter the situation, individuals so assigned need to master the requirements for their specific duties. Note that some communities or states impose higher standards than those of the NFPA.

Candidates for progression under the standards are required to demonstrate an ability to express themselves clearly, orally and in writing. They are also required to demonstrate their ability to interact with the public tactfully and with discretion when performing code enforcement, fire investigation, or fire safety education duties, as well as to demonstrate a knowledge of personal safety practices and procedures.

People aspiring to be fire inspectors at level I must demonstrate knowledge and abilities in many specific areas, including flammable and combustible liquids, compressed and liquefied gases, explosives, heating and cooking equipment, principles of electricity, safety to life, code enforcement procedures, fire cause determination, and miscellaneous provisions. The Fire Inspector II is required to master those subjects in greater complexity and to be able to develop plans relating to fire safety. Fire inspector III must also master administration and management functions.

In states where fire investigators are considered officers of the law, they are required to meet the applicable minimum standards set for law enforcement officers. They must also demonstrate knowledge and abilities in areas including legal procedures, evidence collection, photography, fire scene examination, records, reports and documents, and courtroom procedures.

The public fire and life safety educator must demonstrate knowledge and abilities in general areas and in public information and public education fields. The individual must have the teaching ability to fulfill those requirements successfully. He or she must also understand industrial plant and institutional emergency organization and have the associated instructional abilities. The Public Fire Educator III also must master administration and management functions, including the development of examinations and the evaluation of personnel skills.

There are many reasons for adopting professional qualification requirements for fire safety personnel. The system helps develop professionals and goes a long way toward assuring the city administration and the public that fire prevention, investigation, and education duties are being carried out by qualified individuals. In fact, the necessity for courtroom qualification of fire investigators as expert witnesses makes the system almost imperative for people in this role.

Many states require individuals who perform fire safety inspections to be certified under a state program that they meet adopted standards for job proficiency. Some states also certify fire investigators and public fire and life safety educators as well. In those states, the basis for the certification requirement is generally the NFPA standards and state requirements for law enforcement officers mandated for those with the power to arrest.

Many jurisdictions use fire suppression personnel for fire inspection duties. In states with certification requirements, that means all such firefighters must be certified as fire inspectors. Firefighters carry out the majority of fire inspections, and fire prevention officers serve as resource specialists and handle complex fire safety assignments.

Professional qualification systems have been in effect for years in many other careers, such as for public accountants, insurance underwriters, electricians, barbers, and beauticians. The public can have reasonable assurance that a barber is adequately trained but cannot always have that assurance about a fire inspector. Under the professional qualifications system, a Fire Inspector I in New Hampshire has basically the same abilities as a Fire Inspector I in Arizona or New Jersey.

More than 250 two-year fire science programs may be found in community colleges in the United States. Many of them offer courses in fire prevention and fire investigation. The attendees are primarily from public fire departments. However, building inspectors, architects, and engineers may be found in these classes as well. Most fire science programs offer courses in building and fire prevention codes and standards. In addition, many fire departments are encouraging fire prevention bureau personnel to enroll in fire science programs. Time off to attend classes or special financial stipends may be offered as inducements to enroll.

The 4-year fire protection and safety engineering technology program at Oklahoma State University, Stillwater, has been looked on as a leader in academic fire protection programs. Since the program's inception in 1937, hundreds of graduates have had a substantial impact on fire prevention and fire safety in many countries. Graduates obtain a Bachelor of Science degree in Fire Protection and Safety Technology. But Oklahoma is not the only place in the United States where 4-year fire science/fire protection–related programs are offered. There are over 26 more, and all are affiliated with the U.S. Fire Administration/National Fire Academy's Fire and Emergency Services Higher Education (FESHE) Recognition Program. Note also that the University of Florida[10] has an Internet-based program with courses in fire, EMS, and emergency preparedness management and planning. And the University of Maryland, College Park, is viewed as a leader in offering fire protection as an engineering curriculum.

Summary

Historically, the primary motivation behind the organization of most fire departments has been the suppression of fire in the interest of public safety. Little thought was given to the allied function of fire prevention. Training programs for fire service personnel have concentrated for the most part on fire suppression, with a conspicuous lack of emphasis on the prevention of fire, even in the programs of the major fire service conferences and in the national periodicals that serve the fire protection field.

The current trend is toward recognition of the rightful place of fire prevention as an equal partner with fire protection. Major fire departments are beginning to include fire prevention training as part of their basic training programs. More attention is also being paid to the subject of fire prevention by writers and periodicals in the field, and many city charters specifically name fire prevention as a public responsibility.

Fire prevention training for all fire service personnel can have many beneficial results. It promotes professionalization of the fire service because more knowledge is required than for basic fire suppression duties, and it opens up new fields of specialization for growth on the job for all fire service personnel. It also helps to generate respect for the goals of fire prevention.

A vigorous fire prevention program can also do much to enhance the image of the fire department in the community. If the public recognizes fire service personnel as active public servants who are on the street preventing fires when not engaged in firefighting or emergency medical activities, certainly the overall good is served. The citizen who is favorably impressed with the fire service is more inclined to support efforts to obtain funding for needed apparatus, equipment, new stations, and other fire department improvements.

Specialized training needs for fire prevention inspectors include a thorough comprehension of the local fire code and the procedures used to conduct an effective inspection; an understanding of other related codes, such as building codes and health codes; familiarity with fire investigation procedures; and recognition of the legal aspects of inspection responsibilities. Many of the larger departments have formal training programs for fire prevention inspectors. Smaller departments may employ on-the-job or state training. One of the earliest U.S. training programs for fire inspectors on record is a 1913 program of the Philadelphia Fire Department, which outlined principles that are surprisingly applicable today.

Although the more technical fire inspection duties require more extensive training, basic fire safety education and home inspections can be effective with relatively simple training methods.

An important development was the establishment of a National Professional Qualifications System for state certification of fire service personnel at several levels of professional competence in the classifications of firefighter, fire inspector, fire investigator, public fire and life safety educator, fire service instructor, and fire service officer. Standards were originally developed by committees appointed by the Joint Council of National Fire Service Organizations, with staff support by the National Fire Protection Association. Adoption of those standards is a major step in professionalizing the fire service and improving the nation's fire protection capabilities.

Review Questions

1. Fire prevention has not always been well recognized in the overall fire service picture because:
   a. fire suppression is more exciting.
   b. fire prevention is rewarded with immediate evidence.

c. fire suppression is confrontational.
d. there are more national fire prevention conferences.

2. Fire prevention training for fire suppression personnel opens new fields of:
   a. staffing.
   b. specialization.
   c. enforcement.
   d. governing.
3. Master plans should include:
   a. a survey of resources and personnel.
   b. short- and long-term prevention and control needs.
   c. a plan to meet needs with an estimate of the costs.
   d. all of the above.
4. The purpose of a prefire plan is to:
   a. preplace apparatus.
   b. establish command strategies.
   c. familiarize personnel with building locations and floor plans.
   d. all of the above.
5. Fire inspectors can obtain training:
   a. at the National Fire Academy.
   b. from state fire organizations.
   c. in larger fire departments.
   d. all of the above.
6. NFPA Standard _____ applies to fire inspectors.
   a. 1021
   b. 1031
   c. 1041
   d. 1061
7. Additional NFPA standards for fire prevention include:
   a. 1901 and 1902.
   b. 1033 and 1035.
   c. 1500 and 1561.
   d. 13 and 231.
8. Which organization sponsors the Fire and Emergency Services Higher Education (FESHE) recognized school program?
   a. International Association of Fire Chiefs
   b. National Emergency Response Training Center
   c. National Board of Professional Standards
   d. National Fire Academy
9. Which is *not* considered in the development of a master fire plan?
   a. Fire prevention
   b. Fire and building code enforcement
   c. Emergency medical services
   d. Fire safety education
10. Which fire prevention personnel are likely to be found in larger departments?
    a. Fire inspector
    b. Fire investigator
    c. Public fire educator
    d. All of the above

End Notes

1. U.S. National Commission on Fire Prevention and Control Report, *America Burning* (Washington, DC: U.S. Government Printing Office, 1980), p. 25.
2. Public Law 93-498, Federal Fire Prevention and Control Act, 1974, U.S. Fire Administration.
3. Howard D. Tipton, "Master Planning for Community Fire Protection," in *Annual Meeting of International City Management* (Toronto, Canada, 1976), p. 8.
4. Personal communication to author Love from Thomas Jensen, Chief of Fire Prevention, Fire Department, City of New York, 2013.
5. Russell E. Sanders, "Chief's Briefing," *NFPA Journal*, November/December 1993, p. 14.
6. *The President's Conference on Fire Prevention: Report of the Committee on Fire Fighting Services* (Washington, DC: U.S. Fire Administration, May 6–8, 1947). Accessed November 30, 2013, at http://www.usfa.fema.gov/downloads/pdf/47report/firefighting.pdf
7. Ibid.
8. Powell Evans, comp., *Official Record of the First American National Fire Prevention Convention, 1st Philadelphia, 1913*

(Philadelphia: Merchant and Evans Co., 1914), p. 150.

9. *NFPA 1031: Standard for Professional Qualifications for Fire Inspector and Plan Examiner* (Quincy, MA: National Fire Protection Association, 2014). Accessed November 30, 2013, at http://www.nfpa.org/codes-and-standards/document-information-pages?mode=code&code=1031
10. Personal communication to author Robertson from L. Charles Smeby, Jr., Academic Instructor, University of Florida, 2007.

CHAPTER

7 Organization and Administration of Municipal Fire Prevention Units

Mike Love

KEY TERMS

fire ground, *p. 134*

fire prevention bureau, *p. 128*

SARA Title III, *p. 129*

OBJECTIVES

After reading this chapter, you should be able to:

- Identify the primary components of a fire prevention program.
- Describe the local government's responsibility with respect to fire prevention.
- Explain the role of the fire prevention bureau within the fire service and the community.
- Describe generalists and specialists in fire prevention and where they are likely to be encountered.
- List potential sources of conflict between fire prevention and fire suppression personnel and how an advisory committee resolves such conflicts.
- Explain an intensive inspection program as it is recognized in Cincinnati.
- Explain the relationship between a fire prevention entity and other municipal agencies.

Fire prevention has been recognized as a function of municipal government since the earliest days of this country. In the colonial period, the administrative functions of the town with respect to fire, police, sewage, water, health, and street lighting were often carried out through volunteer citizen service. Some of the early municipal charters required citizens to serve in certain positions of responsibility without pay,

FIGURE 7-1 The fire station in many suburban and rural communities is often a center of activity that includes municipal business such as plan review and permitting for fire safety.

Mike Love

and they usually readily accepted such responsibilities. In thousands of communities in the United States, fire protection and fire prevention services are provided by volunteer fire departments (Figure 7-1).

The Place of Fire Prevention in Municipal Government

By tradition, fire prevention and fire protection in the United States have been primarily local government responsibilities. Fire and fire-related problems are more locally confined than are most other public service concerns, such as crime, health, welfare, and employment. It is quite unusual, for example, for a structural fire to burn in more than one jurisdiction. Furthermore, fire safety code enforcement can usually be most effectively carried out at the local level.

In 2002, the U.S. Fire Administration report *America at Risk, America Burning Reconsidered* addressed this concept:

> The primary responsibility for fire prevention and suppression and action with respect to other hazards dealt with by the fire services properly rests with the state and local governments. Nevertheless, a substantial role exists for the federal government in funding and technical support.[1]

Most communities form a fire department when some event demonstrates the need for fire protection for public safety and welfare. A community typically gets along without a fire department until it becomes obvious to the citizens that they can no longer rely on neighboring communities to provide fire protection or until a large fire involving loss of life or serious property damage brings urgent demands for community action to organize a fire department.

In most areas of the United States, the incorporated municipality is considered the most logical agency to develop and operate a fire department. In some areas, fire protection is not considered a municipal function, and service is provided by an incorporated fire department, an arrangement often used in unincorporated communities.

In municipal government, elected officials are responsible for the overall administration of government. Administrative heads of departments, who are responsible for carrying out the day-to-day functions of the government, may be political appointees or may be appointed through a merit system. Lower ranks of municipal employees are usually selected through a merit system or civil service procedures.

Recognition of the close interrelationships of municipal functions is essential for good city government. The lack of coordination among municipal officials may be a reflection of some weakness of the city administration, or it may be that the individual departments are more concerned about preserving their autonomy than about performing as part of a city government team. Some fire department administrative personnel believe that fire protection is so specialized that it has little relationship to other municipal government functions.

The fire department is an important entity in municipal government, but to function most effectively, it must cooperate with other municipal agencies for the mutual goal of improved public service.

CHARTER RESPONSIBILITIES

The charters of most cities address fire protection as a responsibility of the municipality. Not all mention fire prevention. An example of a charter that specifically includes fire prevention is the charter of the College Park, Maryland, Fire Department, which states the following purpose for existence: "Its object shall be the prevention and extinguishment of fires and the protection of life and property in College Park and vicinity."[2]

Even if fire is not specifically mentioned, the responsibility for the public's safety is clearly stated in the charters of practically all cities. There is no question that the public safety includes protection from fire, so it may be assumed that broad public safety provisions give charter backing for appropriate actions and measures in fire prevention.

Recent years have seen the role of the fire service expanded to include new responsibilities far beyond its original goals. As an example, the stated goals of the Fullerton, California, Fire Department are to provide services designed to protect lives and property of the people in the city of Fullerton from the adverse effects of fires, sudden medical emergencies, and exposure to dangerous conditions created by either man or nature.[3]

Fire Service Administration in Local Government

With thousands of incorporated cities and municipalities in the United States, it is understandable that there is a wide diversity of public fire services, including volunteer or call groups in smaller communities, combination volunteer and career forces, and fire departments with specialized divisions for fire prevention. In the traditional municipal fire department, the fire chief reports directly to the mayor or other top administrator, an arrangement that makes fire safety responsibility equal to other municipal functions.

Fire Prevention Functions

Prevention of fire is a primary goal of a fire department and an important component in the fire defenses of a community. Fire prevention is primarily made up of measures directed toward preventing the inception of fire. Among those measures are fire and life safety education, fire prevention inspection, fire code enforcement, investigation of fires to determine causes, and investigation of suspicious fires. Analysis of information gained from the latter two functions can be of great value in evaluating and improving practices to help prevent fires.

There are many organizational approaches to assigning responsibility for fire prevention functions. Some fire departments do not consider fire prevention a full-time activity and may assign fire prevention responsibilities as a collateral duty. In contrast several cities have a **fire prevention bureau**, which is a separate arm of government with a fire prevention chief or fire marshal reporting directly to the city manager or chief executive.

fire prevention bureau ▪ the organizational element that coordinates all fire prevention activities within the fire department and makes important contributions to fire department public relations through its inspection and fire safety education programs.

The fire prevention bureau is the entity that coordinates all fire prevention activities within the fire department and makes important contributions to fire department public relations through its inspection and fire safety education programs or the fire marshal's office. In some community fire departments (and in a number of federal and industrial fire departments), fire prevention is a line function for suppression forces.

These departments make sure that personnel are trained and are employed not only as firefighters but as fire "preventionists" as well.

The term *fire prevention bureau* is no longer the sole title for the arm of the fire department that carries out fire code enforcement, public education, and fire investigation duties. Fire loss management, risk management, environmental control, community risk management, community risk control, fire safety loss prevention, and a myriad of other titles may be found throughout the country. It is important to recognize that the word *fire* has a clear and distinct meaning to practically everyone, although the rest of some of those titles may be baffling to the average citizen.

In Montgomery County, Maryland, a reorganization of the Fire and Rescue Service created a new division with a diverse range of services. While the traditional services of the fire marshal and fire prevention bureau were core, the new division included other services such as emergency management, master planning, public information, and member recruiting and retention. The term *fire marshal's office* no longer properly described the new scope of services. So Montgomery County determined that the new division would be renamed for its diverse services as the Community Risk Reduction Services. Many other communities have reorganized and consolidated their fire and life safety divisions, bureaus, and sections as staffing changes, downsizing, and organizational consolidation resulted in broader-based services and flatter organizational structures.

SARA Title III is a federal program that requires industry and related occupancies to maintain records on hazardous materials and their characteristics. Some fire departments assign monitoring duties related to SARA Title III plans to the fire prevention bureau. The fact that inspectors are checking on hazardous materials in the community makes this a logical arrangement. Although an actual chemical release may activate fire suppression forces, routine oversight responsibilities can well be carried out by fire prevention personnel.

SARA Title III
■ a federal program that requires industry and related occupancies to maintain records on hazardous materials and their characteristics.

The Fire Corps is a local level program through which citizens can voluntarily assist their local fire departments in nonoperational roles, including many fire prevention–related tasks. For example, the Layton City, Utah, Fire Corps assists through a fire safety education program in the local elementary schools. Fire Corps members from two area high school drama clubs develop props and write and perform their own skits at the elementary schools.[4]

The Fire Prevention Bureau

In fire departments serving communities with populations of 25,000 or more, the most common organizational concept is a fire prevention bureau that functions as a separate arm of the fire department. Personnel are assigned to the bureau full time and usually have no fire suppression responsibilities except in emergencies, when they may be pressed into service in firefighting. In departments with a separate fire investigation bureau, fire prevention personnel are also called to assist in investigations on occasion.

The fire prevention bureau coordinates all fire prevention activities within the fire department and makes important contributions to the department's public relations through its inspection and fire safety education programs. Fire prevention personnel are in daily direct contact with the citizens of the community.

In some fire departments, the attitude prevails that fire prevention work is for those who are physically unable to carry out active firefighting duties. People who have been injured in the line of duty or who are suffering from some physical disability that precludes their effective service in fire suppression are assigned to the fire safety bureau. This practice may be the result of a lack of understanding of the true role of fire prevention in the fire defenses of a community. Under the fire laws and regulations in effect in most jurisdictions, the fire department has the authority and obligation to carry out essential fire prevention measures. Furthermore, fire prevention work has some strenuous aspects. Conducting inspections, for example, may require climbing ladders and other physical effort that may be beyond the capacity of the physically challenged.

THE ONE-MEMBER BUREAU

In a smaller community, only one individual may be assigned to head as well as function as a one-member bureau. Great care should be given to his or her selection because of the range of duties, and because so much of the work involves contact with the public. Clearly, there are challenges in a one-person shop, one of which is the need to plan for absences, such as for vacation and illness. In a one-member bureau efficient work practices such as mobile connectivity and communications and administrative work in the field is a must. But there also needs to be some administrative function or base of operations for files, permit application, and other customer oriented functions.

Many smaller communities have outstanding fire prevention bureaus, though just as in any profession, there are potential pitfalls. After some years on the job and especially after becoming proficient at it, the lone fire prevention officer may become entrenched and possibly sidetracked from promotion in other areas of the fire department. In addition, that officer, because of constant public contact, may become better known in the community than the chief of the fire department, which could create internal problems. Close coordination and understanding between the fire prevention officer and the chief of the department are needed to maintain harmony.

GENERALISTS VERSUS SPECIALISTS IN THE FIRE PREVENTION BUREAU

In a fire department large enough to establish a fire prevention bureau consisting of more than one individual, organizational concepts must be well planned to ensure effective use of personnel and ability to cover all work needs. Because there is a range of challenges from routine to an emergency and from simple to complex, how work is assigned can ultimately impact on the overall performance of the bureau. For example, how personnel will be assigned work, such as by geographic region, needs to be considered because it affects how far resources must be spread to cover the workload. Careful consideration should be given to personnel assignments based on the work that needs to be accomplished.

In a fire prevention bureau consisting of two to seven people, it may be practical to assign all personnel as generalists so that any member of the bureau can handle any assignment for inspection, plan review, fire prevention education, or investigation. Assignments might be made on a geographic basis. For example, personnel assigned to one area of town might be responsible for all inspections within that area, regardless of the type of occupancy. This arrangement has the advantage of providing inspections at the most reasonable cost because less travel is involved. The inspector can go to the assigned section of the town and go from door to door, conducting inspections for an extended period of time, for example, in the business district or in shopping centers located on the outskirts of the town. One disadvantage of this system is the possibility that having inspectors work different parts of town could bring about variations in inspection procedures or in interpretations of code requirements for similar facilities. A nursing home in the northwestern part of town, for example, might not be inspected in the same manner as one in the southeastern part of town.

Larger fire prevention bureaus find it advantageous to develop and assign inspectors as specialists in a particular field. For example, one inspector may specialize in flammable-liquids installations, another in health facilities, and another may handle all industrial occupancies. If the inspection load in a community does not warrant full-time assignments in the inspector's area of expertise, the specialist may spend some time handling general-assignment work.

The disadvantages of the specialist system of inspections are that transportation costs are higher when inspections are not confined to one area and inspectors have less opportunity to establish close contacts in the community than they do when working in a single neighborhood. The advantages and disadvantages to the community and to the department should be thoroughly weighed to determine the best system for a particular organization.

Here, too, as with the one-person bureau, care must be taken to avoid dead-ending the specialist inspector because of expertise in a given field. Personnel assignments must take into consideration the promotional policies of the bureau and the fire department. Promotion to captain, for example, often requires several years' experience in fire suppression. The individual with a specialized background should not be penalized by not qualifying for movement within the department.

BUREAUS IN MAJOR CITIES

A brief description of the structure of the Bureau of Fire Prevention of New York City is offered below as an example of the wide range of services necessary in a major city. Note that the fire investigation and fire safety education functions are rendered by separate divisions in the department.

> The Bureau of Fire Prevention (BFP), staffed by 400 professionals, includes civilian fire protection inspectors, engineers trained in various disciplines, explosive and pyrotechnic experts, and uniformed personnel working out of Fire Department Headquarters. The BFP generates revenue annually in inspection fees, permits, and certificates.
>
> The Bureau is commanded by the Chief of Fire Prevention, who reports directly to the Chief of Department, as mandated by the Administrative Code. That section of the code empowers the Bureau to perform the duties and exercises the powers of the Commissioner in relation to dangerous articles such as combustibles, chemicals, explosives, flammables, compounds, substances, or mixtures. The Bureau also is responsible for the prevention of fires and the management of crowds, obstruction of aisles, passageways and means of egress, standees, fire protection, and fire extinguishing appliances in all buildings and occupancies within the City of New York.[5]

SUITABLE QUARTERS

In the development and organization of a fire prevention bureau, it is important to provide suitable quarters for adequate operation and public contact. Unfortunately, some fire prevention bureaus are tucked away in a remote corner of the central fire station that is poorly located for public access. In some cases it is necessary for the fire prevention bureau "customer" to go through sleeping quarters or dayrooms occupied by firefighting personnel to reach the fire prevention bureau. This situation not only may cause inconvenience for the public but also may result in resentment toward the fire prevention bureau by the fire suppression personnel.

The physical location of the fire prevention bureau in the fire station or city hall may also be a factor in the public's attitude toward the fire prevention program. If the bureau is poorly located, perhaps in a building that is not in compliance with fire or building codes, the implication is that the bureau is of limited importance in the fire department and, in fact, in the overall municipal government. The fire prevention bureau should be located in a section of the fire station or city hall that can be reached easily by the public. Parking facilities should be made available for visitors.

COMMUNICATION WITH THE PUBLIC

The bureau offices should have adequate facilities for receiving telephone calls and electronic communications. Citizens calling to transact business with the fire prevention bureau should be able to reach a person who can supply the needed information without having to talk with several other people who may not be directly concerned with fire prevention. The fire prevention bureau should also have clearly accessible mail and e-mail addresses.

Because of public uncertainty, the chief of the fire prevention bureau, or the fire marshal as he or she may be called, may receive communications, telephone calls, or visits that should properly have gone to the chief of the fire department. Conversely, calls may be referred to the fire chief that should have been referred to the fire prevention bureau. With responsibilities for the overall administration and day-to-day operations of the fire department, the fire chief is not likely to be fully informed of routine fire prevention bureau activities. Establishing an internal method of screening calls and visits to the fire department may provide a simple solution to this problem.

Many fire prevention bureaus have discovered that they can improve customer service and increase efficiency by making useful information accessible on the Internet. The Internet and fire prevention bureau websites are often the first place people seek for information. For example, the Fairfax County, Virginia, fire marshal's website can be found easily through common Internet search engines. Its home page provides seasonal information and also the office business hours, training opportunities, code compliance information sheets, fire safety information sheets, links to comprehensive information of their business process, and a directory of contact information for fire marshal staff. By proving a reliable and up-to-date website, Fairfax County has increased the likelihood that customers can find the information they need without the need for a phone call.

WORKING HOURS

Personnel in the fire prevention bureau are often assigned shorter working hours than other fire department employees. However, there are many exceptions to this rule. At one time, assignment to the fire prevention bureau might have been considered desirable from the standpoint of working hours, but this view has all but disappeared with changes in working conditions for fire suppression personnel.

Practice dictates that fire prevention personnel work a schedule comparable to headquarters office personnel and to city government office staff. The workweek is normally 40 hours, although some cities have reduced this to 37 or fewer hours. Some fire prevention bureaus work four 10-hour days per week.

In comparison, a number of cities have firefighting personnel working 42, 48, or 56 hours a week, but with shifts arranged on a 10-hour, 14-hour, or 24-hour schedule. That gives them daytime hours off to participate in other activities during the normal daytime workweek. Such an arrangement cannot be made for an individual assigned to a 40-hour workweek in the fire prevention bureau. For that reason, some cities are finding it difficult to encourage people to voluntarily enter the fire prevention field. As a result, some cities are upgrading fire prevention personnel and taking other steps to make the job more attractive to firefighters, while still other municipalities are turning to nonuniformed workers to conduct fire marshal work.

Periodic reviews of time allocations are important in a fire prevention bureau. Funding shortfalls and the complexities of business operations have brought about some innovative reassessments of typical fire prevention bureau duty assignments. An example is the Asheville, North Carolina, Fire Department. Its fire prevention bureau is divided into three sections: periodic fire inspections, new construction, and fire investigations.

The periodic inspection section is organized around two battalions (each in a fire station) with three staff each and decentralized from bureau headquarters. Decentralization was an efficiency move that geographically placed the office in the general area of the inspections they performed, which reduced both cost and travel time.

The Asheville deputy fire marshals are unique in that they are shift-based like their firefighter counterparts. They retain their firefighting certifications and assist at the scene of fires, serving as aides to the incident commanders. The benefit of this arrangement is that it reduces callback when there are code issues during nonbusiness hours, and provides for critical safety and strategic support during fire operations. The fire marshals also can visit the occupancies that operate during the nonbusiness hours such as bars, restaurants, theaters, and some public meetings, such as the city council, where crowding issues can become a problem. The shift fire marshals have been able to improve sharing of information and communication with the operational fire staff by their shift presence. Information about the buildings they inspect and complaints they investigate, as well as heads up when they become aware of vacant buildings and other hazards go a long way toward improving the firefighters' situational awareness of the area they protect.

The construction division of the Asheville bureau has four personnel who integrate with the city's development services department, offering a "one-stop-shop" concept. The integrated development services department is an industry-wide best practice that enhances customer service by reducing the time and effort required when seeking permits for work. Time efficiency in building and development is a critical factor that ultimately can impact economic development. The fire inspectors involved in Asheville's construction activities benefit from a close working relationship and proximity to inspectors for the other trades. Asheville has reported that customers are happier for the reduced time it takes to obtain permits.

Another best practice that Asheville employs is a common plan-review software system that allows tracking of all work and comments and increases communication between reviewers and customers. Common plan-review systems increase efficiency by being able to conduct simultaneous reviews by the different disciplines, instead of being a serial process that can stretch out review of plans sometimes by weeks.

The City of Asheville implemented a full cost-recovery-fee system to fund its inspection services. Generally, customers are in favor of cost recovery fees, as long as they see benefits through higher efficiency and customer service that reduce the time and effort needed to carry on business. Asheville has produced some of the more innovative improvements to thrive in the ever-changing government and business environment.[6]

Sources of Conflict Within the Fire Department

One of the major responsibilities facing the fire chief is bringing about a full understanding and appreciation of the fire prevention bureau by other fire department personnel. Misunderstandings are not limited to those connected with hours worked. Many other areas of misunderstanding arise concerning the operation of the bureau. For example, friction may develop between bureau and other department personnel because fire suppression personnel may feel that the fire prevention staff is not exposed to the physical punishment associated with firefighting and emergency medical services.

A conflict also may arise between the two segments because of the procedure followed in many communities of assigning automobiles to fire prevention bureau personnel. The vehicles are assigned for official use only, including investigation of overcrowding complaints and inspection of nightclubs during the inspectors' normal off-duty time. However, an elevated status may be attached to possession of a city car, especially if it may be taken home at night. There are many justifiable reasons for assigning automobiles to fire prevention bureau personnel. For example, it may be necessary for them to respond to fires to carry out fire investigation duties.

Fire suppression personnel may also resent the fact that fire prevention bureau personnel are in the public eye more frequently. Inspectors are in daily contact with business owners, educators, operators of institutions, and other community leaders in the normal course of their duties. Fire suppression personnel have little public contact except on the **fire ground** and on emergency medical calls.

fire ground
■ the area where fire department emergency operations are conducted at the scene of a fire.

Fire Prevention Advisory Committee

A concept that has been found to be quite helpful in creating a better understanding of the role of fire safety is the fire prevention advisory committee, made up of members from all divisions of the fire department (Figure 7-2). One or two representatives might be selected from each of the battalions or districts within the department. The committee formulates the fire prevention program for the department and meets from time to time to evaluate its efforts and suggest changes in direction. Through the establishment of a fire prevention advisory committee,

Dotshock/Shutterstock

FIGURE 7-2 Fire prevention advisory committees and task forces made up of stakeholders can offer expertise and valuable input on community fire safety issues.

personnel throughout the department can be given an appreciation of fire prevention duties and responsibilities.

Intensive Inspection Program

The Cincinnati, Ohio, Fire Department has been recognized as a leader in fire prevention. The Cincinnati Fire Department has maintained an outstanding inspection program for many years. Procedures used are worthy of study and emulation by other cities. The Cincinnati Fire Department inspects all buildings, including homes in the city. The department's inspection program includes block inspection, a tactical inspection unit, and focused area inspections. Each is described in the following paragraphs.

The *block inspection* came about when the Cincinnati Fire Prevention Program underwent revisions to reflect the constraints on working hours available and fire company personnel levels in a modern urban fire department. Today, the fire department has a four-person minimum requirement for each fire company. This minimum seldom allows the luxury for an individual to be available to leave the fire company for the purpose of inspecting in the field. As a result, the need to "block-inspect" or "unit-inspect" almost exclusively is utilized at the company level.

Cincinnati's *tactical inspection unit* is called on when the frequency of fires has increased critically in certain areas. The unit, composed of a task group of fire inspectors, is deployed to reduce hazards. They are detailed from various companies throughout the city under the command of a superior officer and concentrate on fire hazards within the critical area until they are rectified.

Current activities also involve *focused area inspections* in defined communities within the city of Cincinnati. Members of the Cincinnati fire, health, police, and building departments concentrate their efforts for a 90-day period to address blight, crime, and fire hazards in the identified area of the city. The overall effect is to address issues that need to be corrected and to halt deterioration that can lead to an increase in hazardous buildings.[7]

Cincinnati's fire prevention code-enforcement efforts are directed toward serious violations where legal action is indicated. Also responsible for preventing environmental hazards, the fire department utilizes an environmental crimes unit. This unit is organizationally placed within the Cincinnati Fire Prevention Division.

Cincinnati also has changed its emphasis from fire suppression and fire prevention activities to emergency medical service (EMS) responses. This change in emphasis is typical of a majority of fire departments nationwide. Information regarding the effect this change has had on the department's long-standing fire prevention efforts is related here as an example of this phenomenon:

> EMS responses were a major factor in the change in the Cincinnati Fire Department's inspection procedures. In 1971 a "busy" fire company completed 1,000 responses a year. Today a "busy" fire company completes almost 4,000 responses a year. Originally, the goal of the fire department was to inspect *all* dwellings within the city each year and most businesses at least twice a year. As emergency responses of all sorts increased, it became obvious that the fire department could not meet those inspection goals. New methods of setting goals were defined, and now the local fire company commander sets the goals. One of the parameters for those goals is that one, two, and three-family residential inspections could be conducted on an as-called-for basis. Certainly, the commanders can implement a yearly or biannual inspection program for a troubled area where an outbreak of fires occurs. This method has been used successfully for a number of years.
>
> But what we found was that the greatest intervention we could affect in residential fire loss was to make sure early fire warning was achieved. This has reaped many positive results and has decreased civilian fire deaths significantly. This success is an incentive for the Cincinnati Fire Department to continue its efforts to reach its goal of zero fire deaths.[8]

Fees for Services

Some communities collect permit fees for certain fire prevention services. Others have historically offered all fire safety services at no charge. But due to the increased cost of government services and community growth, communities are looking more often to a variety of fees to assist with fire safety resources. Using the model that building departments have followed, fire services are starting to adopt a variety of procedures to acquire the needed resources, including the use of a fee structure, permits, licenses, and charges for capacity certificates.

Although an increasing number of jurisdictions are collecting fees for their fire safety services, few are covering the entire cost of their operations with this revenue stream. The fire administrator is still in the position of needing to justify expenditures for fire safety services. In fact, spending limitations imposed by state initiatives have forced several municipalities to look to the private sector for financial support to continue certain fire prevention activities, including public fire safety education programs. This financial need may place the administrator in an uncomfortable position when private sector benefactors seek relief from fire code edicts, perhaps initiated by the same department employee who is using private sector funds to carry out fire safety programs.

Personnel Trends

A number of cities are using direct-entry personnel for fire prevention duties. Dallas, Texas, for example, recruits persons with at least 2 years of college directly into the fire prevention bureau. The career ladder is limited to the fire prevention bureau supervisory positions, however. These individuals receive 710 hours of specialized training in subjects related to fire prevention.

Proponents for using people with fire suppression backgrounds as fire inspectors point out that such individuals have a greater appreciation for fire spread characteristics based on their personal experiences and are better at describing fire spread to building occupants.

Program Goals

Goals need to be set for the various fire prevention functions within the fire service, including public education, fire prevention, fire investigation, and hazardous materials. The Santa Clara County, California, Fire Department has devised a set of annual performance goals for each of those categories, and it measures the effectiveness of the department's service to its citizens accordingly. The importance of each element of activity is explained in a publicly distributed flyer.[9]

The Santa Clara County comprehensive fire prevention program includes fire and life safety inspections, plan review, and public education. Because Santa Clara County found 90 percent of businesses that experienced a fire closed within a year of the fire, its 2011 goal was to limit commercial fires to 4 per 1,000 occupancies inspected annually[10]. The aggressive goal resulted in greater improvement than expected: 2.1 fires per 1,000 occupancies inspected, so observers may anticipate even more aggressive goals for the future.

Every year Santa Clara County's goal is to limit reportable releases of hazardous materials to 1.5 per permitted occupancies. Public education goals are to make contact with 15 percent or more of the population served by the department during the year. Thorough fire investigation is given strong support because its results are valuable from a public education standpoint. There is also value from an arson suppression angle, because additional arsons may be averted by arrests and convictions. No specific performance goals are set for that activity.

The Santa Clara County Fire Department's goal statements summary can be a model approach for any fire department wishing to enhance community fire safety:

> Public education and fire prevention are where we can make the biggest difference in our communities. This is where we truly save lives and spare citizens the

pain and anguish fires cause. In addition to standard programs, employees are encouraged to offer "value added" services including such simple things as checking a smoke alarm or providing fire safety tips when called to someone's home.[11]

State Responsibilities

State agencies may exercise control over local fire-code-enforcement activities. As an example, Wisconsin law provides for local enforcement of the state's fire codes. However, the state provides funding through an insurance premium tax for the local jurisdictions to perform the necessary inspections and other related activities. The state employs several inspectors who verify that the local municipalities are actually conducting required inspections. Funding can be terminated if inaction is found.

The Relationship of a Fire Prevention Entity to Other Municipal Agencies

Certainly, the fire department is the major municipal agency with an interest in fire prevention, and it is the one responsible for the overall control and enforcement of fire safety measures. A number of other municipal agencies have a corollary interest in fire prevention as part of their major functions. To understand the place of fire prevention in municipal government, it is helpful to discuss the ways in which the entire municipal government is involved.

The *municipal water department* has a clear-cut tie-in to fire protection. Any improvements in the municipal water supply and distribution system will result in improvement in the overall fire protection in the community. For this reason, the fire chief or fire marshal and the director of the water department should work closely together. Plans for expansion of the water system should always include fire protection considerations. As an example, a 6-inch main may be adequate for domestic service in a given section of town, but an 8-inch or 10-inch main may be far more desirable from a fire protection standpoint. On the other hand, it has been mentioned that economic incentives have become more prevalent where the voluntary installation of residential fire sprinklers in one- and two-family homes and multifamily structures results in a municipality allowing smaller water mains and even longer distances between fire hydrants. The sprinklers reduce fire risk and the reduced materials of larger mains and hydrants reduce the cost to developers. If the fire department cooperates with the water department, fire protection needs can be taken into consideration at the time of original development. Plans also should be coordinated if the water system is privately owned. This may require extra effort to establish a good working relationship with the private water authority.

Another area of cooperation between the fire department and the water department is in testing of hydrants. Coupled with this activity may be the maintenance of fire hydrants, including painting and color coding.

Coordination with the *police department* is essential for effective fire prevention. In some cities, there has been real or imagined friction between police and fire personnel, which may stem from rivalry or from misunderstanding of duties and responsibilities. Regardless of the reason, certainly the public loses when misunderstandings affect services rendered. On the other hand, the public stands to benefit greatly when police and fire personnel cooperate closely in the interest of fire prevention.

In the course of their normal duties, police officers may observe conditions that should be reported to the fire prevention bureau for appropriate action. Overcrowding in places of assembly, for example, is more likely to come to the attention of the police than to fire service personnel. The police department can also be of great assistance in enforcing regulations covering the transportation of hazardous materials. Police officers are in a better position than members of the fire department to check on vehicles transporting dangerous materials to ensure the use of the required markings, the use of designated parking facilities, and the observance of other safeguards. On the fire ground, police and fire suppression forces traditionally work well together. The same kind of cooperation in other areas can advance the cause of fire prevention in the community.

The ultimate coordination of police and fire services is found in the combination police–fire department. Regardless of the pros and cons of the overall plan, it may safely be said that it is a mistake to expect personnel who are looked on primarily as police officers to carry out fire prevention responsibilities. The fire service has depended on a low-key public image to help in implementing fire prevention programs. The psychological impact of a person uniformed and armed as a police officer may create the opposite effect. One of the major advantages of the combined police–fire program is said to be the reduction of idle time. It can be argued that a fire department that is carrying out a wide variety of fire prevention and emergency medical programs in a busy organization has little time for police duties.

The local *health department* will probably have some responsibility for the inspection and approval of facilities from a sanitation standpoint. In the field of restaurant inspection, there is room for a great deal of cooperation between health and fire department personnel. The same may be true in inspection of day-care centers, boarding houses, hotels, and numerous other facilities where health department inspection and approval are required.

Coordination with the *building department* lies in review of plans and specifications, as well as cooperation in inspection for occupancy of structures within the community. Such coordination is extremely important. It should begin with coordinated reviews of preliminary plans and follow through to include cooperative inspections of occupancies. There are many examples of close cooperation between the two agencies. Traditionally, there have been misunderstandings between the building and fire departments in some communities. Such misunderstandings and conflicts can result in compromising public safety—with dire results.

Fire department responsibility for building code enforcement was addressed in *Municipal Fire Administration,* a publication of the International City/County Management Association. This text states, "As a matter of strict logic, since most features of building regulation are for fire safety, the building code and fire prevention code could both be made a fire department responsibility."[12] This idea has trended in the other direction, with an increasing number of building departments being assigned the responsibility of most or all of the fire and building regulations.

Building codes, however, contain a number of requirements not directly related to fire safety. A fire chief who assumes responsibility for enforcing the building code must have adequate knowledge to carry out the duties. In some cities the building official's duties are assigned to the fire department.

Some communities, counties, and states have found it desirable to place all responsibility for enforcement of the fire-related sections of the building code in the hands of the fire department or fire marshal's office. The building department retains full responsibility for enforcement of all other sections of the code. This arrangement seems to work well. An obligation is imposed on the fire department or fire marshal's office to provide adequate service with competent employees. Recent trends in funding state and local government, however, have resulted in consolidation of services that can be seen as redundant. While this is a difficult process, such consolidation can result in services being more customer-service-oriented. For example, with consolidation of plan reviews, developers, builders, and other building trades have fewer stops and less bureaucracy to deal with in obtaining permits and approval for projects.

In some cities, the building department enforces the fire prevention code. All responsibilities related to building inspection are thus concentrated in one agency. This procedure has also been abandoned in several cities as the proficiency of fire services in this activity has increased.

There should also be coordination with the housing enforcement division of the municipal government. Housing codes cover many features of fire prevention, so coordination is mutually beneficial. The desirability of joint inspections may be reviewed on a local basis. However, in single-family occupancies, it would not be legal to conduct joint inspections without permission. The fire department may inspect such facilities on a nonobligatory, voluntary basis, although the housing section inspections are obligatory. It would not be in the best interest of either agency to have fire personnel enter on the strength of the housing inspector's right of entry. Both programs might suffer under such an arrangement.

Planning and zoning functions have considerable bearing on fire safety in any community. There should be no hesitation in coordinating fire prevention activities with those of planning and zoning to ensure recognition of necessary storage locations for hazardous materials and to ensure adequate clearances between buildings in the interest of preventing conflagrations. In some communities, the fire prevention chief or fire marshal actually sits as an ex officio member of the planning and zoning board. Regardless of how participation is afforded, planning and zoning always provide the earliest opportunity to influence community design with fire safety in mind. For example, trends in "new-urbanism," "sustainable growth," and "smart growth" all have core values that highlight more compact or denser planning that promotes walkable and transit-oriented communities. In areas that have begun to implement such community blueprints, residential construction includes little spacing between combustible structures and very narrow streets. These and other features influence the need for fire protection. If a coordinated review occurs early in the planning process, there is more likely to be understanding from all involved as to how to achieve quality of life balanced with safety and reduced fire risk.

The *city purchasing bureau*, which is generally responsible for the purchase of all supplies for the municipal government, offers an avenue through which the fire prevention bureau can work to have certain fire safety requirements established in city purchasing policies. For example, curtains and draperies and other materials used in public buildings can be specified to meet certain flame-retardant requirements. The purchasing agencies are usually willing to comply with such requests.

The *sanitation department* has a role in fire prevention. Disposal of trash and other combustible waste is, of course, a part of the fire problem in any city. The sanitation department can minimize hazards through uniform pickup procedures. The fire prevention bureau and the sanitation department should coordinate their efforts to ensure fire safety during extreme conditions. An example is a special trash collection during times of civil unrest to reduce the possibility that loaded trash cans will be set on fire by disorderly persons. Other public emergencies, such as storms, may produce a temporary need for storage of large amounts of damaged vegetation and building materials that can create a significant fuel load for an unwanted fire. Fire officials are generally involved in local emergency management and disaster planning and can keep the public aware of the increased hazards of improperly stored solid waste.

Public utilities may be provided by private or public agencies. Regardless of ownership, fire prevention personnel should deal with these agencies just as they would deal with a municipal department. Areas of coordination include fire code enforcement, cooperation in responding to alarms, and other functions in which mutual interest is indicated. Cooperation with electric power, telephone, cable television, and gas companies is especially important.

The *public school system* is a government agency with great potential for cooperation in fire prevention endeavors. Cooperation may extend to arranging for fire department staging of fire safety demonstrations, classroom instructions, and cooperation with fire drills and inspections. This type of work has been most rewarding, but it can be carried out successfully only if there is complete understanding between fire prevention and education personnel.

Another area of coordination is the school construction program. With the pressure to make physical improvements to aging facilities, school districts have numerous renovation and new construction projects under way. Again, close coordination can ensure that fire protection systems are maintained during construction in schools so that students are not at risk. Early planning also can help to identify and correct fire protection deficiencies (for example, retrofitting older schools with automatic sprinkler systems when a renovation is planned).

The *city library*, although not thought of as an agency having a great deal of involvement in fire prevention, can be of value by providing space for displays and exhibits and by making available a wide variety of publications in the fire field. Books are available that may instill an interest in fire safety in the reader. Fire drill procedures in libraries are important, too.

The *judicial system* has a bearing on fire safety. In most states, the judicial system is primarily a state operation, although many states do have municipal courts, which are responsible for hearing cases involving housing, building, zoning, and fire code regulations. Thus the fire prevention inspector may engage in enforcement activities involving cooperation with the courts and the city attorney.

Public welfare programs, which are generally carried out at the state level, offer some opportunities for coordination in fire prevention. A high percentage of public expenditures for fire protection, especially as related to responses to fire alarms and emergency medical calls, go to areas with low-income residents, indicating a need for education in fire safety matters among welfare recipients. Very few jurisdictions have attempted to tie in fire prevention with welfare or human services programs. However, avenues for development should be explored. Partnering property management agencies with the fire prevention bureau can offer an opportunity to impart fire safety information and procedures to new residents. Additionally, welfare systems often operate low-cost multifamily buildings that house elderly residents and other individuals with mobility challenges. Discussions with the property management officials concerning future renovations should include safety upgrades for automatic sprinkler retrofits and improved fire detection and notification systems. It can help to have recommended fire protection upgrades in high-risk buildings identified in the community's fire protection master plan.

Cooperation on the fire ground reflects on fire prevention and protection in several different ways. Fires in the community can have a tremendous impact on fire prevention efforts. They draw public attention to the subject of fire and demonstrate the need for fire prevention in a way that should not be underestimated. It is easy for people working in the fire safety field on a full-time basis to get the idea that everyone else thinks about fire prevention most of the time. This, of course, is not true, and many people give no thought to the subject until fire breaks out in their neighborhood. Awareness of the improper conditions firefighters see when reporting to an incident, even when it is not a significant fire, may lead to code enforcement that would result in a safer building. Most fire prevention bureaus have an internal (for fire department use) and a public fire code complaint system to report problems such as locked exits, improperly operating alarms and suppression systems, and unsafe buildup of waste. Such complaints can trigger an investigation by a fire inspector, even to the extent of an emergent visit during nonbusiness hours.

Full cooperation between police and fire departments at the scene of a fire can go a long way toward smoothing out rough edges in public reactions. A motorist stopped by a hose line across a highway is usually irritated at the fire department, especially if the delay is lengthy. Under those conditions the police department, rather than the fire department, can probably do a better job of handling the public. By treating people in and around the fire scene courteously, they can preserve good public relations and allow the fire team to concentrate on their emergency operations. A planned effort coordinated by fire departments, police, and transportation departments can trigger rerouting and marked detours to reduce the inconvenience to motorists around a fire or incident scene.

Cooperation between the fire department and the *water department* is also essential during major fires and can benefit both agencies. Deficiencies in water supply that may affect automatic sprinkler protection may show up during the suppression of a major fire or even during training where a water supply is used for pumping operations. A fire emergency may also show the need for up-to-date information on water supplies, maps, and methods of designating hydrant locations and sizes, water main sizes, and the like. It may also show the need for augmenting water supplies during major fires and for providing additional hydrants in a given location. All these factors can be used to press for public support for necessary water supply improvements. This activism will

Billy Morris Director of Fire Science Cisco College

FIGURE 7-3 Emergency communication centers, or 9-1-1 centers, where fire emergencies are received and dispatched fill a vital role in rapid fire control.

benefit the community in another way, since water supply facilities are recognized as a major factor in the classification of a municipality under insurance rating schedules.

The *municipal streets* or *public works department* can lend support to the fire department at the fire ground. This support may include the provision of barricades at the scene of a major fire to assist in crowd control.

As important as the fire service and fire prevention functions are to the community, other municipal departments are clearly responsible for the safety and welfare of the public (Figure 7-3).

Case Study: Morgantown, West Virginia, Fire Prevention Program

Morgantown, the home of West Virginia University, has 31,000 residents and a fully career fire department of 48 personnel. Approximately 30,000 additional people live in Hagerstown during the school year, mostly students at the university. Commercial development in the city is robust, with the evolution of high-density rental properties and plans for 10 new high-rise buildings that will support the university campus. The daytime population for Morgantown is approximately 100,000 and the city has a Class 3 Insurance Services Office (ISO) rating. At one time the city had the reputation of being the "sofa-burning capital" of the United States, owing to post athletic game activities of students. It no longer has this reputation.

Faced with a funding request for additional fire suppression personnel and the memory of the 2001 deaths of two university students in substandard lodging, a city council committee recommended in 2005 the hiring of two additional fire inspectors, bringing the total in the Morgantown Fire Department to four. Those individuals were selected for their interest in fire prevention rather than their rank in the department. The council committee felt that aggressive code enforcement plus fire safety education, plan review, and thorough fire investigation would probably save more lives and property than would additional fire suppression personnel.

An evaluation of the effectiveness of this program was prepared 3 years after the program was implemented in 2005. Not only did Morgantown surpass all other cities in a similar population category in West Virginia in code enforcement activities, but no fire deaths or injuries had occurred during the 3 years included in the survey. An average of more than 8,000 inspections were made per year at that time, with more than 3,000 violations detected, most of which were corrected. Officials in Morgantown have found that as the inspection program matured, inspection numbers declined. For example, in the 3-year period of 2010 to 2012, there was an average of 4,000 inspections per year. Approximately two-thirds of the inspections were focused on commercial and multifamily residential occupancies and one-third on assembly occupancies.

As they began to achieve better control of the risk in the downtown district, the officials were able to spend less inspection time, especially in assembly occupancies like bars, nightclubs, and restaurants, which tend to have in the range of 10,000–15,000 patrons over a weekend. The ability to reduce the number of assembly inspections offered the city an opportunity to evaluate its overall risk environment for planning and resource allocation purposes. City officials have initiated the collection of data and have compiled the data in an occupancy vulnerability assessment profile (OVAP) as related in personal communication from the City Fire Marshal, Captain Ken Tennant. The profile resulted in a score that helped define risk, based on factors that include, for example, occupancy type, construction classification, water supply, fire protection systems, number of occupants, and time period building is occupied. The profile is set up to interface with the department's computer programs and helps the department allocate resources to control risk, including plans for scheduling staff, patterns of fire incidents, and the need for additional prefire plans.

During the early years of implementing the fire safety program, Morgantown was averaging over 250 plan reviews per year, but the recent challenging economy has seen the number drop to an average of 120 per year. The department conducts an average of 5,225 events through its public education programs, reaching all demographic groups in the city. These programs range from preschools to offices, industrial plants, and centers for older adults.

Training and certification requirements are quite stringent for fire inspectors; all have fire suppression backgrounds and are subject to fire ground duties during major emergencies. Inspectors issue citations that can result in the imposition of fines. A major strength of the inspection program is the goal of obtaining compliance with applicable codes through detailed knowledge of applicable code provisions. For example, domestic sprinkler heads are permitted in lieu of 1-hour fire separation for heating plants in rooming houses.

The city averaged 43 reported structure fires a year between 2010 and 2012, with an average of 12 per year being fires that actually caused a recordable loss. Property damage from fires in Morgantown was on average below $1 million per year between 2010 and 2012, and the department feels this is a result of its fire prevention efforts. The department protects property with an estimated value of several billion dollars.

Inspectors are assigned inspections based on the occupancy rather than the territory in the city. This arrangement is felt to give the inspector a higher degree of expertise than with the geographic arrangement that is used in some jurisdictions.

In June 2013 the Minger Foundation was awarded a fire prevention and safety grant by the U.S. Department of Homeland Security to create a campus community service project that offers fire safety outreach to at-risk communities near the campus. West Virginia University was one of the five schools included in the community service project.

Minger Foundation was established in the memory of a college student killed in a dormitory fire in another state. The Minger Foundation is working in partnership with Campus Firewatch, a college campus fire information service, and the City of Morgantown. The grant teams students with firefighters from Morgantown Fire Department to conduct fire prevention and education programs in high-risk sections of the city. In addition to the education and outreach activities, the partnered groups distribute smoke alarms, carbon monoxide alarms, and fire extinguishers to the residents without charge. The effort is intended to help students become more involved and responsible for their own fire safety and to help them become aware of hazards affecting some of the most at-risk people in a community and how community activity can have a positive impact. The students help by increasing the staff available to the fire department while learning and in some cases experiencing firsthand the potential effects of fire on the community while coming away with a greater respect for fire and its disastrous outcome.

What has been seen, as a tradition around many college campuses—nuisance fires set on the streets and in dumpsters—has been a fire safety challenge for Morgantown for over 40 years. Data collected by Morgantown identified approximately 3,200 street fires since 1979. Morgantown implemented a positive approach to dealing with these malicious fires that included an arson task force with the city fire marshal division and police, campus police, and state police. This approach formally integrated all stakeholders into the process, including the university, since students set many of the fires. Since 2001, 124 people have been charged and prosecuted criminally for setting fires. West Virginia University has expelled or suspended a number of the students involved in intentionally setting fires.

In a short time, Morgantown's focus on fire safety as a priority in the whole community has shown positive results that should over time help the city improve and sustain an effective level of safety. This work did not go unnoticed and was rewarded by further resources under the grant with the Minger Foundation. The continued efforts to analyze risk and a current project to qualify the department for accreditation is evidence of the commitment to the best possible service for the community.

CHAPTER **REVIEW**

Summary

By tradition, fire prevention and fire protection in the United States have been primarily local government responsibilities. In the charters of most cities, public fire protection, which can easily be interpreted to include fire prevention, is named as a responsibility of the municipality. Court decisions have clearly indicated that the legal responsibility of a municipality to protect its citizens encompasses fire protection.

With thousands of incorporated cities and municipalities in the United States, the public fire services vary widely and include volunteer and call groups in the smaller communities, combination volunteer and career forces, and large, fully career fire departments with specialized divisions for fire prevention and training. In the traditional form of the municipal fire department, the fire chief reports directly to the mayor or other top city or local government administrator. In some municipalities, the fire department is placed under a director of public safety, who is also responsible for the police department. Another variation is the combination police–fire department, often known as a *public safety department*.

The term *fire prevention* refers primarily to measures directed toward preventing the inception of fire, among which are fire prevention education, fire safety inspection, fire code enforcement, investigation of fire to determine causes, and investigation of suspicious fires. In fire departments serving communities with populations of 25,000 or more, a fire prevention bureau generally coordinates all fire prevention activities. The bureau may range from a one-person operation to a relatively large operation with full-time personnel assigned to specialized functions, such as educational programs and technical inspections. Fire prevention personnel should be able to deal well with the public, and they should also have suitable quarters, easily accessible to the public, with adequate communication facilities to carry on their work.

Conflicts may arise between fire suppression personnel and the fire prevention bureau for several reasons: lack of understanding on the part of fire suppression personnel of the nature and importance of fire prevention work, differences in hours, social status associated with vehicles assigned to fire prevention personnel, and a general feeling that the fire prevention people have it easy and do not have to take the physical punishment that firefighters do. Some departments require all fire service personnel to spend some time in fire prevention work; others make fire prevention training a part of their basic training; still others make fire prevention assignments a requisite of the rank of lieutenant or captain.

Another innovation that advances the cause of fire prevention is the fire prevention advisory committee. Such committees are made up of members from all divisions of the fire department—and sometimes also of people from outside the fire department.

Fire prevention is a concern of other municipal agencies, too, and the fire prevention bureau should cultivate opportunities to work with them in every possible way, to get the message of fire prevention across to the whole community.

Review Questions

1. Fire prevention and fire protection are primarily the responsibility of:
   a. the federal government.
   b. state government.
   c. local government.
   d. private enterprise.
2. Responsibility for fire protection in most cities is usually established by:
   a. charter.
   b. state law.
   c. county protocol.
   d. the fire department.
3. The fire prevention bureau may also be called:
   a. fire loss management.
   b. risk management.
   c. community risk control.
   d. all the above.
4. A generalist in fire prevention could be assigned to:
   a. inspection and investigation.
   b. plan review.
   c. fire prevention education.
   d. all the above.
5. A source of conflict between suppression and prevention personnel is:
   a. titles and status.
   b. chain of command.
   c. training opportunities.
   d. assignment of automobiles.
6. The department most likely to assist fire prevention with overcrowding in public assemblies is the _____ department.
   a. building
   b. police
   c. zoning
   d. water
7. Plan review is best coordinated with the _____ department.
   a. building
   b. police
   c. engineering
   d. planning
8. The term *fire prevention* refers to:
   a. fire prevention education.
   b. fire safety inspection and code enforcement.
   c. origin and cause investigations.
   d. all the above.
9. Fire prevention bureaus of two to seven people often find it practical to assign people as:
   a. specialists.
   b. generalists.
   c. engineers.
   d. investigators.
10. Larger fire prevention bureaus find it advantageous to:
    a. employ full-time fire prevention engineers.
    b. assign personnel as specialists.
    c. assign personnel as generalists.
    d. hire civilian nonsworn employees as code enforcement personnel.

End Notes

1. U.S. Fire Administration, *America at Risk, America Burning Reconsidered* (Washington, DC, June 2002), p. 15.
2. College Park, MD, Volunteer Fire Department *Constitution, Article* I, p. 8.
3. *Orange County Fireman* (Anaheim, CA, Summer 2003), p. 10.
4. Fire Corps, Program Profiles, Layton City, UT. Accessed April 30, 2013, at http://www.firecorps.org/news/program-profiles/965-layton-city-ut-fire-department
5. Personal communication to author Love from Chief of the Bureau of Fire Prevention, Thomas Jensen, New York City Fire Department, 2013.

6. Personal communication to author Love from Fire Marshal Wayne Hamilton, Asheville, NC, Fire Department, 2013.
7. Personal communication to author Robertson from Frederick Prather, District Chief, Fire Prevention, Cincinnati, OH, Fire Department, 2008.
8. Ibid.
9. Fire Department, Santa Clara County, CA, *2011 Annual Report*. Accessed March 31, 2014, at http://www.sccfd.org/forms/annual_report_2011.pdfannual_report_2011.pdf
10. Ibid.
11. Ibid.
12. International City Management Association, *Municipal Fire Administration* (Washington, DC, 1967), p. 232.

CHAPTER 

8

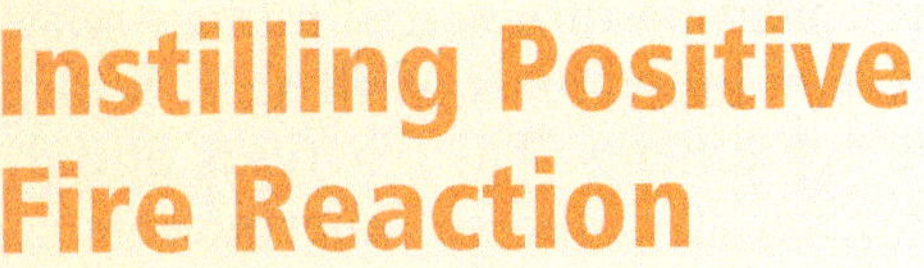

Instilling Positive Fire Reaction

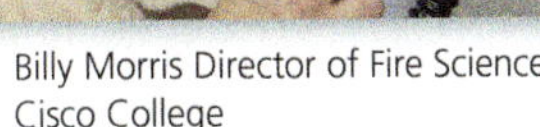

Billy Morris Director of Fire Science
Cisco College

KEY TERMS

fire reaction, *p. 150*

fire safety manual, *p. 159*

flashover, *p. 150*

OBJECTIVES

After reading this chapter, you should be able to:

- Explain the importance of fire reporting as it relates to fire prevention.
- Identify what the emphasis on home fire drills should be.
- Explain the difference in focus of institutional fire drills and fire drills in occupancies where residents are ambulatory.
- Explain the careful application of fire extinguisher use in certain occupancies.
- List some of the building safety devices or concepts that can lead to successful evacuations of a structure during a fire.
- List some behaviors researchers have discovered often demonstrated by people during fires.
- Identify fire reaction in specific occupancies, including high-rises, special occupancies, and housing for older adults.

The concept of life safety in buildings holds that occupants will be safe under fire conditions as long as they are able to successfully escape threatened areas before the fire makes things untenable. But there can be precious little time, and evidence is mounting that the window of time for escape closes steadily. The reaction of the individual in a fire emergency is a critical element in life safety considerations, and the human behavior factor should be carefully analyzed in the development of fire safety codes. Public **fire reaction** is an integral part of fire prevention planning and education.

fire reaction
■ fire-related human behavior that must be planned for in designing fire prevention training and education that may include, for example, leaving by the same door a person entered from, taking in cues of what is going on before deciding to evacuate, helping others escape a fire, having a narrowed focus while under stress.

Any assessment of anticipated human reactions in the event of a fire must take into account individuals of all ages, health categories, physiques, occupations, and temperaments. The reactions of individuals to fires vary in relation to these characteristics and to other factors such as the severity and location of the fire, the manner in which it started, the presence of other individuals within the fire area, and the specific condition of the individual at the time. The latter consideration includes factors such as whether the individual is awake or asleep, is intoxicated or under the influence of drugs, or is fully alert and capable. It also includes any other conditions that may have an effect on the individual's reaction to the specific situation with which he or she must cope.

Individual Decisions When Fire Occurs

Bernard M. Levin, Ph.D., formerly of the National Institute of Standards and Technology, studied fire reaction for many years. He concluded that people usually make rational decisions during a fire. However, they may make inappropriate decisions that increase their danger because they do not understand the nature of fire, its escalation, and the elementary principles of fire protection.

Levin suggested that fire safety education programs address the occurrence of **flashover**, the speed with which a fire that seems small can grow to flashover, the concept of margin of safety, the size and nature of fires that building occupants can extinguish, the value of closed doors, and the danger of opening doors. He explained the concept of *margin of safety* as the time between when everyone has left the area and when anyone in the room can be injured or killed (Figure 8-1).[1]

flashover
■ the point when a room fire becomes fully developed with all contents completely involved in fire. People intimate to a fire when it reaches flashover are not likely to survive.

A 2007 nationwide study of 1,000 adults in the United States conducted by the Society for Fire Protection Engineers (SFPE) revealed that respondents' first reactions to fire could place them in danger. When asked what they would do if there was a fire in their building, only 28 percent answered that they would evacuate, whereas 39 percent said they would call the fire department. Although calling 9-1-1 seems like a sensible response, people should first exit the building and then call the fire department once in a safe location, according to Chris Jelenewicz, SFPE's engineering program manager.

Other responses in this study were notifying others (24 percent), fighting the fire (4 percent), and searching for the source of the fire (3 percent). Delays before deciding to evacuate, time spent searching for the fire, gathering belongings, and trying to fight the fire are observed repeatedly in real fire situations.

The survey also revealed that the elderly, who are statistically a high-risk group, generally respond to a fire situation incorrectly. More than half of people 65 years or older would first call the fire department; only 30 percent would exit the building first.[2]

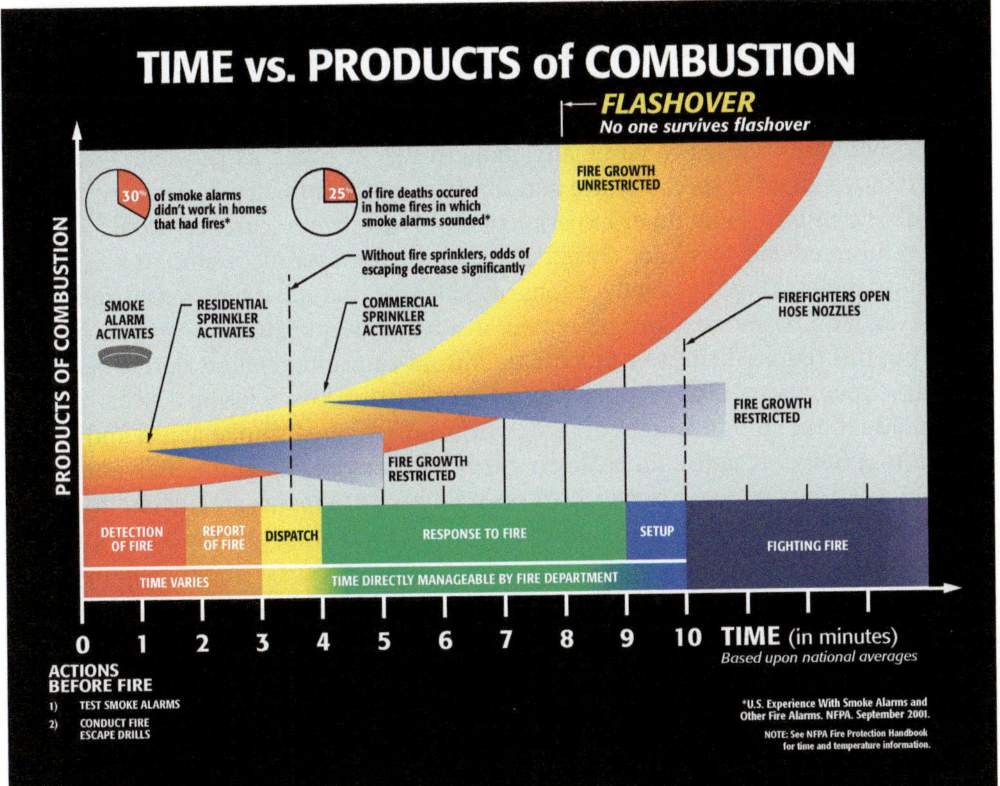

U.S. Fire Administration

FIGURE 8-1 For demonstration purposes this diagram shows functions that occur to alert residents and control fire so residents can react in time to escape.

Fire Reporting Procedures

An example of fire reaction training is the ongoing campaign to inform the public of how to properly summon the fire department. In addition to signs and special printed plastic wraps applied to city fleet vehicles and public transportation vehicles like buses and subway cars, some cities have resorted to roadway banners in an effort to increase public awareness of fire-reporting procedures. The effort is being made to train individuals so that they will immediately report a fire as soon as they have safely evacuated and use the national emergency number of 9-1-1.

Home Fire Drills

Another example of fire reaction training is the home fire drill program. This outstanding program, which has been under way for a number of years, was given considerable impetus by the Ohio State Fire Marshal's Office in the late 1950s. At the same time, a major home fire drill project undertaken by the Kansas City, Missouri, Fire Department, resulted in widespread adoption in the greater Kansas City area. That program gained national stature and is being carried out in some communities as part of the annual Fire Prevention Week in October. Members of the public are encouraged to conduct a home fire drill at a predetermined time during Fire Prevention Week. In many communities, the program is known by the acronym EDITH, which stands for "exit drills in the home."

The program encourages individuals to locate potential escape routes and to make plans to leave the dwelling place without delay in the event of an emergency. This means determining all possible avenues of emergency egress and training occupants to leave by those routes. The program also emphasizes the importance of leaving doors to occupied bedrooms closed to eliminate smoke travel as much as possible. Dramatic evidence shows that the spread of fire and smoke is restricted when doors are closed; a closed door can buy some time. Experiments by NIST fire researchers in old college dorms reinforced the fact that closed doors limit the spread of smoke and gases throughout the building.[3] Recent versions of NFPA 72, National Fire Alarm and Signaling Code, require smoke alarm placement inside the sleeping area as well as the corridors, but this requirement was added a significant time after smoke alarms began to be used, so many homes have smoke alarms only in corridors. Closing a sleeping-area door could reduce an individual's ability to hear an activated smoke alarm and possibly lead to a longer reaction time. Under the home fire drill program, individuals are taught to feel doors before attempting to open them. Participants are advised to check for smoke and, if it appears safe, to enter the corridor crawling on hands and knees, if necessary, to stay in the portion of the corridor that has the least smoke buildup.

Another important part of the home fire drill program is a consideration of the need for all occupants to assemble at a predetermined place once they are outside. This measure gives the fire department personnel who arrive on the scene a way to know which rooms might still be occupied and thus to make any necessary rescue operations more successful.

The home fire drill program emphasizes the need to call in an alarm without delay, as soon as the occupant notifies other occupants of the fire and safely evacuates.

Home fire drill training is a logical sequel to smoke alarm publicity. The tremendous popularity of smoke alarms has resulted in the production of countless pamphlets, instruction sheets, and guides relating to home fire drills. Promoting home fire drills and providing actual step-by-step instructions for a home escape plan may be the local fire department's and safety group's most useful activities.

Information is readily available on the benefits of home fire drills. The effective yet simple website homefiredrill.org offers resources for, instruction in, and helpful explanations of home fire drills.

Closely related is training in what to do when clothing catches fire, that is, to "stop, drop, and roll." These steps have saved many lives and minimized injuries in countless incidents.

The babysitter training program is an adjunct to training in home fire drills. It includes other facets of safety; however, the fire prevention and fire reaction functions are of major importance. Training is given in measures to reduce the possibility of fire as well as in procedures to follow should a fire occur. Babysitters are instructed in means of evacuating infants and small children, in reporting fires, and in other measures relating to fire reaction.

School Fire Drills

Probably the most widespread program to condition public fire reaction is the fire drill program carried out in schools throughout the United States and Canada. Practically every state and province requires that fire drills be held periodically in public, private, charter, and faith-based schools. The drills are designed to ensure, so far as possible, the immediate evacuation of students from a school building in the event of a fire or other emergency. Most schools accept the fire drill program as a routine part of their operation and have shown little hesitation in actively requiring drills.

Many hidden benefits have come from the school fire drill program. Undoubtedly, individuals have reacted properly to fires in places other than schools as a result of training they received while they were in school. Fire drills in schools have probably helped individuals to more readily accept fire drill programs in institutions, industries, and other locations. Reactions to fires in theaters and other public assembly locations are also more positive as a result of conditioning received in school fire drill programs.

Institutional Fire Drills

In an institution, especially where nonambulatory individuals reside, reaction is usually expected only from the institutional staff and visitors. The patients most often are not counted on for reaction of any kind, and institutional personnel initiate all evacuation measures. Complete evacuation, which is usually achieved in educational occupancy drill programs, is not generally expected in institutions. The drill may include horizontal evacuation to a place of safety within the structure rather than to a location outside the building. In some cases "defending in place" is recommended.

Obviously, typical evacuation procedures are not practiced in places of incarceration. In such institutions, measures are taken to eliminate the chance of escape as far as possible. Therefore evacuation is, at best, most difficult and must be planned to fit the particular conditions in each facility. Although inmates in a prison are most likely physically capable of departing from a burning structure, they must first be released from their cells by correctional officers.

Industrial Fire Drills

Years ago little emphasis was given to fire drill programs in industry. Long hours coupled with arduous tasks left little time for them. Two high-profile, significant examples of workplace risk "gone bad" are the Triangle Shirtwaist Factory fire in 1911 and the Hamlet, North Carolina, poultry plant fire in 1991. Both indicated the need for comprehensive fire reaction programs in this class of occupancy. Workplace deaths from fires and explosions continue to occur, underscoring the importance of the continued need for fire reaction training, workplace fire inspection, and workplace management's responsibility to bring the work environment into fire and life safety compliance.

In arranging fire reaction programs for industry, officials must remember that shutdown time is expensive and that many machines cannot be safely left running unattended. Personnel can be familiarized with the location of fire exits, the location of fire alarm stations, and procedures to follow in the event of fire. As in the case of institutions, employees in industry may have duties to perform in reaction to fire that involve more than merely evacuating the structure. They may be responsible for obtaining and operating fire-extinguishing equipment and calling the fire department. These duties should be described in an emergency action plan. Illegal removal of products during fire drills is of concern in some industries.

Training in escape procedures may be one of the best ways to reduce the stress related to an actual emergency. When facing an actual emergency, increased stress levels heighten the sense of danger and thus tend to result in a more immediate decision to escape. The late Guylène Proulx, senior researcher at the National Research Council of Canada, described how stress can enhance what a person learned in training or drills. Proulx explained that training and exercise of a well-developed decision making plan ahead of time is more likely to be successful when a person is under stress. She further explained that when faced with a stressful situation a person's decision-making will become more focused and consider only a few options and that under heightened stress new solutions may not even be considered.[4]

Proper Use of Fire Extinguishers

Fire extinguishers have put out thousands of fires in the incipient stage. However, a major weakness in fire reaction programs is failing to adequately train people in their use and limitations. Records are replete with incidents of individuals who attempted to control fires with extinguishers that were inadequate or improper for the conditions encountered. In cases where the individual also delays reporting the fire to the fire department while using the extinguisher, large losses may ensue.

A great deal of thought has been given to the relative role of the portable fire extinguisher in the fire protection field. This device depends entirely on proper human fire reaction for its effectiveness. Directions printed on the fire extinguisher are sometimes detailed and may be somewhat intimidating.

Programs for education in the use of fire extinguishers are widely available. The all-class fire extinguisher, which may be used on Class A, B, or C fires, has simplified training procedures considerably.

Traditionally, training in the use of fire extinguishers has been confined to individuals who have a direct responsibility for protecting specific locations, such as nurses in hospitals, custodians in schools, crowd control managers in a nightclub, or maintenance employees in industrial plants. Even in industry, training of this type is not usually given to individuals who have little likelihood of ever using an extinguisher. Likewise, office workers in industry seldom are given such training.

Large Residential Occupancies

Another place where fire reaction training can enhance fire safety is the multi-family residential occupancy. Apartment houses and condominiums with a number of units and floors generally have interior fire alarm systems as well as fire extinguishers mounted at various locations in the structure. Still, there are very few apartment houses or condominiums in which drills are carried out or in which fire extinguisher training is given.

Individuals living in apartments generally give little thought to the fire safety features of the structure. If a fire drill were staged periodically, tenants might feel more secure. For example, fire drills are fairly common in college dormitories. The procedure has saved lives on many occasions.

Measures to Facilitate Successful Evacuation

A number of positive steps can be taken to increase the success of escape from a fire. The importance of every second of time saved cannot be overemphasized. On several occasions, an individual has taken control of an emergency by immediate action. For example, in a theater fire some years ago in Florence, South Carolina, a young usher went to the front of the theater, announced a fire drill, and asked everyone to leave in an orderly manner. Many people thought it was in fact only a drill. They reacted calmly and departed the structure only to realize that there was a major fire in the building. Quick thinking on the part of an individual can be quite important, especially if the individual can be identified as having some authority.

POSTING OF MAXIMUM OCCUPANCY SIGNS

Posting maximum occupant capacity for structures may help prevent crowding, but not exceeding the capacity needs to be a managed process. Facility management is responsible for keeping the capacity within the established limits. This responsibility means using some means of counting patrons. Counting can be accomplished by use of turnstiles or other means of queuing lines with someone standing by with a hand tally counter clicking each person going through the queue. Selling tickets is another way to control capacity prior to an event.

Sometimes being turned away can create a gate-crashing incident. A contingency plan for overflow, or for people wanting to enter when capacity has already been reached, is beneficial. Ticket sales or other preevent control plans plus a procedure at the gate itself can help prevent crashing. Adequate notice in advertisements should explain the policy for capacity as well.

The fire marshal can help by providing education to event venue management and staff well in advance of events to increase awareness. It is also helpful to provide staff training in fire safety, and if event attendee thresholds meet the requirements of the local code, then crowd managers as well as training may be in order. Posting of allowable attendee numbers on occupant capacity certificates should be based on measurable standards, so the number of people inside the structure at no time exceeds the exit capacities and predetermined limitations based on a square-footage formula.

Overcrowding is a very difficult condition to control, especially where people are moving in and out randomly. It is one of the key contributing factors to high-fatality fires and is especially a problem with assembly occupancies. It takes time to move a large number of people, even with adequate egress, so they need to be made aware of an emergency as soon as possible. Occupancy overcrowding contributed to the high number of deaths in the fire at The Station, a nightclub in Rhode Island, on February 20, 2003,[5] one of a number of similar incidents that occurred over a short period of time and led to adoption of new procedures in model codes to help facilitate safe management of crowds.

NFPA 101, Life Safety Code, and both model fire codes from NFPA and ICC include crowd manager requirements. One key responsibility is to manage the crowd by keeping it within the structure's designed occupancy capacity. Within the code requirements, crowd managers receive specific crowd management training, including procedures for facilitating safe evacuation, evaluating the size of crowds, and reducing crowding when occupancy has exceeded capacity.

EMERGENCY LIGHTING

Emergency lighting of the exit way is a significant safety measure. When general lighting fails or is obscured by smoke, emergency lighting helps to facilitate safe evacuation.

EXIT SIGNS

A combination exit and emergency light is on the market. It provides both directed lighting under the fixture for exit visibility and lighting of the exit sign.

Much study has been given to choosing the proper color for exit signs. Some codes require red; others require green. That is because some people associate red with fire and exit lights, while others automatically go toward green to reach a place of safety. There are valid arguments on both sides.

MAINTENANCE OF EXIT FACILITIES

An extremely important measure in successful evacuation is maintenance of exits. No exit can be considered usable if opening it requires a key or the removal of obstructions. Individuals react differently under the stress of even a small amount of smoke or heat, so although having to clear an exit may be a minor problem, in a time of stress even moving a chair away from a door may be a monumental task.

Any blockage of an exit could cause death. It is not difficult to imagine a number of people moving quickly to an exit only to trip over an obstruction and then have more people tripping behind them, soon completely blocking the exit. A 2013 fire in the Kiss Nightclub in Santa Maria, Brazil, claimed the lives of over 200 people. Critical contributors to the loss of life were overcrowding by nearly twice the designed capacity and only one known exit, the front entrance.

In venues where crowds are expected, crowd managers can help. Their training makes them aware that all exits need to be free and clear, unlocked, and ready for any emergency evacuation. A trained staff also can help by facilitating the movement of people toward less-used exits to reduce waiting time.

Exits must be considered in the original construction of a building. Often, occupancy changes bring changes in exit requirements. Through periodic inspections, appropriate fire officials can note those changes, and proper remedial steps can be taken.

VOICE COMMUNICATION SYSTEMS

Voice communication systems are extremely helpful in immediately notifying occupants of any emergency condition that arises. In some cases, a voice recording operated by automatic control is used instead of the more common fire bell or horn. Fire drills can help differentiate the emergency evacuation alarms from other notification systems, such as the telephone system and any audible shift-change indicators. Public assembly occupancies may be equipped with both fire alarm and voice communication systems. *Note:* Emergency power is essential to the successful use of a public address system for fire evacuation.

A voice communication system also is the desirable choice because the best way to encourage rational and adaptive responses to fire situations is to provide occupants with information about the incident. This information is helpful not only in promoting successful evacuation but also in discouraging neglect or apathy in response to fire alarms. If a fire alarm goes off and information about its true nature is not provided (Is it an actual emergency, a minor nuisance, a false alarm?), people are likely to regard every fire alarm as false because they rarely experience one that signals an actual fire.[6]

In an effort to prepare people for the possibility of a fire or other emergency, some jurisdictions require advance notice to patrons, encouraging them to look for exits within the structure. Jurisdictions may mandate the use of a printed program, printed sign, oral announcement, or instructions on a screen to the effect: "For your own safety, look for an exit now. In case of emergency, walk, do not run, to that exit." This kind of instruction is especially prevalent in movie and other theaters, where it can be projected onto the screen or printed on the program (Figure 8-2).

Billy Morris Director of Fire Science Cisco College

FIGURE 8-2 Mass notification systems can provide building or area occupants with timely and carefully directed instructions or information.

Human Behavior in Fire

John L. Bryan studied and wrote extensively about human behavior related to fire. He was the author of a chapter called "Human Behavior and Fire" in the *Fire Protection Handbook*.[7] In the handbook, Dr. Bryan provided an overview of the complex area of human behavior and fire. As Bryan was beginning his career as a fire protection engineer in the 1940s, the United States was regularly experiencing mass fatality fires, including, for example, the fires in the Cocoanut Grove Nightclub in Boston, Massachusetts (492 killed); the Winecoff Hotel in Atlanta, Georgia (119 killed); the Rhythm Club in Natchez, Mississippi (209 killed); and the La Salle Hotel in Chicago, Illinois (61 killed). There was a complete lack of interest in the study of human behavior related to fires in the 1940s and 1950s. An exception was Bryan's involvement in an interview study of some survivors of a fire in Arundel Park, Maryland, in 1956. The fire occurred in a social hall where a fund-raiser was held for the St. Rose of Lima Roman Catholic Church. Nearly 1,200 people attended the event; 11 were killed and hundreds injured. The interview study verified that when family members reentered the hall trying to reach loved ones, they contributed to the confusion and blocked the escape of others.

Beginning with the comprehensive study of the U.S. fire problem in *America Burning*,[8] the 1970s and 1980s were very productive in the study of fire and human behavior. In reference to that time, Bryan wrote:

> The emphasis of the studies in human behavior in fire during this time period was on defining the behavioral actions of the occupants in fire situations, the examination of the then popular concept of "panic behavior," and an emphasis on the study of the evacuation process as it occurred in high-rise building fires.[9]

His choice of words in framing the study appears to support the emerging evidence that would begin to build a new norm for human behavior in fires. For decades the word *panic* was used to describe the behavior of fire victims, survivors, and casualties alike. However, evidence and data do not support the assertion that panic is any more than a rare occurrence. In fact, investigation by behavioral scientists finds that people intimately involved in an escape from fire actually tend to be calm, determined, and concerned for the welfare of others. Again, we can look to John Bryan's reflections:

> Thus, the behavior of the individuals intimately involved with the initiation of the fire is critical not only for themselves but often for the other occupants of the building. It should be recognized that the altruistic behavior observed in most fires with the interaction of the occupants and the fire environment in a deliberate, purposeful manner appears to be the general mode of reaction. The nonadaptive flight or panic-type of behavioral reaction is apparently unusual in fires.[10]

The frightening reality of fire can be seen in the real-time video recording of the pyrotechnics-ignited foam insulation in the overcrowded Station nightclub in 2003. One behavior that night was people watching the fire before deciding on their personal danger. Though recognizing the danger, a number of them could be seen calmly making gestures indicating the need for others to head for the exit. The delay on the part of many is a human trait: the need to collect cues or investigate a situation before deciding flight is necessary. Dr. Erika Kuligowski, a Fire Protection Engineer and scientist with NIST, described this behavior in the NIST report The Process of Human Behavior in Fires. Kuligowski explained that decisions are made only after the interpretation of the situation and the consideration of any risk based cues presented and that actions taken are based on this process. Kuligowski presented that actions are not random, but the result of a predictable decision-making process.

What could be seen in The Station nightclub video was order and calm as people walked toward one of the exits surrounded by smoke. There was no running or trampling to get out. Instead, there was a steady stream of people leaving. Left to reflect on their nightmare escape once outside, hundreds stood scared, shaken, and worried about friends, but they were not panicked. Many of the people who had already escaped went to other possible exits and could be seen trying to help people still trapped inside with the fire. None of this suggests that there was no sign of chaos or terror in the nightmare that unfolded on the video. But there was evidence of the best kind of human behavior offered in an emergency.

Panic behavior in fire is overstated. In a 2009 paper Fahy, Proulx, and Aiman said, "Over several decades, studies specifically looking at panic behaviour in fires have consistently shown that non-adaptive and irrational behaviours are actually a rare occurrence."[11] According to the article, "Fighting Fire with Psychology,"[12] researchers discovered a few things about people's behavior during fires: People generally do not panic, they are often altruistic, they try to exit through the door they entered, and they will move through smoke when necessary.

Articles and papers presenting findings about human behavior during fires say that panic is a rare occurrence. An article by the Newhouse News Service that appeared in the *Montreal Gazette* in November 1982[13] briefly described actions of how people, despite being trapped in the 1980 fire at the MGM Grand Hotel in Las Vegas, took the time to patiently help others who were not able to self-evacuate due to disabilities. The article went on to describe this as one of the surprising behaviors discovered in, at that time, new and fast-evolving research in human behavior that was dispelling myths about how people react to fire. The article quotes Dr. Bryan: "If you're not scared (during a fire), you're not intelligent. But the point is that people will describe their feelings as panic, but what they are describing is a high degree of anxiety and tension."[14]

Fire Safety Manuals

fire safety manual ■ a document that forms the basis of fire drill procedures and is tailored to a specific occupancy.

In many occupancies, a formal fire brigade is impractical. In those locations, a **fire safety manual**—a document that forms the basis of fire drill procedures and is tailored to its specific occupancy—may be used to promote proper fire reaction.

Academic residence halls and hotels often have posted fire safety instructions for occupants. Sometimes this information is included in a guest information folder.

Fire Reaction in Housing for Older Adults

In a 2013 report to Congress on the projected impact of aging Americans on Fire and Emergency Medical Services, the Federal Emergency Management Agency (FEMA) said: "Older adults represent one of the highest fire-risk populations in the United States."[15] They are two and one-half times more likely to die in a fire than the general population. In 2002, 34 percent of all fire deaths in residential structures and 14 percent of all fire-related injuries involved people 65 years and older.[16] This problem is likely to get worse with the rapid increase in the older adult demographic. According to the U.S. Census, between 2000 and 2010 the 65 and older segment grew at its fastest rate ever (15.1 percent) compared to the growth rate of the general population (9.7 percent).[17]

An outstanding example of guidance for instilling the proper reaction to fire is the publication *Life Safety from Fire: A Guide for Housing for the Elderly*, a report prepared for the Architectural Standards Division, Federal Housing Administration, of the U.S. Department of Housing and Urban Development (HUD).[18] The report was specifically designed to pinpoint problems in housing for older adults (age 65 and over). It was based on a study of thousands of fires and then reduced to a detailed study of 1,000 fires. It points out that many older adults are housed in buildings that were not specifically designed for them. New occupancy classifications are generally developed through experience with disastrous fires.

Few, if any, jurisdictions provide specific requirements for apartment houses occupied by older adults. The report suggests that a series of fires may need to occur before such requirements are imposed. The report emphasizes that older adults do not react to fire in the same manner as younger people. A code requirement for 100-foot travel distances to exits is mentioned as being satisfactory for younger people, who might be able to run that far under smoky conditions. However, this is an almost impossible feat for a person over 80. A 21-year-old could easily escape from a second-floor window; this would be very difficult for a 78-year-old. An individual's reaction time is considered most critical. This is the time it takes the occupant to react to the fire and then to attempt to escape. Reaction times for older adults are usually almost twice as long as those for young people.

Older adults who have aged in place, creating the so-called Naturally Occurring Retirement Community (NORC), may be in an environment that is not well suited for the physical and cognitive changes occurring with the aging process. Their residential structures in most cases have changed little while the needs of the residents have changed considerably. Many of the buildings, especially residential high-rises, were built between the 1940s and 1970s, before building codes required fire sprinkler protection for apartments and more sophisticated evacuation alarms that can be readily heard in the apartments. When fires do occur, there is a greater risk for the elderly residents due to losses in their hearing, vision, and mobility. Some cities such as Louisville, Kentucky, have planned for this situation and have required retrofitting of fire sprinklers in residential high-rises. Other jurisdictions have attempted to adopt similar retrofit requirements but have not been able to overcome the resistance from more powerful stakeholders.

There is no building code classification that provides specifically for the special accommodation safety needs of older adults still able to live independently. One recommendation for future planning and design in regard to aging is to champion a building code classification that provides a higher level of safety in multistory residential buildings built exclusively for older adults. Ideas to consider include areas of refuge and robust elevator hoist ways with positive air pressurization that could accommodate safe evacuation.

The California State Fire Marshal recognizes that now it can take as little as 3 minutes to escape a home fire.[19] However, the increasing severity and danger of home fires resulting in a significantly limited time to escape plus a rapidly increasing number of older adults with a list of physical and cognitive limitations affecting their survival of a fire should raise concern about even aggressive actions to increase home fire safety.

The Montgomery County, Maryland, Senior Citizen Fire Safety Taskforce set out to reduce the fire risk to older adults in its community. The task force found the key to success in preventing an increased fire fatality rate for older adults is changing sociocultural attitudes regarding fire safety. The attitudinal shifts are detailed in the task force's report and focus on several approaches as follows: (1) Educate older adults, particularly those living alone and aging in place, about the vulnerabilities common among older adults and the necessity of responding appropriately in the event of an emergency; (2) orient families, friends, and neighbors of older adults to be responsible for installation and routine maintenance of fire safety devices such as smoke alarms; (3) educate older adults and their families about appropriate residential placement options, given potential impairments and disabilities that reduce full functional independence for response in emergencies; (4) instill in all citizens, regardless of age, background, and national origin, that fire safety is an individual responsibility of all members of society and that negligent behavior in causing fire fatalities and injuries may have legal consequences; and (5) improve the capacity of emergency responders to better understand the potential functional impairments associated with aging and to communicate more effectively in assisting older adults in emergency situations.[20]

The National Fire Protection Association has focused significantly on all aspects of fire safety for older adults with its trademarked "Remembering When" services. "Remembering When" is a cooperative effort with the Centers for Disease Control (CDC) and is offered to promote the ability for older adults to live independently and safely for as long as they can. The program has a dual purpose: fire prevention and fall prevention. The two subjects are addressed in 16 safety messages, 8 for each risk area.

Fire Reaction in High-Rise Structures

High-rise buildings can present serious fire reaction problems. A number of factors make it difficult for individuals to get out of these buildings in the event of fire. One of them is the common use of the elevator. People occupying high-rise buildings seldom use stairways, and they, as well as their visitors, may not be familiar with the location of stairways.

An example of this problem is a fire that occurred in a large southwestern city in a nine-story modern office building. All windows were of the fixed-sash type so that the balance of the air-conditioning system would not be disturbed, a common feature in structures of this type. The fire occurred during working hours. It was relatively minor and was confined to the basement of the structure. The problem was that smoke from the fire rapidly permeated the upper floors of the building through the air-conditioning system. The fire was brought under control, and firefighting equipment was being prepared for return to service, when a large executive chair came hurtling down from an upper floor, hitting the hose bed of one of the fire pumpers.

Personnel in the street looked up. Occupants had gathered around the window they had broken by throwing the chair and were calling for help. An aerial ladder, which was still on the scene, was raised to the floor level and used to rescue some of the people. Fire department personnel also went up the stairways to the top floor and assisted others to safety.

All the people assembled on the top floor were in a near state of panic and had not given thought to using either of the stairways, which were readily accessible nearby. For some reason, not one of these people thought of trying any means of egress other than the elevator. Countless high-rise buildings throughout North America present similar evacuation problems. Normal fire reactions must often be modified in such structures.

Many studies have been carried out to develop improved fire safety procedures for high-rise structures. One of the most promising developments is the voice communication system (already described earlier in this chapter) for advising people about fire conditions. Protection for high-rise structures also should include automatic sprinkler protection. With this equipment, normal fire reactions might be entirely suitable because fires would be held to such minimal proportions as to enable egress by normal methods.

Lighting is another special factor for consideration in the high-rise occupancy. A power failure at a time of darkness can be extremely critical with or without a fire. The high-rise structure should be equipped with emergency lighting units or emergency generators to give occupants the opportunity to reach a place of safety without appreciable delay. They should not have to stumble down many levels in the dark to reach the outside of the building. This problem was reemphasized in the World Trade Center bombing in New York in 1993. More than 100,000 people were evacuated in the incident, 6 employees were killed, and more than 1,000 persons were injured.

A study of behavior during that explosion revealed some interesting findings. The study verified that persons evacuating by stairway in high-rise buildings will continue their downward travel even as smoke conditions intensify. The need for evacuation training for all occupants, not just fire wardens, was amplified. Occupants need to have an understanding of smoke movement, the stack effect, and the dangers of falling glass to persons on the street below.[21]

Terrorist Activities

An errant (nonterrorist) airplane may strike a major building adjacent to an airport much the same as a plane struck a hotel adjacent to the Indianapolis airport years ago. But the events of September 11, 2001, were incomprehensible to most Americans.

The 2001 attacks on the World Trade Center buildings in New York City gave impetus to studies relating to overall safety factors in high-rise buildings. Among the most prominent studies was one conducted by the National Institute of Standards and Technology (NIST) with the assistance of the National Institute of Building Sciences. The process of investigation into the collapse of the World Trade Center towers was the result of the October 2002 National Construction Safety Team Act initiated by President George W. Bush to perform critical building safety investigations similar to activity by the National Transportation Safety Board investigations of critical transportation crashes and disasters. The act gives NIST complete investigative authority, including that needed to subpoena evidence, to have access to building sites, and to record, document, preserve, and move evidence. The act does not empower NIST to mandate reforms or establish preventive regulation or standards as a result of its findings.

An investigation into the collapse of the World Trade Center was announced by NIST in August 2002. The investigation was fully funded by the U.S. Congress through the Federal Emergency Management Agency and was conducted under the authority of the 2002 National Construction Safety Team Act. The investigation was initiated to improve high-rise building safety. The goal of the investigation focused on the contribution of construction, materials, and technical conditions to the collapse. It was intended to improve how tall buildings are designed, constructed, maintained, and ultimately used. A goal was also to identify recommendations for revising existing codes and developing new codes, standards, and practices related to tall buildings.[22]

The *Final Report on the Collapse of the World Trade Center Towers* offered 30 recommendations for the model codes, standards industry, design community, and the emergency response.[23] An ad hoc committee of the International Code Council (ICC) reviewed the findings and developed code change proposals, which were considered by the ICC membership. Though not all of the code change proposals were approved by the ICC members, many were. Some of the proposals that were not approved were strongly considered and have influenced structural design. A range of expense (from very high expense to no cost) was related to some of the code changes. Opposition during the testimony phase of the code considerations by some stakeholders cautioned that the cost would negatively impact the profitability of real estate. Although this was a narrow stream of opposition, it shows that even with the loss of life from the September 2001 terrorist attacks, there is still a call for a cost-balanced approach to safety.[24]

The World Trade Center disaster ultimately led to extensive changes in the New York City building and fire codes, which included requirements for automatic sprinklers in older commercial high-rise buildings and requirements to reduce the potential for building collapse. Recommendations of the *Final Report on the Collapse of the World Trade Center Towers*[25] also addressed basic code changes needed to facilitate speedy emergency evacuation (fire reaction) from high-rise structures. Several examples follow:

- An additional exit stairway for buildings more than 420 feet in height
- A minimum of one fire service access elevator for buildings more than 120 feet in height
- Luminous markings delineating the exit path (including vertical exit enclosures and passageways) in buildings more than 75 feet in height to facilitate rapid egress and full building evacuation
- Significantly limiting the length of horizontal transfer corridors in stairways and requiring signage when such corridors change direction
- Increasing the physical separation distance (remoteness) between exit stairway enclosures beyond that currently required to protect the egress system from accidental structural loads in addition to fires
- Enhancing the functional integrity and survivability of exit stairway wall enclosures for buildings greater than 420 feet in height by requiring wall surfaces to possess a minimum level of structural robustness through an ability to withstand a sudden increase in pressure[26]

In 2008 researchers from NIST released their findings that the September 11, 2001, collapse of the 47-story World Trade Center Building 7 was caused by fire, with the loss of water supply for sprinklers as a factor. They believed that this was the first time a fire caused the failure of a modern skyscraper.[27]

An example of a training program that addresses occupancy evacuation is the Hotel Employee Life-Safety Program (H.E.L.P.) taught by the Las Vegas, Nevada, Fire Department in the hotels of that city. The program trains hotel security officers, management, and staff on what to do during an emergency prior to arrival of the fire department and how to assist firefighters after they arrive on the scene. It has proved to be quite successful.

Summary

An integral part of fire prevention planning and education is instilling awareness in the public of how to act and what to do in the event of a fire, at home or in a public place. Any assessment of human reactions that might be anticipated in a fire emergency must take into account individuals of all ages, health classifications, physiques, occupations, and temperaments. A fire's severity and location, how it started, and the presence of people in the fire area must also be considered.

Escaping a building on fire is the highest priority for an occupant. After that is achieved, the next thing to do is call for the fire department. Public marketing of the universal emergency telephone number 9-1-1 on public transportation, municipal vehicles, and the sides of fire department vehicles can keep the fire department telephone number constantly before the public.

Fire drill practices in homes, schools, institutions, and industry teach people how to respond in a fire emergency in a planned, orderly way. Fire drill programs in schools throughout the United States have probably had a more lasting effect in conditioning the public to proper fire reaction than any other single effort.

There is a need for more widespread public education in the proper use of portable fire extinguishers and particularly in recognizing their limitations, both in the home and in public buildings. Although fire extinguishers have been used successfully on thousands of fires, many fires involving large losses have ensued because of failure to use fire extinguishers properly or delays in reporting a fire while attempting to use an extinguisher.

In planning measures for improved public fire reaction, the planner must consider conditions in buildings that may impede evacuation. The required fire detection and alarm systems should be provided to give occupants plenty of time to consider cues to emergencies and escape without delay, especially in public places. Typically, crowded occupancies such as places of assembly require the most robust means available to facilitate evacuation, including trained crowd managers, enforcement of maximum occupancy limits, emergency lighting, clear designation and maintenance of exits, use of public address systems for emergency instructions, and advance notice to patrons to look for the exits nearest them. Closely allied to those measures is posting of signs near public elevators warning passengers not to use the elevators in the event of a fire and to use marked exit stairways instead.

Fire reaction safeguards in housing for older adults and in high-rise structures have received special emphasis as a result of several serious fires. Studies are being made of the special problems in both types of occupancies. The essential links of life safety are measures for fire prevention, detection, alarm, confinement, control and extinguishment of the fire, and escape or refuge from the fire.

Not to be overlooked is the role of security personnel in fire protection of the premises in institutions, mercantile establishments, and industrial plants. Responsible people should be hired for this job and should be given adequate training to ensure proper fire reaction.

Review Questions

1. In an assessment of anticipated human reactions in the event of a fire, one must consider:
   a. height.
   b. weight.
   c. age.
   d. sex.
2. The condition of the individual at the time of the fire includes factors such as:
   a. intoxication.
   b. marital status.
   c. occupation.
   d. temperament.
3. Which national fire safety program encourages individuals to locate potential escape routes and leave the dwelling place without delay in the event of an emergency?
   a. National Emergency Number Association
   b. Fire Wise Program
   c. Exit Drills In The Home
   d. Home Fire Sprinkler Coalition
4. In the article *Fighting Fire with Psychology*, researchers discovered that:
   a. people do not panic.
   b. people try to exit through the door they entered.
   c. people will move through smoke when necessary.
   d. all the above.
5. Institutional drills may include all of the following *except*:
   a. complete evacuation.
   b. horizontal evacuation.
   c. staff actions.
   d. inmate simulation.
6. In the 2013 Kiss Nightclub in Santa Maria, Brazil, one critical contributor to the loss of life was:
   a. dark stairways.
   b. overcrowding.
   c. slow fire department response.
   d. use of elevators.
7. Successful evacuation can be increased by use of:
   a. emergency lighting.
   b. public address systems.
   c. exit signs and maintenance of exits.
   d. all of the above.
8. Older adults are at risk from fire owing to:
   a. limited mobility.
   b. reduced hearing and sense of smell.
   c. poor vision.
   d. all of the above.

End Notes

1. Bernard M. Levin, unpublished paper presented at National Fire Protection Association Fall Meeting, Nashville, TN, 1988.
2. "Americans 'React Incorrectly' to Fire," *Fire Chief*, August 2007, Vol. 51, Issue 8, p. 17.
3. NIST Engineering Laboratory, *Smoke Alarms + Sprinklers + Closed Doors = Lives Saved in Dorm Fires* (Gaithersburg, MD: NIST Engineering Laboratory, March 30, 2010). Accessed May 30, 2013, at http://www.nist.gov/el/fire_research/dorms_033010.cfm
4. Guylène Proulx, "Cool Under Fire," *Fire Protection Engineering Magazine*, No. 16, Fall 2002, pp. 33–35. National Research Council Canada document, NRCC-45404. Accessed May 23, 2013, at http://archive.nrc-cnrc.gc.ca/obj/irc/doc/pubs/nrcc45404/nrcc45404.pdf
5. National Institute of Standards and Technology, *Recommendations—NIST Investigation of The Station Nightclub Fire* (Gaithersburg, MD: National Institute of Standards and Technology, June 28, 2005). Accessed December 1, 2013, at http://www.nist.

gov/public_affairs/factsheet/ri_recomm_factsheet.cfm

6. Personal communication to author Robertson from Mark Chubb, Southeastern Association of Fire Chiefs.
7. J. L. Bryan, "Human Behavior and Fire," Chapter 4.1 in *Fire Protection Handbook*, 19th ed (Quincy, MA: National Fire Protection Association, 2003).
8. National Commission on Fire Prevention and Control, *America Burning* (Washington, DC: National Commission on Fire Prevention and Control, 1973).
9. Bryan, pp. 4-3.
10. Ibid., pp. 4-3 to 4-4.
11. Rita F. Fahy, Guylène Proulx, and Lata Aiman, *Panic and Human Behaviour in Fire*, NRCC-51384 (Ottawa, Ontario: National Research Council of Canada, July 13, 2009). Accessed May 30, 2013, at http://archive.nrc-cnrc.gc.ca/obj/irc/doc/pubs/nrcc51384.pdf
12. Lea Winerman, "Fighting Fire with Psychology," *Monitor on Psychology*, September 2004, Vol. 35, No. 8 (American Psychological Association), p. 28. Accessed May 30, 2013, at http://www.apa.org/monitor/sep04/fighting.aspx
13. Newhouse News Service, "Fire Study's Surprise: Most People Don't Panic," *Montreal Gazette*, November 17, 1982 (Google News). Accessed December 2, 2013, at http://news.google.com/newspapers?id=XokxAAAAIBAJ&sjid=NqUFAAAAIBAJ&pg=1084%2C3047651
14. Ibid.
15. Federal Emergency Management Agency, *Impact of an Aging Population on the Fire and Emergency Medical Services: Fiscal Year 2013 Report to Congress* (FEMA, May 23, 2013).
16. U.S. Fire Administration/National Fire Data Center, *Fire and the Older Adult,* FA-300 (Washington, DC: Department of Homeland Security, January 2006). Accessed May 24, 2013, at http://www.usfa.fema.gov/downloads/pdf/publications/fa-300.pdf
17. Carrie A. Werner, *2010 Census Briefs: The Older Population: 2010*, C2010BR-09 (Washington, DC: U.S. Census Bureau, November 2011). Accessed May 24, 2013, at http://www.census.gov/prod/cen2010/briefs/c2010br-09.pdf
18. Raymond D. Caravaty and David S. Haviland, *Life Safety from Fire: A Guide for Housing for the Elderly*. Report prepared for the Architectural Standards Division, Federal Housing Administration (Washington, DC: U.S. Government Printing Office, 1968).
19. CAL FIRE–Office of the California State Fire Marshal, *California State Fire Marshal Smoke Alarm Task Force Final Report Analysis and Recommendations: Understanding, Utilization, and Effectiveness of Smoke Detection Technology including Ionization, Photoelectric, and Other Technologies* (August 2011). Accessed December 1, 2013, at http://osfm.fire.ca.gov/firelifesafety/pdf/Smoke%20Alarm%20Task%20Force/CSFM%20SATF%20Report%208-16-11.pdf
20. Montgomery County Fire and Rescue Service, *Seniors at Risk: Creating a Culture of Fire Safety: Senior Citizen Fire Safety Task Force Final Report* (Rockville, MD: Montgomery County Fire and Rescue Service, September 2008). Accessed on May 24, 2013, at http://www6.montgomerycountymd.gov/content/frs-safe/downloads/older/scfst080905.pdf
21. Rita F. Fahy and Guylène Proulx, "A Study of Human Behavior during World Trade Center Evacuation," *NFPA Journal,* National Fire Protection Association (March/April 1995), pp. 59–67.
22. National Institute for Standards and Technology, *World Trade Center Disaster Study*. Accessed November 2, 2013, at http://www.nist.gov/el/disasterstudies/wtc/
23. National Institute for Standards and Technology, *Final Report on the Collapse of the World*

Trade Center Towers (Gaithersburg, MD: Author, September 2005), p. 203. Accessed December 2, 2013, at http://www.nist.gov/customcf/get_pdf.cfm?pub_id=909017

24. Eric Lipton, "Agency Fights Building Code Born of 9/11," *New York Times,* September 8, 2008, p. A1. Accessed May 27, 2013, at http://www.nytimes.com/2008/09/08/washington/08codes.html
25. National Institute for Standards and Technology, *Final Report on the Collapse of the World Trade Center Towers* (Gaithersburg, MD: National Institute of Standards and Technology, September 2005), p. 203. Accessed December 2, 2013, at http://www.nist.gov/customcf/get_pdf.cfm?pub_id=909017
26. NIST news release, www.nist.gov, July 6, 2007.
27. *Star Telegram* (Fort Worth, TX: August 21, 2008), p. 2A.

CHAPTER 9

Fire Prevention Efforts of the Private Sector

AA World Travel Library/Alamy

KEY TERMS

Congressional Fire Services Institute, *p. 173*
Fire Protection Research Foundation, *p. 171*
FM Global Group, *p. 177*
International Association of Arson Investigators, *p. 172*
International Association of Fire Chiefs, *p. 172*
International Association of Fire Fighters (IAFF), *p. 172*
International Fire Service Training Association, *p. 174*
National Fire Protection Association (NFPA), *p. 170*
National Safety Council, *p. 173*
National Volunteer Fire Council, *p. 174*
Society of Fire Protection Engineers (SFPE), *p. 173*
Underwriters Laboratories, Inc. (UL), *p. 179*

OBJECTIVES

After reading this chapter, you should be able to:

- Describe the primary role of the National Fire Protection Association.
- List some organizations with primary fire prevention functions.
- List some organizations with an allied interest in fire prevention.
- Describe the relationship between fire prevention and the insurance industry.
- Explain the role of the Insurance Services Office as it relates to the fire service.

Visitors from other countries who are in the United States studying fire protection activities are often amazed at the important role played by nonpublic fire safety agencies. In many countries, the private sector is of little significance in the development and implementation of fire protection procedures. By tradition, fire prevention activities in the United States have been motivated to a great extent by private organizations.

Fire safety activities of the private sector may be divided into three categories: those in which fire prevention is the primary function of the nonpublic organization; those in which fire prevention is a secondary effect of activities in the promotion of trade; and those in which fire prevention activities are carried out for the internal protection and well-being of the organization.

Organizations with Primary Fire Prevention Functions

NATIONAL FIRE PROTECTION ASSOCIATION

National Fire Protection Association (NFPA)
■ an organization that aims to reduce the risk of fire and other hazards by offering consensus codes and standards, educational and training materials, and fire safety advocacy.

In the category of organizations having the prevention of fire as a primary responsibility, there is little doubt that the **National Fire Protection Association (NFPA)**, headquartered in Quincy, Massachusetts, holds the top position. The mission of NFPA is to reduce the worldwide burden of fire and other hazards on the quality of life by providing and advocating consensus about codes and standards, research, training, and education.

The NFPA was founded in 1896. It is a nonprofit educational and technical organization devoted entirely to preventing loss of life and property by fire and other hazards. Membership is open to any individual, company, or organization having an interest in its objectives. NFPA is recognized worldwide for its impact in the fire prevention field. It has members from more than 100 countries.

Activities of the NFPA include the developing and publishing of technical standards, codes, recommended practices, and model ordinances. More than 300 such publications have been developed, prepared, and updated under the guidance of approximately 250 technical committees. More than 5,500 individuals serve on the committees, all on a volunteer basis. They usually include representatives from industry, public fire services, equipment manufacturers, the insurance field, governmental agencies at all levels, and other qualified individuals. Committees are arranged to preclude their domination by any represented groups that may have special interests in the subject at hand.

Standards are adopted only after detailed study and vote by the committee. Built into the process of developing and adopting standards are calls for proposals to amend existing documents or to shape the content of new documents. The public proposals, together with committee action on each proposal, as well as committee-generated proposals, are published in the *Report on Proposals* for public review and comment. Public comments, along with committee action on each comment, are then published in the *Report on Comments.* Only after this public review and comment cycle has been completed is the final committee report brought before the membership for action. Once adopted by the NFPA membership at either an annual or a fall meeting and issued by the Standards Council, standards are published and made available for voluntary adoption. Many of the codes and standards have had great influence and are used as the basis for legislation and regulation at all levels of government.

The NFPA has for many years published educational materials in the fire prevention field. They include school curricula (K–12), pamphlets, posters, occupancy studies, fire records, films, videos, and a wide variety of materials suitable for use in local fire prevention promotional efforts. More than 800 publications are available from the NFPA. The NFPA also maintains the world's most comprehensive fire protection library at its headquarters. The NFPA provides technical and other topical fire safety information in its membership magazine, *NFPA Journal,* which is published six times a year. Other periodical publications for members include *NFPA Update,* a bimonthly newsletter containing current fire protection information and standards-related articles.

Among the many books published by the NFPA, the one best known is undoubtedly the *Fire Protection Handbook*. The first edition appeared in 1896, and through the years, it has become the bible of the fire protection field. Other handbooks include *Industrial Fire Hazards, Automatic Sprinkler Systems, National Electrical Code, Life Safety Code,* and *Flammable and Combustible Liquids Code.*

The NFPA is a leader in public education programs concerning fire protection and prevention. Risk Watch® is NFPA's injury prevention curriculum for grade school students. "Learn Not to Burn" is the theme and focus of the NFPA's comprehensive public fire safety education program, which includes a national media campaign, a fire safety curriculum for schoolchildren, and an extensive outreach program supported by a regional field network of NFPA representatives. NFPA's Remembering When® program is designed to be implemented as a community coalition to help older adults live safely. Remembering When was developed by NFPA and the Centers for Disease Control with a combined strategy that addresses both injury and fire safety, including eight fire and eight injury safety messages. Fire Prevention Week is sponsored annually by the NFPA as a major national event designed to promote fire safety awareness. Sparky the Fire Dog® is the trademark creation of the NFPA.

The NFPA's Engineering Services Division has several field service projects working in specific phases of fire protection, including electrical, life safety, flammable liquids, gases, and marine fire protection. These special field service groups provide guidance in the application of pertinent standards in specialized areas.

The association annually publishes statistics on the fire problem in the United States and maintains one of the world's most extensive fire experience databases. Combining its annual survey of fire departments and Fire Incidence Data Organization (FIDO) files on large fires and others of major technical interest, the NFPA publishes two dozen major annual overview reports on the nation's fire problem. It also provides numerous special topical reports and responds to thousands of individual and organization requests for data.

The Fire Investigations Department conducts investigations of major fires of technical or educational interest. Important lessons learned from the fires investigated provide input to NFPA technical committees and are vital in understanding how fires originate and how similar fires can be avoided in the future.

In 1982, the NFPA formed the **Fire Protection Research Foundation**, an organization whose specific goal is identifying fire problem areas and developing solutions to those problems. Foundation programs help meet the urgent need to begin or continue research into hazardous materials, improved safety equipment for firefighters, better and more realistic code enforcement, improved public education, and a safer environment.

Fire Protection Research Foundation
■ a nonprofit organization that provides fire and life safety research that supports NFPA's mission and provides specific guidance to NFPA technical committees.

The NFPA has more than 80,000 members. Many of them are affiliated with 1 of the 14 membership sections. These include the International Fire Marshals Association, the Industrial Fire Protection Section, the Fire Service Section, the Education Section, the Electrical Section, and the Railroad Section. Each section has elected officers and carries out functions relating to fire prevention within its sphere of interest. Section business meetings are usually held during the NFPA annual meeting.

In an effort to provide better service to members, the NFPA has regional managers assigned to the New England, Mid-Atlantic, Southern, Central, Denver, Western, and Canadian areas. The Washington, D.C., office serves as a liaison with all federal agencies on many matters in the fire prevention field. With the ever-increasing interest of the federal government in fire protection, coordination between federal and private agencies has become imperative. NFPA offices have been valuable in coordinating efforts and gaining recognition of the part played by nongovernmental interests in fire protection.[1]

The NFPA's Risk Watch program offers a primer on how to build community child safety programs through coalition building. Risk Watch publishes a paper outlining and promoting the effectiveness of coalitions. NFPA has been one of the more active groups in coalition building, what might be considered a ground-up approach to impacting safety. One of the most successful fire safety efforts seen in recent years involved the Coalition for Fire Safe Cigarettes. This effort, coordinated by NFPA, involved building a network of individuals and groups by getting them to sign on as members, and in return the coalition members received well-developed materials that educate and spread the word about the so-called fire safe cigarette. The coalition website, which is produced by NFPA, offers an impressive list of coalition members and also provides keys to effective implementation of laws, especially model legislation. The Coalition for Fire Safe Cigarettes can be seen as an effective model strategy for any fire safety effort at the community level.

International Association of Fire Chiefs ■ a worldwide organization with a primary objective of leadership and advocacy that provides positions on key issues important to the fire service and, in particular, fire chiefs. Its Fire and Life Safety Section leads efforts in fire prevention and is active in codes and standards development.

International Association of Arson Investigators ■ an organization whose primary objective is to improve the professional development of fire and explosion investigators through its availability as a global resource for investigation, technology, and resources.

International Association of Fire Fighters (IAFF) ■ an organized-labor representative for the majority of the career firefighters in the United States and Canada, active in fire prevention through participation in standards development, code change, and fire and life safety forums.

OTHER FIRE PREVENTION—ORIENTED ORGANIZATIONS

International Association of Fire Chiefs

The **International Association of Fire Chiefs** has as a primary objective: the encouragement of fire prevention activities by local fire departments. The association is composed primarily of fire chiefs and their deputies in the United States and Canada.

International Association of Arson Investigators

The **International Association of Arson Investigators** is an organization of professional fire investigators. Its primary objective is the prevention of fire through the suppression of arson. The association encourages adequate training to enable its membership to fulfill responsibilities in this important field.

International Association of Fire Fighters

The **International Association of Fire Fighters (IAFF)**, an affiliate of the AFL-CIO, represents a large percentage of career fire department personnel in the United States and Canada. Local members have been active in fire prevention and have on occasion directly sponsored fire prevention displays at public gatherings. The

IAFF has become very active in areas of model code development and fire research related to building construction and materials.

International Association of Black Professional Fire Fighters

The International Association of Black Professional Fire Fighters (IABPFF) has an interest in the promotion of fire prevention in high fire-risk areas. It is active in national efforts to reduce the risk of fires and offers community-focused fire safety programs with educational and promotional materials, generally at no charge.

The IABPFF's Stop Fire Campaign focuses on home fire safety through its message, "Fire-Safe Cooking: A Recipe for Saving Lives." In 2009 the organization launched a campaign targeting grilling and outside cooking. Videos on safety can be viewed on the YouTube social media site.

IABPFF also has a program that addresses children at home alone. The program, No Child Left Alone, was launched in October 2010. It focuses on providing parent and caregiver awareness of the dangers of leaving children alone and gives tips on making children "fire-safe." An educational video, information pamphlets, and a community poster are available at no charge.

Congressional Fire Services Institute

The **Congressional Fire Services Institute** represents fire protection interests at the federal legislative level. The institute provides liaison between the Congressional Fire Caucus and the fire protection community. Members of the Senate and the House of Representatives belong to this caucus.

The institute has a national advisory committee that reviews legislation relating to fire safety. Its work has been quite successful in coordinating fire service legislative initiatives.

Congressional Fire Services Institute ■ an organization that provides a liaison between the Congressional Fire Caucus and the fire protection community.

National Safety Council

The **National Safety Council**, an organization with a major impact in the field of safety, includes fire prevention as an overall safety objective. The annual National Safety Congress is probably the largest gathering in the world of people with an interest in safety. Many of those active in the National Safety Council have fire prevention responsibilities as part of their overall safety duties. The Canada Safety Council has a similar role.

National Safety Council ■ an organization whose major impact is in the field of safety, including fire prevention as an overall safety objective.

Society of Fire Protection Engineers

The **Society of Fire Protection Engineers (SFPE)** was established in 1950 and was incorporated as an independent organization in 1971. The society has approximately 4,500 members in the United States and abroad, and 57 regional chapters. SFPE conducts an annual educational conference as well as periodic seminars and workshops. It has also been active in delivering free educational Internet webinars. One recent free webinar presented the latest information on concealing methods (enclosing pipe in walls and ceiling) for chlorinated polyvinyl chloride (CPVC) plastic. The SFPE also publishes books and technical guides, a quarterly newsletter called *SFPE Today*, a peer-reviewed quarterly journal, and a quarterly technical magazine called *Fire Protection Engineering*.

Society of Fire Protection Engineers (SFPE) ■ a professional society representing those practicing in the field of fire protection engineering.

International Fire Service Training Association

International Fire Service Training Association
■ an organization that identifies needed areas of training materials and supports those needs through the development and validation of training materials for the fire service and other related services.

The **International Fire Service Training Association**, formed in 1943, develops and publishes training material for fire service personnel. Manuals on fire prevention subjects are included. Committee members from the United States, Canada, and other countries gather annually under the auspices of Oklahoma State University to update the manuals and develop new material. A majority of states and provinces use the manuals for their fire service training programs.

International Society of Fire Service Instructors

The International Society of Fire Service Instructors was organized in 1960 by a group of state and provincial fire service instructors. The society has since expanded to include local training officers and other fire service personnel. The society has strived for improved instructor capabilities and has generally represented fire service training interests at the national level. Improved training in fire prevention is a goal of the organization.

National Volunteer Fire Council

National Volunteer Fire Council
■ an organization that represents the interests of the nation's volunteer fire service.

The **National Volunteer Fire Council** represents the interests of the nation's volunteer fire service. Many volunteer fire departments have aggressive fire prevention programs.

National Association of State Fire Marshals

The National Association of State Fire Marshals represents state fire marshals at the national level. The organization has actively supported fire safety education and smoke alarm installation and has assisted in enhancing the effectiveness of state-level fire safety agencies.

Organizations with Allied Interests in Fire Prevention

Private sector organizations with fire prevention as an adjunct of their operations may be divided into categories. They include professional associations that issue standards having some bearing on fire protection, trade associations that deal in areas having some relationship to fire protection, organizations that develop model building codes, and the insurance industry.

PROFESSIONAL ASSOCIATIONS

Examples of associations that issue standards having a bearing on fire protection include the Air Conditioning and Refrigeration Institute, the American Industrial Hygiene Association, the American Institute of Architects, the American Society for Testing and Materials, the American Society of Agricultural and Biological Engineers, the American Society of Mechanical Engineers, the American Society of Safety Engineers, the American Water Works Association, the American Welding Society, the Institute of Electrical and Electronics Engineers, the American Society of Heating, Refrigerating, and Air Conditioning Engineers, Inc., and the NFPA. All of these groups prepare standards and procedures that have some bearing on fire protection.

The American National Standards Institute (ANSI) does not prepare or write standards; rather, some 250 separate standards developers, including those listed in the previous paragraph, prepare standards. ANSI enables the development of standards by accrediting the procedures of standards-developing organizations. Many of those standards are then submitted for acceptance as American National Standards (ANS). Standards are accepted through a collaborative and consensus agreement process from representatives of the participating stakeholders of a candidate standard. Vetting of the standard and input of the stakeholders, as well as public opportunity to comment and provide input, makes for a highly open process that leads to acceptance of the standards. There are more than 13,000 ANSI-approved American National Standards. The American National Standards Institute is the coordinator and issuer of ANS designations.

TRADE ASSOCIATIONS

Trade associations that have some interests bearing on fire prevention include the American Fire Sprinkler Association, the American Forest and Paper Association, the American Hospital Association, the American Health Care Association, the American Gas Association, the American Iron and Steel Institute, the American Petroleum Institute, the Automatic Fire Alarm Association, the Compressed Gas Association, the Fire Equipment Manufacturers Association, the Gypsum Association, the Institute of Makers of Explosives, the Lightning Protection Institute, the Manufacturing Chemists Association, the Manufactured Housing Institute, the National Association of Fire Equipment Distributors, the National Association of Home Builders, the National Electrical Manufacturers Association, the National Fire Sprinkler Association, the National Propane Gas Association, and the Society of the Plastics Industry. These trade associations, as well as a number of others, have an interest in products and services that relate in some manner to fire safety. They often include fire prevention subjects in their publications and in programs presented for their membership.

To illustrate how a trade association may be concerned with fire prevention, consider the example of the National Propane Gas Association. The industry represented by this association markets a product that is inherently hazardous. To be marketable, the material must have the ability to burn—a characteristic that creates a problem in handling the material. To minimize dangers, the association has taken steps to be sure that all individuals who handle the material are properly indoctrinated in safety measures.

The National Propane Gas Association, with headquarters in Chicago, has maintained active representation on the Liquefied Petroleum Gas Committee of the NFPA. Together, they have participated in the development of standards for the safe use and handling of propane and have actively promoted employee training for member organizations. Liquefied petroleum gas associations at the state level are primarily concerned, as is the national organization, with the development of industrial and domestic uses of liquefied petroleum gas, but they also place considerable emphasis on safety programs of member companies. All these associations have actively worked to promulgate and enforce liquefied petroleum gas safety codes in many jurisdictions.

The American Gas Association, in addition to performing trade association services, operates testing laboratories in Cleveland and in Los Angeles for the purpose of testing all types of gas appliances. Safety testing is a major part of this function.

BUILDING CODES

The International Code Council has a major impact on construction practices in the United States. The council's model construction-related codes, including building and fire codes, are widely used and are replacing locally developed building codes to a great extent.

INSURANCE INDUSTRY

As a result of the major fires that ravaged many American cities in the late 1800s and early 1900s, the property insurance industry played a major leadership role in the development and encouragement of fire prevention and fire suppression measures in the United States. Many advances came about as a result of the direct relationship between the fire insurance rate structure and the consequent implementation of local fire protection efforts.

In the 1980s, the insurance industry's emphasis on fire protection began to shift. The progression of insurance policies from covering only fire and lightning to package policies covering several property perils, as well as liability, lessened the insurance companies' centralized emphasis on the risk from fire alone.

In addition, strong competition developed among the insurance companies, which resulted in independent postures, lessening their joint efforts. The "mix" of business written by insurers has changed dramatically. Companies known as "direct writers" have concentrated their sales efforts on personal lines, whereas the large-agency companies have emphasized commercial risks. Rather than supporting broad-based industry efforts, more and more companies maintaining loss control departments are specializing in the accounts each individual company wishes to insure.

There also have been major expense pressures on the insurance companies. As a result, many companies have reduced their contributions to trade organizations. That reduction, in turn, has shifted those organizations' emphasis from public service activities, such as providing fire prevention activities and publications, to primarily dealing with federal and state government relations. Loss control services are aimed at specific occupancies. With the reduction in fire protection activities by the insurance industry, more and more of the functions they formerly performed are now conducted by governmental, educational, and private nonprofit organizations.

Two of the insurance organizations currently performing major activities in the fire protection area are the FM Global Group and HSB Professional Loss Control (Figure 9-1).

FM Global Group
■ a significant leader in research to reduce property loss, this organization specializes in insuring major corporations' commercial property.

FM Global Group

The **FM Global Group** is an organization that specializes in insuring major corporations' commercial property and is a leader in its market. From year to year, FM Global Group ranks in the top five of those insuring large commercial properties.

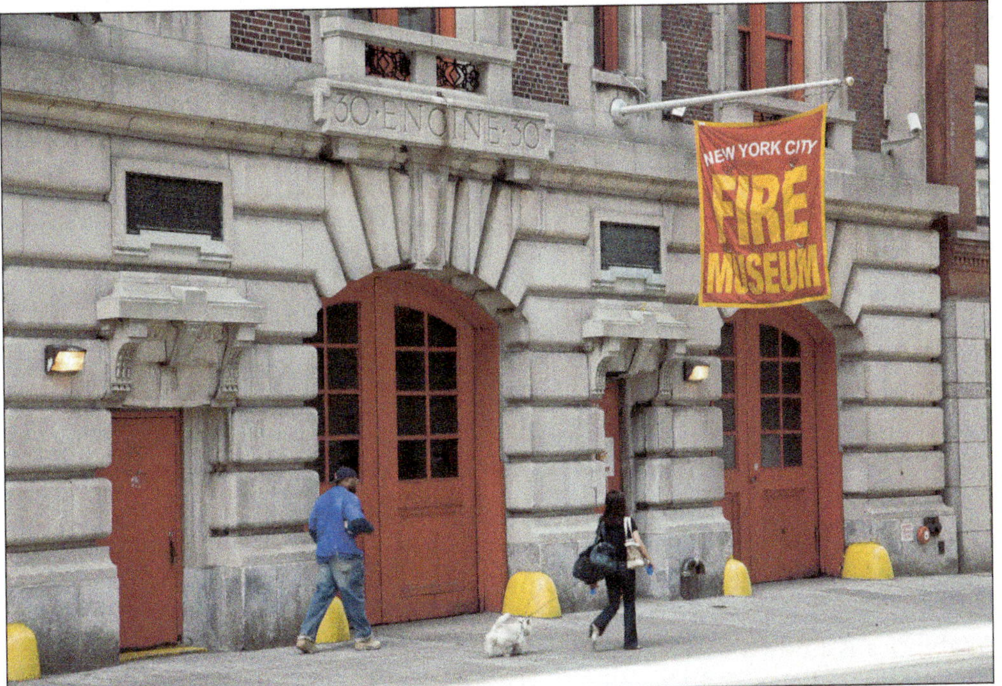

AA World Travel Library/Alamy

FIGURE 9-1 Municipal government and industry cooperated to establish a fire museum that includes expansive home safety working displays.

The FM Global Group is today's outgrowth of what was termed the Factory Mutual System. As is common in the insurance industry today, FM Global Group has evolved through mergers and acquisitions of other Factory Mutual System carriers whose operations were similar.

FM Global Group predominantly insures only those large commercial property accounts considered highly protected risks (HPR) or certain preferred property risks that do not meet the HPR standards but that meet the FM Global Group's rigorous underwriting requirements.

FM Global Group is best known for its risk management and loss control services, which are provided by its approximately 1,400 engineers. It maintains what has been called the largest fire and natural hazards testing center in the world. It researches the effectiveness and reliability of industrial firefighting equipment. It also tests electrical equipment and building materials. The issuance of the FM seal of approval signifies that the equipment bearing the seal meets FM Global Group's acceptability standards. It is the promulgation of its acceptability standards, the inspections of insured risks, and the research performed by FM Global that closely interrelates with the fire prevention and protection efforts of public and private authorities.

HSB Professional Loss Control

A subsidiary of the American International Group (AIG), HSB Professional Loss Control insures well-protected industrial plants and institutions. The organization provides combined engineering and loss control survey services to insured factories, hospitals, and other large occupancies. The services of its nationally recognized fire safety laboratory are available for training of personnel in the maintenance and use of various types of fire protection equipment usually found in a major industrial plant.

Insurance Services Office, Inc. (ISO Inc.)

The Insurance Services Office, Inc. performs the functions that before 1971 were performed by the regional and single-state fire insurance rating bureaus. Those functions include:

- Classifying, surveying, and assigning public protection classifications to municipalities and fire protection districts
- Promulgating loss costs (rather than rates) from which insurance companies themselves make rates
- Issuing rules for determining program eligibility and classifications
- Developing insurance contracts, policy forms, and endorsements
- Gathering premium and loss information
- Rating commercial buildings and their contents based on the buildings' construction, occupancy, and fire protection features

ISO Inc. performs a wide variety of services for the insurance industry. It also combines risk factors and satellite imagery to pinpoint potential wildfire exposures, conducts on-site surveys of commercial and residential properties to determine their exposure to loss. The process of classifying a community's suppression capabilities is known as ISO's Public Protection Classification, which helps the community evaluate its fire protection services. The system uses a scale from Class 1 to Class 10, with Class 1 being the highest rating and Class 10 considered no recognized protection.

ISO Inc. recently developed a schedule, similar to its Fire Suppression Rating Schedule, that grades municipal and county building code and enforcement measures for each community. This schedule is named the Building Code Effectiveness Grading Schedule. Communities that have up-to-date building codes with an effective and timely enforcement process in place generally have evidence of a more effective loss experience. This reduces the risk, but more important, ISO's classification of the jurisdiction can lead to lower insurance rates for property owners.

ISO's specific commercial Property Evaluation Schedule reflects specific building loss control and fire prevention measures for underwriting purposes, thereby recognizing fire safety factors. In one state alone, more than 140,000 properties are evaluated under this schedule (Figure 9-2).[2]

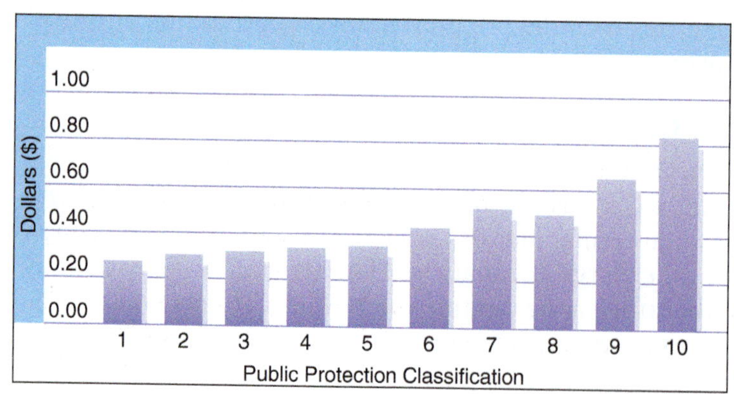

FIGURE 9-2 Communities with superior fire protection ratings (Public Protection Classifications) have lower fire losses than those with lesser services. *(ISO Chart included in 2008 Action Report of Mountain View Fire District, Colorado)*

Risk Management

Risk management is the development of a systematic approach to the handling of pure risk (not speculative risk) by individuals and businesses. Fire prevention is part of a loss control technique. Many colleges and universities offer programs and degrees in risk management and insurance.

A professional approach to insurance buying and loss control was developed in the 1950s through a series of buyer organizations that led to the organization known as the Risk and Insurance Management Society (RIMS). RIMS has been very significant in the promotion of loss control, including fire prevention, as part of the overall risk management process. This organization publishes the well-known magazine *Risk Management* as well as many reports and studies dealing with loss prevention.[3]

Underwriters Laboratories Inc.

Underwriters Laboratories Inc. (UL) is an independent not-for-profit organization having a comprehensive program of testing, listing, and reexamining equipment, devices, materials, and assemblies as they relate to fire protection and safety.

Underwriters Laboratories Inc. (UL)
■ an independent not-for-profit organization that has a comprehensive program of testing, listing, and reexamining equipment, devices, materials, and assemblies as they relate to fire protection and safety.

The laboratories, which were first established in 1894, test building materials of all types, fire protection equipment, gas and oil equipment, accident and burglary prevention equipment, as well as electrical appliances and equipment. The electrical testing program of UL, as an example, has had a major impact on fire protection, especially through the reduction of fires caused by the faulty design and manufacture of electrical devices and appurtenances. The UL label is internationally recognized as indicating that the product has met minimum standards for safety. Comprehensive tests have been devised to reduce the possibility of fire, shock, or injury of any type through the use of a product listed by UL.

An important feature of UL is its follow-up service program. UL's field representatives visit the factories where products are produced under UL service at least four times annually. The public is able to discern which items have been tested because of the use of a printed designation on the product. More than 21 billion labels are applied annually.

Underwriters Laboratories writes most of the standards used to investigate products. Those standards are regularly reviewed and revised on the basis of such considerations as field experience that indicates a product is unsatisfactory from a fire protection or safety standpoint. Product directories issued annually show names of manufacturers of products that are listed by UL. Inspection centers are maintained worldwide in more than 70 countries.[4]

As an example of public utilization of a privately sponsored and developed activity, several jurisdictions require that all electrical equipment sold within the state or municipality be listed by UL or another recognized testing laboratory. For a number of years, Maine has had such a requirement, which came about because the state, in the far northeastern corner of the United States, was becoming a dumping ground for inferior or faulty electrical equipment from all parts of the country. North Carolina enacted a similar requirement in 1934, and Maryland did the same in 1973.

Private Organization Efforts in Fire Prevention

Fire prevention activities carried out by private organizations for the improvement and protection of their own properties are many and varied. In the majority of business establishments, industries, and privately operated institutions and schools, the primary concern with fire prevention is brought about by the local fire department, fire marshal's office, and insurance carriers. Some of the more progressive businesses maintain extensive internal fire prevention programs.

Some industries handle fire prevention at a local level, assigning a person or persons to such duties in each plant. In other cases, a floating fire prevention person may service a number of plants out of a central location.

SELF-INSURANCE PROGRAMS

An example of a program in which there is a direct relationship between fire prevention efforts of private enterprise and expenditures is the self-insurance program. Under a self-insurance program, a business organization or government body accepts the risk connected with fire, either partially or completely. The industry or government body elects to handle fire losses with its own resources rather than depend on commercially available fire insurance and/or comprehensive liability insurance.

Whether the program is operated by a governmental agency or by a private organization, a self-insurance program places a direct responsibility on management to be sure that all proper precautions are taken to prevent fire. Because there is a chance that losses might not be fully replaced or that any reserve fund might be depleted, an obligation exists to ensure minimal losses.

The most effective self-insurance plans are those in agencies or businesses that have a spread-of-risk arrangement, so that the possibility of a total loss at one location is avoided. A state government, for example, has hundreds of fixed installations; the likelihood that a number of them will have a major fire during a given year is highly remote. As long as the reserve to handle the anticipated losses can be maintained, the plan should function satisfactorily.

One unfortunate feature of the self-insurance plan is the possibility the reserve fund will be used for other purposes. If, for example, a company with a self-insurance program decided to open a new operation and start-up funding was required, there would be a real temptation to tap the fire-loss reserve for this purpose.

A self-insurance program must consider how to enforce the requirements necessary to maintain low fire losses. The usual procedure is to give someone or some group the responsibility to determine what measures must be taken to make the place reasonably safe from the standpoint of loss. In industry, a consultant firm is often used to provide outside inspections for the company.

Controls needed to ensure continued satisfactory operation include plan-review and inspection functions and detailed analysis of all special hazards within the industry or institution. It is also necessary to have a soundly developed program for training maintenance personnel and others directly engaged in the day-to-day operation of the structures in question.

EDUCATIONAL INSTITUTIONS

Several private nongovernmental universities offer programs related to fire safety. The Worcester Polytechnic Institute, with its Center for Fire Safety Studies in Worcester, Massachusetts, has carried out research in a number of fire safety matters and offers a master's and a doctorate in fire protection engineering.

FIRE DEPARTMENT ASSOCIATIONS

Although made up primarily of members of public agencies, regional, state, and local fire associations are actually private organizations. They exist in practically every state and province. Many have both career and volunteer fire department membership. In some states, career and volunteer fire service groups have organized separate associations functioning within the same state. In either case, fire prevention may well be considered a major activity of an association. Most states have fire chief and fire inspector associations as well.

Generally, a fire prevention committee is one of the standing committees of an association. Its function is to encourage the promotion of fire prevention by the member departments. The committee may hold meetings for the purpose of distributing fire safety information, and in some states it conducts an annual fire prevention contest.

An example of a fire prevention group within a state fire association is the Fire Prevention Committee of the Maryland State Firemen's Association. This committee carries out an active program of fire prevention within the state and conducts several fire prevention contests each year. The contests include fire prevention activities by individuals as well as the fire companies within the state.

A number of county fire associations also sponsor active fire prevention programs. The programs may directly reach the public or may merely be designed to encourage fire prevention activities on the part of member companies.

CHAPTER **REVIEW**

Summary

Fire prevention efforts in the United States have been motivated predominantly by private organizations, rather than by public agencies, as in most other countries.

Among organizations with a primary function in fire prevention, the unquestioned leader is the National Fire Protection Association, which has members in more than 100 countries. It is a nonprofit educational and technical association devoted to promoting scientific methods of fire protection and prevention to guard against loss of life and property by fire. Among the many publications of the NFPA, probably the best known is the *Fire Protection Handbook,* which has become the bible of the fire prevention field. Others include educational material and various periodicals, records of fires and fire loss statistics, and over 300 publications on technical standards, codes, recommended practices, and model ordinances, which are developed, prepared, and updated under the guidance of more than 200 technical committees.

Among other fire prevention–oriented organizations that have contributed greatly to the cause are the International Association of Fire Chiefs, the International Association of Arson Investigators, the International Association of Fire Fighters, the Society of Fire Protection Engineers, and the National Safety Council.

Organizations with allied interests in fire prevention are noncommercial organizations, professional associations and institutes in special fields that prepare standards and procedures having some bearing on fire protection, and a long list of trade associations devoted to the promotion of trade in products and services that relate in some manner to fire prevention.

Building code development groups have made a major impact on fire prevention. Operating at the national level is the International Code Council.

The insurance industry has been a dominant influence in the development and encouragement of fire prevention measures in the United States, and there is no doubt that many improvements have come about as a result of their direct relationship to insurance rate structures. Here, too, the associations and mutual systems representing a number of insurance companies have made significant contributions.

Insurance Services Office, Inc., establishes insurance rates and is responsible for the grading of municipal fire defenses. In all states, insurance rates are subject to review by the state insurance commissioner.

Underwriters Laboratories Inc. has long been an important force in fire prevention through its comprehensive testing programs for equipment, materials, and assemblies that relate to fire protection and safety. It has earned an international reputation for the UL label.

Some private business organizations maintain extensive internal fire prevention programs for their own protection. Others elect to operate under self-insurance coverage. The federal government might be considered a leading self-insurer. A self-insurance program puts direct responsibility on management to make sure that all proper precautions are taken to prevent fire.

And last, the contributions of educational institutions that conduct courses in fire science and administration, and of firefighters' associations that are active in fire prevention education, must also be counted in the monumental effort that has been made by the private sector in this country toward the cause of fire prevention.

Review Questions

1. The National Fire Protection Association was founded in:
   a. 1853.
   b. 1896.
   c. 1925.
   d. 1947.
2. Activities of the NFPA include:
   a. code enforcement.
   b. developing technical standards.
   c. mandating standard operating procedures.
   d. establishing laws.
3. Which one of the following is not an organization that promotes fire prevention?
   a. International Association of Fire Chiefs
   b. International Association of Arson Investigators
   c. International Association of Fire Fighters
   d. International Municipal Fire Fighters
4. The American National Standards Institute:
   a. prepares standards.
   b. accredits standards-developing organizations.
   c. approves standards.
   d. supervises standards-developing organizations.
5. Associations that have an interest in fire prevention include:
   a. Central Station Alarm Association.
   b. Compressed Gas Association.
   c. National Propane Gas Association.
   d. all of the above.
6. The Insurance Services Office, Inc.,:
   a. sets insurance rates.
   b. rates fire departments.
   c. investigates fires.
   d. both a and b.
7. The primary mission of the NFPA is to:
   a. decrease insurance costs.
   b. decrease the likelihood of fire in industrial complexes.
   c. prevent accidents.
   d. prevent the loss of life and property by fire and other hazards.
8. How are NFPA standards adopted?
   a. After third-party testing by Underwriters Laboratories
   b. After careful review by the National Fire Academy technical services division
   c. After detailed study and vote by committee
   d. NFPA standards are not adopted.
9. What is the highest rating a fire department can receive from the Insurance Services Office?
   a. Class 1
   b. Class 2
   c. Class 3
   d. Class 4
10. The Congressional Fire Services Institute represents:
    a. fire prevention efforts at the local level.
    b. protection of federal assets such as the National Fire Academy.
    c. protection of collective bargaining rights for union firefighters.
    d. fire prevention at the federal legislative level.

End Notes

1. Communication to author Robertson from NFPA, 2008.
2. Communication to author Robertson from Edward Straw, Regional Senior Technical Coordinator, ISO, 1996.
3. Communication to author Robertson from J. Frank Hodges, Ph.D., 2007.
4. Communication to author Robertson from John Bender, PE, Underwriters Laboratories Inc., 2007.

10

CHAPTER

Fire Prevention Responsibilities of the Public Sector

Billy Morris Director of Fire Science Cisco College

KEY TERMS

Bureau of Alcohol, Tobacco, Firearms, and Explosives (ATF), *p. 187*

Consumer Product Safety Commission, *p. 196*

Federal Aviation Administration, *p. 192*

Federal Bureau of Investigation (FBI), *p. 187*

Federal Emergency Management Agency (FEMA), *p. 195*

Federal Housing Administration, *p. 192*

Mine Safety and Health Administration (MSHA), *p. 191*

National Park Service, *p. 187*

Occupational Safety and Health Administration (OSHA), *p. 190*

U.S. Coast Guard, *p. 185*

U.S. Department of Health and Human Services, *p. 191*

U.S. Department of Homeland Security, *p. 193*

U.S. Department of Housing and Urban Development, *p. 192*

U.S. Forest Service, *p. 188*

OBJECTIVES

After reading this chapter, you should be able to:

- Identify federal resources that may be called to assist at the local level.
- Describe the role of the Bureau of Alcohol, Tobacco, Firearms, and Explosives in fire investigation.
- Compare the role of the federal government in fire prevention in the United States with its counterparts in other countries.
- Identify which federal agency has had the greatest impact on fire prevention.

Governmental agencies in the United States have played a less prominent role in fire prevention than have their counterparts in other countries. However, the trend toward heavily urbanized areas has made greater governmental responsibility in the field of fire prevention necessary, and the influence of governmental agencies at all levels is steadily increasing. In this chapter, the role of the federal government as well as that of state and local governments is outlined.

Rapid changes are taking place in the sphere of governmental responsibilities, and activities by the federal government that were unheard of 40 years ago are commonplace today. The major change has taken place at the federal level. Not many years ago, federal responsibility was confined primarily to direct protection of federal properties and fire safety enforcement related to a rather narrow group of activities involving interstate commerce. Current interpretations of the term *interstate commerce* have broadened the scope of federal responsibility and operation in that field. The findings and recommendations of the National Commission on Fire Prevention and Control, established by an act of Congress in 1968, are also having a strong influence on fire prevention responsibility in the public sector.[1]

Within the federal government, fire prevention efforts are broad in scope and encompass practically all federal agencies in some manner. A number of larger federal government installations maintain their own fire departments with personnel that are responsible for fire safety inspections, enforcement of fire prevention regulations, and public fire prevention education, in addition to their fire suppression duties.

Generally, standards for fire safety within the federal government are set by individual agencies, which maintain guidelines and regulations that must be adhered to by their employees. Regulations and standards of construction vary among agencies, although the advent of the General Services Administration as a property management arm of the federal government has probably brought about more uniformity.

The responsibilities of certain federal agencies, especially the armed services, demand a high degree of attention to fire prevention because of the nature of the equipment they employ. The seagoing services—the U.S. Navy and **U.S. Coast Guard**—place a great deal of emphasis on fire prevention because of the grave problems related to fires aboard ship. The same condition prevails with aircraft assigned to the military services. The Coast Guard, a military branch of the Department of Homeland Security, has a long history of legal responsibility in the field of fire prevention, both aboard vessels and at U.S. ports and waterways.

U.S. Coast Guard
■ one of five armed forces branches and the only military branch within the Department of Homeland Security, it safeguards U.S. maritime and environmental interests and has a long history of legal responsibility in the field of fire prevention, both aboard vessels and in U.S. ports.

Although primarily aimed at enhancing fire prevention within federal facilities, government efforts in internal fire prevention have had an effect on the fire prevention behavior of the public. There is no question that fire safety training and actual fire experiences while on active military duty have made a lasting impression on significant segments of the present civilian population.

In a similar way, military reservists and National Guard members are exposed to fire prevention education through weekly or monthly drills that may include an occasional fire prevention lecture. Training programs offered to federal employees can also be of value in home fire prevention and fire response conditioning.

Federal agencies with fire prevention responsibilities that have an impact on members of the general public include most cabinet departments and a number of independent agencies. Examples follow of the fire prevention work being done by the some of the federal departments and agencies.

U.S. Department of Defense

The Department of Defense includes the U.S. Army, U.S. Navy and Marine Corps, and U.S. Air Force, in addition to several other agencies.

DEPARTMENT OF THE ARMY

An army activity having a bearing on fire prevention and protection is the maintenance of explosives ordnance disposal units stationed throughout the country. The highly trained units are available at the request of local fire or police personnel to remove and render harmless military explosive devices that may be found in a community.

Several research projects carried out by the army have likewise contributed to the field of fire prevention. They include the development of fire-retardant materials.

DEPARTMENT OF THE NAVY

The Department of the Navy, which includes the U.S. Navy and U.S. Marine Corps, has made a substantial contribution to the fire prevention field through extensive research activities. Although much of the research relates to the development of fire suppression devices and equipment, some effort has been directed toward fire prevention. Fire suppression research is, in fact, an effort to bring about more effective control of fire, which contributes to fire prevention through reduction of fire spread.

DEPARTMENT OF THE AIR FORCE

The Department of the Air Force has devoted a great deal of research to the development of specialized equipment for fire suppression. Funding and technology have been available to permit extensive testing of new approaches in fire suppression before they are put to widespread use.

As with the navy, a number of air force fire suppression advancements have had a spin-off effect in fire prevention where innovation is introduced to the business and civilian world. The very nature of fire suppression devices and procedures has the indirect effect of calling attention to the subject of fire.

U.S. Department of Justice

The U.S. Department of Justice is the enforcement agency for the Americans with Disabilities Act, which has a strong influence on building and fire codes. An agency of the Department of Justice, the Federal Bureau of Investigation (FBI), pursues activities that contribute to the cause of fire prevention, as does the Bureau of Alcohol, Tobacco, Firearms, and Explosives (ATF).

DRUG ENFORCEMENT ADMINISTRATION

The Drug Enforcement Administration has an increasing role in cooperation with the nation's fire services in connection with illegal drug laboratories. Routine fire inspections have uncovered some of those illicit operations.

FEDERAL BUREAU OF INVESTIGATION

The **Federal Bureau of Investigation (FBI)** has for many years provided laboratory services to local law enforcement agencies, including examination of materials found in and about a fire scene. Many individuals have been prosecuted for arson and unlawful burning based on the FBI's examination of evidence.

The FBI has been of considerable assistance in arson investigation through the National Crime Information Center and its extensive fingerprint identification program. With authority over fugitive felons crossing state lines, the FBI has been of assistance in apprehending wanted arsonists.

Federal Bureau of Investigation (FBI) ■ the top law enforcement agency in the Department of Justice, offering intelligence services and threat-focused national security; supports state and local law enforcement on crimes such as arson by facilitating access to national crime data and specialized forensic management of evidence.

BUREAU OF ALCOHOL, TOBACCO, FIREARMS, AND EXPLOSIVES

The **Bureau of Alcohol, Tobacco, Firearms, and Explosives (ATF)** of the U.S. Department of Justice enforces federal statutes related to explosives. The agency regularly inspects and licenses explosives dealers, manufacturers, and users when any aspect of the operation may involve interstate commerce. Public display fireworks, for example, that are purchased in one state and transported to another require possession of a license from the Bureau of Alcohol, Tobacco, Firearms, and Explosives.

The bureau's personnel are empowered to investigate certain cases of arson, explosions, use of firebombs, and careless activities relating to explosives. This function often brings agents in contact with local and state fire department and law enforcement personnel.

The ATF operates a highly sophisticated laboratory designed to aid in suppression of arson. It also facilitates the database, Bomb Arson Tracking System (BATS). The BATS database offers a Web-based intelligence resource that allows federal, state, and local law enforcement officials to share information about bomb and arson incidents in the United States.

Bureau of Alcohol, Tobacco, Firearms, and Explosives (ATF) ■ a law enforcement agency in the Department of Justice that regularly inspects and licenses explosives dealers, manufacturers, and users when any aspect of the operation may involve interstate commerce; investigates certain cases of arson, explosions, use of firebombs, and careless activities relating to explosives.

U.S. Department of the Interior

The Department of the Interior maintains two bureaus whose functions directly and indirectly contribute to public safety and fire prevention education.

NATIONAL PARK SERVICE

The **National Park Service** maintains a program to encourage conservation and environmental awareness in the national parks. The park service is charged with maintaining and protecting millions of historical artifact, archives, structures, and places. Guests in the parks account for at least some of the potential risk to the legacy through use and sometimes abuse. To reduce the risk the National Park Service promotes a high state of conservation and environmental protection. A keen focus on the education and awareness of fire prevention in this role of protecting the legacy property not only helps reduce risk within the national parks but also can have a residual effect outside the parks.

National Park Service ■ a division of the Department of the Interior that promotes conservation and environmental awareness in the national parks, including countless historical structures and icons, through administration of an aggressive fire prevention program.

BUREAU OF LAND MANAGEMENT

A good portion of the nation's federally owned land in the western states is administered by the Bureau of Land Management. This agency provides fire prevention and fire suppression support in connection with the lands it manages.

BUREAU OF INDIAN AFFAIRS

The Bureau of Indian Affairs is responsible for oversight of Native American reservations throughout the country. A number of fire departments are sponsored as part of this responsibility.

U.S. Department of Agriculture

The federal department that has traditionally had the greatest fire prevention impact on the general public is probably the Department of Agriculture. The publicity campaigns of several of its agencies—the Forest Service, Natural Resources Conservation Service, and Extension Service—reach most citizens directly.

U.S. FOREST SERVICE

U.S. Forest Service ■ a division of the Department of Agriculture that maintains a widespread fire suppression force and conducts an extensive nationwide fire prevention promotional program.

The **U.S. Forest Service** is a part of the Department of Agriculture. It maintains a widespread fire suppression force and conducts an extensive nationwide fire prevention promotional program. The program, which made Smokey Bear a national figure, reaches a high percentage of the U.S. population through television, radio, and newspaper advertising, as well as through signs and billboards.

Smokey observed his 70th birthday in 2014. His message, "Only YOU can prevent forest fires," is recognized by 95 percent of adults and 85 percent of children aged 8–12. The number of accidental wildfires caused by humans has been cut in half since Smokey arrived, even though the number of visitors to forests has grown 10-fold during this period. The original Smokey was a cub that had survived a New Mexico wildfire.[2]

The Forest Service fire prevention program has been widely accepted by the public, as demonstrated by a number of polls. State forest fire personnel have cooperated in federal efforts and have on many occasions promoted forest fire prevention to school groups and civic organizations.

NATURAL RESOURCES CONSERVATION SERVICES

Closely related to fire protection is the program of the Natural Resources Conservation Service of the Department of Agriculture. This agency has been responsible for establishing countless farm ponds, which are put to use for fire protection by rural fire departments. Extension service programs include an emphasis on fire prevention. The National 4-H program, for example, includes farm fire prevention themes.

U.S. Department of Commerce

BUREAU OF THE CENSUS

A major function of the Bureau of the Census, an arm of the U.S. Department of Commerce, is to take a population census every 10 years. Data collected serve a variety of needs in the fire prevention field. For example, the census records may be used to identify the varieties of fire experiences within different types of communities and income levels. They may also be used to compare effectiveness based on population densities.

NATIONAL INSTITUTE OF STANDARDS AND TECHNOLOGY

The National Institute of Standards and Technology (NIST) has increased its role in fire prevention as a result of the enactment by Congress of the Fire Research and Safety Act of 1968. The institute, formerly known as the National Bureau of Standards, has a long history of cooperation with state and local governments in establishing standards of measurement and performance. As an example, the program of weights and measures, which greatly affects the marketing of all items sold on the basis of weight or other measurement, has long been under its purview. Although specific weight and measurement control programs are enforced at the state or local level, the overall standardization among states is based on procedures established by the National Institute of Standards and Technology.

The institute's Fire Research Division, formerly the Building and Fire Research Laboratory, is organized with responsibilities that include the measurement, prediction, and dynamics of fire and how it interacts with the built and natural environments. The division leads research in fire science, fire engineering, and fire safety. The Federal Fire Prevention and Control Act of 1974 strengthened programs previously operated by the bureau in those fields.

The Fire Research Division is made up of five groups of focused research activities: Firefighting Technology, Engineered Fire Safety, Flammability Reduction, Wildland–Urban Interface Fire, and the National Fire Research Laboratory. In 2012 a comprehensive strategic planning process conducted by the Fire Research Division resulted in the articulation of a long-term vision to remove unwanted fire "as a limitation to life safety, technical innovation and economic prosperity in the United States."[3] In order to achieve the vision, the Fire Research Division will be guided by long-term goals meant to establish a reduction of the impact of fire on communities, structures and their occupants, the fire service, and the economy. For example, one goal is accomplishing in a generation a one-third reduction in the nation's preventable fire burden. NIST's key to achieving the vision and goals is in the development and demonstration of what the planners call measurement science. In general, measurement science in any field of studies is the method for quantifying the knowledge gained from research and study. It is the dimensions, quantities, and capacities associated with the investigative work, an accumulation of facts and observations that lead to procedures and inventions and breakthrough discoveries of why a behavior can be observed and what it means. In the measurement science of fire, NIST identifies what it strives to document through its efforts:

- Understanding of the phenomena responsible for an observed effect (e.g., ignition)
- Development of performance metrics, measurement methods, predictive tools, and protocols as well as reference materials, data, and artifacts
- Conduct of inter-laboratory comparison studies and calibrations
- Evaluation and/or assessment of technologies, systems, and practices
- Development and dissemination of technical guidelines and the basis for standards, codes, and practices—in many instances via testbeds, consortia, or partnerships with the private sector[4]

Over decades of testing and research, NIST and its Fire Research Division (in its many different configurations) have accomplished many helpful breakthroughs in fire safety by documenting their measurement science. One discovery or understanding of a metric leads to another. For example, the invention of computer technology that can model fire behavior and then use video and graphics to visualize how fire flows through a structure—all can lead a researcher to recognize a fire dynamic that possibly was never before observed. As in the past, the result of NIST's contributions to reducing America's fire problem will be significant.

In addition to conducting fire-related research, the National Construction Safety Team (NCST) Act of 2002 authorized the institute to establish an NCST team for deployment after events causing the failure of a building or buildings that resulted in substantial loss of life or posed significant potential for substantial loss of life. The institute leads teams of public and private sector fire and safety experts in conducting fact-finding investigations of building-related failures.[5]

U.S. Department of Labor

Under the Occupational Safety and Health Act of 1970, the Department of Labor embarked on an enforcement program that brings it in close contact with a high percentage of places of employment in the United States. This act has probably had more direct impact on fire prevention enforcement at the local level than any federal legislation previously enacted.

OCCUPATIONAL SAFETY AND HEALTH ADMINISTRATION

Occupational Safety and Health Administration (OSHA) ▪ a division of the Department of Labor that is charged with developing safe and healthy working conditions through standards and regulations, including those covering fire protection.

The regulations of the Department of Labor's **Occupational Safety and Health Administration (OSHA)** are applicable to employees, businesses, industries, institutions, farms, and other places of employment that engage in interstate commerce. The term *interstate commerce* includes all places that, for example, obtain supplies from, or market to, businesses in other states. For all practical purposes, few places of employment are not covered by this broad definition.

OSHA is charged with developing occupational safety and health standards and regulations and has therefore promulgated a comprehensive set of regulations, including a number that incorporate National Fire Protection Association standards. Flammable liquids, liquefied petroleum gases, building exits, and electrical safety are among the fire protection subjects covered.

The legislation that established OSHA also established a procedure by which state governments can, if adequately prepared, enforce the provisions of the act within their jurisdiction. A number of states are utilizing this arrangement. In those that are not, the Department of Labor provides direct enforcement with its own personnel. Federal matching funds assist states that provide their own enforcement personnel. Local fire prevention inspections, as well as those by state agencies, are not preempted by this federal program, according to procedures set forth by the agency.

OSHA also oversees one of the most extensive indexes of product testing and listing agencies anywhere. The OSHA Nationally Recognized Testing Laboratories (NRTLs) Accreditation Program came about as a result of frequent requirements or implied reference in OSHA procedures calling specifically for certification of products listed by Underwriters Labs (UL) and Factory Mutual Research Corporation (FMRC). Other testing organizations may have been impacted economically by exclusion and have worked to make changes in the federal regulations so that they include any appropriately qualified labs. The accreditation recognizes mostly private sector organizations that provide safety testing and certification services. The testing and certifications are conducted based on U.S. consensus safety standards. The standards are not developed by OSHA; instead, they are issued by U.S. standards-developing organizations such as NFPA and ANSI.

MINE SAFETY AND HEALTH ADMINISTRATION

The **Mine Safety and Health Administration (MSHA)** has a direct responsibility for fire prevention and safety within coal, metallic, and nonmetallic mines, often sharing this responsibility with a state bureau of mines. Records show that fires and explosions in coal mines have resulted in major tragic losses of life. Inspection and fire prevention procedures in mines follow basic principles of fire prevention, but special consideration must be given to the unusual structural and atmospheric conditions inherent in mining operations.

Mine Safety and Health Administration (MSHA)
▪ an agency of the Department of Labor that has direct responsibility for fire prevention and safety within coal, metallic, and nonmetallic mines, often sharing this responsibility with a state bureau of mines.

U.S. Department of Health and Human Services

The **U.S. Department of Health and Human Services** has had a major impact on fire safety. The department, which was formed in 1953, has a number of agencies with direct interests in fire prevention. One is the Centers for Medicare and Medicaid Services.

Responsibilities for ensuring the provision of proper fire safeguards in health-care facilities receiving funds through Medicare (Title 18) and Medicaid (Title 19) rest with the Centers for Medicare and Medicaid Services. Fire prevention standards have been established for facilities housing patients in either program. Thousands of patients in long-term care facilities and hospitals are funded under these programs.

U.S. Department of Health and Human Services
▪ formed in 1953, this department has had a major impact on fire safety through a number of agencies with direct interests in fire prevention.

State agencies assist the Department of Health and Human Services in ensuring that fire safety requirements are met. Funding may be discontinued for lack of compliance with the requirements, which encompass the Life Safety Code of the National Fire Protection Association.

U.S. Department of Housing and Urban Development

U.S. Department of Housing and Urban Development ■ devoting considerable attention to fire protection for lower-income housing units, this department has funded several research projects in the fire protection field; it also enforces requirements for manufactured housing.

Federal Housing Administration ■ an arm of the Department of Housing and Urban Development that has set certain standards for structures insured under the National Housing Act and other federal laws, which include minimum fire safety standards.

One of the responsibilities of the **U.S. Department of Housing and Urban Development** is the creation and direction of programs to bring about lower-cost housing. The department has devoted considerable attention to fire protection for lower-income housing units and has funded several research projects in the fire protection field. Requirements for manufactured housing are also enforced by the department.

The **Federal Housing Administration**, an arm of the Department of Housing and Urban Development, has set certain standards for structures insured under the National Housing Act and other federal laws. These include minimum fire safety standards. Funding is provided by the agency for the construction of housing for older adults, nursing homes, intermediate-care facilities, and nonprofit hospitals. Minimum fire protection standards also are included in their mortgage insurance programs.

U.S. Department of Transportation

The Department of Transportation, formed in 1966, includes a number of agencies that have an interest in fire prevention. Several of them operated independently before the Department of Transportation was established.

FEDERAL AVIATION ADMINISTRATION

Federal Aviation Administration ■ the agency under the U.S. Department of Transportation that is responsible for the control of all aspects of aviation, including safety features incorporated within aircraft.

The **Federal Aviation Administration** is responsible for the control of all aspects of aviation, including safety features incorporated within aircraft. The agency ensures, for example, that carpeting in the cabin of passenger aircraft does not have an excessive flame spread. The agency is also concerned about the provision of adequate fire rescue forces and equipment at commercial aviation facilities.

FEDERAL MOTOR CARRIER SAFETY ADMINISTRATION

The Federal Motor Carrier Safety Administration controls safe transportation by motor carriers in interstate commerce. It conducts inspections to ensure compliance with regulations covering safeguards for such transportation. All commercial carriers employing motor vehicles come under this program when their operations extend to interstate commerce. The motor carrier and highway safety sections are responsible for the investigation of collisions involving controlled carriers.

FEDERAL RAILROAD ADMINISTRATION

The Federal Railroad Administration is responsible for the regulation of railroads. The Bureau of Railroad Safety has specific responsibility for enforcing regulations on safety and fire prevention on interstate railroads.

PIPELINE AND HAZARDOUS MATERIALS SAFETY ADMINISTRATION

The Pipeline and Hazardous Materials Safety Administration (PHMSA) coordinates Department of Transportation responsibilities pertaining to pipeline safety and the transportation of hazardous materials. Responsibilities include the administration of safety regulations, enforcement, and issuance of exemptions and interpretations.

The Office of Pipeline Safety establishes and provides for enforcement of safety standards for the transportation of hazardous and gaseous materials by pipeline. In the pipeline safety program, states may be awarded up to 50 percent of their costs in carrying out enforcement and inspection programs.

The Research and Special Programs Administration (RSPA) is the Department of Transportation's focal point for regulations, exemptions, research, coordination with the states, and the emergency response information published in its guidebook. Cooperation with state agencies is especially important if the state itself incorporates Department of Transportation regulations within its motor vehicles code, which a number of states do.

The *Emergency Response Guidebook* is the most widely distributed technical guide for initial response actions in the event of incidents (spill, explosion, fire) involving hazardous materials. This publication is distributed to emergency responders through a network of key state agencies. It is cosponsored by the Canadian and Mexican governments.

U.S. Department of Homeland Security

The **U.S. Department of Homeland Security (DHS)** became a cabinet agency in 2002. It was established as a direct result of the events of September 11, 2001, bringing under one cabinet secretary a number of responsibilities of critical interest to the fire service. A major funding initiative for fire prevention and safety comes under USFA and is administered by FEMA, an agency assigned to USFA. The Assistance to Firefighter Grant program provides a variety of funding projects such as firefighting staff, apparatus, communications, and so on. They also address funding of projects through the Fire Prevention and Safety grant for selected projects, such as installation of stovetop fire extinguishers, fire safety training for college students, and smoke alarm distribution for high-risk populations. Communities applying for grants are required to provide various levels of matching funds for the projects.

U.S. Department of Homeland Security
■ established as a result of the attacks on September 11, 2001, this department provides funds through matching grants to local fire departments and public safety agencies.

Legislation enacted to create the Department of Homeland Security and subsequent presidential directives have enhanced the role of the federal government in preparation for and response to acts of terrorism and other emergencies.

The department also is responsible for the Federal Law Enforcement Training Center, with facilities at Glynco, Georgia; Charleston, South Carolina; and Artesia, Colorado. The U.S. Secret Service is housed in this department as well.

U.S. COAST GUARD

The **U.S. Coast Guard**, a part of the new Department of Homeland Security, has long had a legal responsibility in the field of fire prevention. Under the port security program of the Coast Guard, regulations pertaining to fire prevention and other security features are enforced in the ports of the United States. The Coast Guard has responsibility for inspection and regulatory enforcement in all waterfront facilities. That means a Coast Guard inspector has concurrent jurisdiction with municipal and state fire prevention agencies. The regulations enforced include requirements for fire-extinguishing equipment, marking of extinguishers, maintenance of access for firefighting equipment, control of cutting and welding operations, and control of smoking.

The Coast Guard also has a fire safety function in connection with vessels operating in U.S. waters. Vessels built in American shipyards are under the constant inspection of the Coast Guard during construction. A major portion of the obligatory inspection relates to fire safety. Once the vessel is commissioned, the Coast Guard continues to inspect and ensure proper maintenance of firefighting equipment and training of crews.

The loading and storage of cargo, especially of hazardous materials, is also under the purview of the Coast Guard. Coast Guard personnel may be detailed to directly monitor the loading and stowage aboard ship of explosives and other dangerous commodities. Comprehensive regulations relating to the transportation and the storage of dangerous articles of all descriptions are enforced by the Coast Guard.

Under the provisions of the Federal Boating Act of 1958, Coast Guard boating safety teams are required to inspect small boats to ensure compliance with required safety measures, including fire safety. The Coast Guard also has responsibilities in the field of abatement of oil pollution and other environmental issues.

In addition to surveillance over vessels, the Coast Guard is responsible for ensuring that merchant marine personnel have qualified through training. Licenses are issued to such personnel for the performance of specific duties after satisfactory completion of comprehensive examinations. The examinations, both written and practical, include a number of fire safety problems.

The Coast Guard also conducts investigations into causes of marine disasters, including fires. Appropriate sanctions are imposed where indicated.

U.S. CUSTOMS AND BORDER PROTECTION

The U.S. Customs and Border Protection agency, which is also part of the Department of Homeland Security, maintains control over merchandise, including fireworks, imported from other countries. It maintains representatives in all ports of entry. Agents determine that markings for flammability, for example, appear on imported textile products as required by the federal labeling provisions. They are also active in the prevention of smuggling.

FEDERAL EMERGENCY MANAGEMENT AGENCY

The **Federal Emergency Management Agency (FEMA)**, a part of the Department of Homeland Security, is the federal government's focal point for emergencies. In addition to the U.S. Fire Administration, which includes the National Fire Academy, the agency is responsible for emergency management and other coordinating programs relating to major disasters.

Federal Emergency Management Agency (FEMA)
■ the agency under the Department of Homeland Security responsible for emergency management and other coordinating programs relating to major disasters.

The Fire Research and Safety Act of 1968 provided for fire research activities and for the establishment of the National Commission on Fire Prevention and Control. The commission had the task of recommending to the president and to Congress procedures for reducing fire losses in the United States.

After extensive study, the commission submitted a report entitled *America Burning* to the president and the Congress. This report, issued in 1973, contained as a major recommendation the establishment of "an entity in the Federal Government where the Nation's fire problem is viewed in its entirety, and which encourages attention to aspects of the problem which have been neglected."[6]

Needs that the commission felt could be addressed by such a federal agency included more emphasis on fire prevention, better training and education in the fire service, more education of the public about fire safety, and recognition of the hazardous environment to which Americans are exposed from a fire safety standpoint. The commission also addressed the need for improvements in building fire protection features and in research activities related to fire safety.

Congress subsequently enacted legislation in 1974 establishing the National Fire Prevention and Control Administration as an agency of the Department of Commerce. Under the provisions of President Jimmy Carter's 1979 reorganization plan, the National Fire Prevention and Control Administration was transferred basically intact to the new FEMA. The National Fire Prevention and Control Administration became the U.S. Fire Administration under Public Law 95-422 in October 1978.

The National Fire Academy, which is housed with the Emergency Management Institute at the National Emergency Training Center in Emmitsburg, Maryland, is responsible for advanced-level fire and emergency services education activities. A wide variety of programs are offered on campus and in field classes offered throughout the country. The academy has gained international stature and is recognized as having a major impact on fire service activities in the United States. Both volunteer and career fire service personnel can take advantage of course offerings. A number of programs relate to public fire education, code enforcement, arson suppression, and fire prevention administration.

The U.S. Fire Administration has responsibilities in the implementation of the Hotel and Motel Fire Safety Act of 1990 under which federal employees are to stay in properties meeting the law's fire protection requirements.

The U.S. Fire Administration also operates the National Fire Incident Reporting System (NFIRS). NFIRS has enabled fire safety personnel to study the causes of fires and to develop more effective public education and legislative programs.

Juvenile fire setters have been the focus of another program of the U.S. Fire Administration. Fire service personnel have been trained in methods aimed at reducing such fires through effective counseling and education.

From time to time, the U.S. Fire Administration has conducted special studies of major fires and has looked at specific fire problems in high-risk areas. Efforts have been made to disseminate information on these matters to bring about improved community fire safety.

Independent U.S. Government Agencies

CHEMICAL SAFETY AND HAZARDS INVESTIGATION BOARD

The Chemical Safety and Hazards Investigation Board (also known as the Chemical Safety Board) has responsibilities very similar to those delineated later in this chapter for the National Transportation Safety Board. It participates in investigations of significant incidents relating to chemical safety and hazards.

CONSUMER PRODUCT SAFETY COMMISSION

Consumer Product Safety Commission
▪ the agency that endeavors to eliminate hazards associated with consumer products, including toys, hazardous substances, and flammable fabrics.

The **Consumer Product Safety Commission** has endeavored to eliminate hazards associated with consumer products, including toys, hazardous substances, and flammable fabrics. The commission has an enforcement and educational program that strives to reduce injuries resulting from the use of common consumer products, most of which are transported through interstate commerce. It enforces regulations requiring proper labeling of hazardous substances and is legally authorized to ban from the market commodities that pose severe fire or safety hazards. It enforces the Flammable Fabrics Act as well. That law prohibits interstate marketing of wearing apparel and other materials that do not conform to flammability standards.

The Consumer Product Safety Commission has been active in the field of fireworks control and has banned the manufacturing and distribution of the more explosive fireworks. Flammable liquids for household use have also received the commission's scrutiny. Its enforcement officers seek out items having unusual characteristics from a safety standpoint. The commission's activities are having an increasing impact on fire safety. In 2007 it issued new standards for mattress fire safety.

ENVIRONMENTAL PROTECTION AGENCY

The Environmental Protection Agency has the responsibility of ensuring the protection of the environment through the abatement and control of pollution. The comprehensive charge to this agency, organized in 1970, includes development of standards for open-air burning and a number of other fire-related pollution problems.

GENERAL SERVICES ADMINISTRATION

The General Services Administration serves as a management agency for the construction and operation of government-used buildings, the procurement of supplies, and the control and disposal of records. In carrying out those responsibilities, fire protection standards have been established by the Public Buildings Service, an arm of the General Services Administration.

NUCLEAR REGULATORY COMMISSION

The Nuclear Regulatory Commission regulates and licenses the civilian use of nuclear material and generally has a responsibility in connection with the safety of this material. Fire prevention procedures are a factor in those regulations.

NATIONAL TRANSPORTATION SAFETY BOARD

The National Transportation Safety Board has specific statutory responsibility for the investigation of accidents in civil aviation as well as for highway, rail, marine, and pipeline accidents. Investigation is mandatory in aviation accidents; other accident investigations are made on the initiative of the board.

Members of the National Transportation Safety Board, who are appointed by the U.S. president, have executive authority in conducting their investigations and also have the authority to conduct formal proceedings for review of appeal on the suspension, amendment, modification, revoking, or denial of any certificate or license issued by the secretary of transportation or by an administrator operating with the department. The National Transportation Safety Board's reports of its investigations of major accidents have resulted in changes in regulations and in changes in procedures in the U.S. Department of Transportation.

State Agencies

All state governments have agencies with responsibilities in the field of fire prevention. Code enforcement and inspections at the state level have been responsible for a multitude of improvements in fire safety measures and fire prevention practices.

STATE FIRE MARSHAL

The state agency that usually has fire prevention as a major responsibility is the state fire marshal's office. Every state has an organization for carrying out fire safety functions at the state level. All provinces and territories in Canada also have such offices (Figure 10-1).

State fire marshal offices were first established shortly before 1900. Since that time they have grown in number at a rather steady rate. Some states have established the office of fire marshal as a direct result of some major tragedy. Other states have done so because of a realization that effective fire prevention must be carried out as a function of the state government.

In several states, the fire marshal's office is an arm of the state police. It carries out a full range of fire marshal duties, including fire prevention, code enforcement, and fire investigation. However, responsibilities in the inspection and code enforcement field may be limited, with other agencies assigned the remaining inspection responsibilities in some of the states.

Because of the past close association of fire prevention with the insurance industry, the office of the fire marshal in some states is assigned to the office of the insurance commissioner. Another reason for this close relationship with the insurance department is that in some states the combined office of the insurance commissioner and fire marshal is funded by a tax on insurance premiums.

Tom Grill/Corbis

FIGURE 10-1 Decommissioned fire stations can provide fire and life safety learning when sustained and used for museums or safety zones.

In an increasing number of states, the office of the fire marshal has become a part of the state department of public safety, an arrangement that makes possible closer cooperation with other state law enforcement agencies, especially in the field of fire investigation.

Personnel Appointments

Considerable diversity exists in the ways fire marshals are selected. In some states, the position is filled by an elected official or by gubernatorial appointment. In other states, appointment is made by the state fireboard or fire prevention commission, or by the attorney general, state comptroller, or other person under whom the fire marshal's office operates. Length of tenure varies from an indefinite period to 4, 5, or 6 years. In states in which the fire marshal's office is an arm of the state police, the appointment is usually based on departmental assignment procedures. Alabama's fire marshal is a merit system position.

Appointment methods for deputies and other personnel within the fire marshal's office are also varied. In some states, all deputies are merit system or civil service employees. In other states, the deputies are appointed to serve at the pleasure of the fire marshal. In the latter situation, suggestions on appointments may come from someone at a higher level in the state government, with possibilities of political involvement in the appointments. This was a fairly common process in past years. Now, however, in a majority of the states, public employees are selected under a merit system or other nonpatronage employment program.

Fire Marshal's Responsibilities

Although the responsibilities of the office of the fire marshal vary from state to state, its primary function is usually to enforce state fire prevention codes and to investigate suspicious fires. Usually, the office also is the coordinating agency for all fire prevention activities within the state government. The state fire marshal strives to obtain compliance with fire prevention regulations in connection with the licensing program of other state agencies, such as the health department and department of education. The fire marshal may recommend the inclusion of fire safety requirements in licensing and accreditation procedures and in other state controls imposed on facilities within the state.

Fire Code Enforcement

In a number of states, the office of the state fire marshal is the primary fire code enforcement agency. In some states, a locally designated fire marshal or fire chief may serve as an ex officio deputy to the state fire marshal for the purpose of code enforcement within the local jurisdiction.

Fire Investigation

The state fire marshal's office has the responsibility in most states to conduct investigations of all fires to which it is called. Notification procedures vary; however, local fire and police departments are usually the source of calls.

Under the Model Fire Marshal Law,[7] which is enacted in many states, the state fire marshal or a representative of the office has right of entry for the investigation of fires and responsibility for control and suppression of arson within the state. A number of states invest the fire marshal with powers of subpoena and arrest; under those legal conditions, the office is considered a law enforcement agency within the framework of state government. The state fire marshal's office may provide investigators for suspicious fires. The agents are usually qualified as expert witnesses in the courts and can play a major part in the successful prosecution of a case. In many states, fire marshal's office personnel are considered peace officers and are provided with firearms. Those individuals have the power of arrest in arson cases.

Fire Statistics

Another duty carried out by most fire marshals' offices is the compilation of fire statistics. An effort is being made to encourage every state to engage in fire recording and to develop statewide statistical data for use in fire safety programs and legislative hearings.

Fire and Life Safety Education

A number of state fire marshals operate public fire and life safety education programs. In some cases, state efforts are directed toward the development of local capabilities in this field. In other cases, fire marshal's office personnel provide programs directly to the public.

Other Duties

Explosives control activities are assigned to the fire marshal in some states. Control may include licensing programs for people possessing and using explosives, as well as a capability to deactivate clandestine devices, bombs, and other explosive devices.

The advent of greater federal participation in fire safety has brought about an interest in improved coordination of services at the state level. Among concepts advanced is a state coordinating commission or council whose purpose is to influence fire safety efforts in the state. The council or commission may or may not have direct authority over any operating agencies, such as the fire marshal's office

FIGURE 10-2 A fire station is one community facility where residents can go for information about fire safety. A well-planned and well-executed program to train firefighters in the most common topics of fire safety ensures that fire station personnel are prepared to provide this service.

Billy Morris Director of Fire Science Cisco College

or state fire training program. Such a commission would likely have as representatives fire chiefs, career and volunteer firefighters, building officials' associations, the state fire marshal and director of fire service training, the state forester, and members of other statewide agencies and organizations that have responsibilities and interests in fire protection (Figure 10-2).

HEALTH DEPARTMENT

All 50 state governments have a department of health. The health agency has an interest in fire safety as a part of its responsibility to safeguard citizens from health hazards and, more specifically, to ensure proper safeguards for facilities under the direct control of the agency: hospitals, extended-care facilities, and other such institutions. The health department may also have an interest in fire prevention through accident prevention programs, which in most states include certain fire safety information.

EDUCATION DEPARTMENT

Each of the 50 states has a department of education. This agency has an interest in ensuring the safety of all schools within the state, including those that are privately operated but come under state accreditation control. As a rule, state department of education regulations require periodic fire inspection and plan reviews for new construction, evacuation plans, fire drills, and other procedures to strengthen fire protection and fire prevention. In some states, school fire safety regulations are issued and enforced in cooperation with the office of the state fire marshal. Departments of education may also be responsible for including fire prevention education in the required curriculum.

DEPARTMENT OF LABOR AND INDUSTRY

State governments have a division for regulation of labor and industry. Although primarily responsible for industrial safety, these agencies often include fire prevention in their enforcement and educational programs. With the advent of the Occupational Safety and Health Act of 1970, the responsibilities of the department of labor and industry have increased in many states. In a number of states, the department of labor and industry has been designated by the governor as the primary enforcement and coordinating agency for state enforcement of the provisions of OSHA.

In several states in which the fire marshal's duties do not include a full range of fire code enforcement—Pennsylvania and Wisconsin, for example—inspections relating to fire prevention are the responsibility of the agency that regulates industrial safety. The control of explosives may be in the same agency.

Industrial safety classes organized and sponsored by departments of labor and industry may include emphasis on fire prevention. North Carolina, for example, has a comprehensive industrial safety education program that stresses fire prevention as an element of safety.

A number of states have enacted "right to know" laws. Under these laws employers are required to give their employees information on toxic substances in the workplace and are required to provide training on the safe handling of such substances in emergency situations. Employers must notify local fire departments of the location and characteristics of all toxic substances regularly present in the workplace. This legislation may affect fire prevention bureau activities because much of the information relates to that sought during code enforcement visits.

Federal legislation likewise requires that information relating to toxic materials be given to employees and to the local fire services under the Superfund Act of 1987. Fire safety efforts are enhanced by greater distribution of information relating to hazardous materials.

PUBLIC UTILITIES COMMISSION

The public service or public utilities commission is another state agency that has an interest in fire prevention. Such commissions usually regulate public utilities, including gas and electrical distribution. Both utilities have a bearing on fire safety.

In several states, the public service commission has a statutory responsibility for conducting electrical inspections within the state or for seeing that such inspections are conducted properly. Electrical inspections are essential in any comprehensive fire prevention program.

BUREAU OF MINES

States with operating mines generally have a state agency responsible for control of the industry. The state bureau of mines usually includes fire prevention as a part of its legally constituted duties. Fire prevention and protection in underground mines is a specialized field.

FORESTRY DIVISION

Forest fire prevention is a major function of divisions of forestry in all states and provinces. These agencies are generally responsible for fire prevention and control in their areas of jurisdiction. They are usually established in such a way as to permit assignment of personnel to fire prevention or suppression, depending on immediate demands. During fire seasons, for example, all personnel are assigned to fire suppression duties, whereas when fires are less frequent, employees may be used for other duties.

Fire prevention efforts of state forestry departments have undoubtedly been responsible for the dissemination of fire safety information to countless individuals who would not otherwise have been contacted.

OTHER STATE AGENCIES WITH RELATED FUNCTIONS

The functions of several other state agencies in this category are described in the following paragraphs.

State Medical Examiner

The state medical examiner or postmortem examiner performs autopsies to determine the cause of death and other related factors. All fire deaths should be investigated, and an autopsy should be performed. Through this means, a factual picture of fire prevention problems as related to life safety can be more readily obtained.

In one state, for example, where such a program was vigorously implemented, it was learned that a number of deaths from causes other than fire were being reported as fire deaths. Individuals in some cases had been killed by weapons before the setting of the fire. Had autopsies not been performed, the deaths would have been carried as fire fatalities. However, a body damaged by fire is not pleasant to examine, and in communities where coroners are not medical doctors, there may be a tendency to automatically write the death off as "smoking in bed" or some other fire-related cause, when in fact the victim died from a cause unrelated to the fire.

Another responsibility is determining whether the victim died from carbon monoxide poisoning before being burned in the fire. Detection of hydrogen cyanide, a product of combustion, is an emerging science and should be included among the tests a medical examiner uses in investigating a death associated with fire.

State Institutions

State institutions with responsibilities for housing patients, inmates, and other wards of the state are likewise responsible for fire prevention measures in connection with the operation of their facilities. Through the cooperation of inspecting agencies, such as the fire marshal's office, fire safety practices in those institutions can be maintained at a high level.

Mental health agencies often have responsibility for licensing structures housing people suffering from various mental disorders. The fire safety requirements have a bearing on the safe operation of these facilities.

Building Code Agencies

Many states have enacted a state building code on either a mandatory or voluntary basis, with functional responsibility in a variety of state agencies. Under the provisions of some state laws, a local jurisdiction may, if it deems it desirable, adopt the state building code for enforcement within the jurisdiction.

Educational Institutions

The college and university system operated by state governments has a very definite role in fire prevention in two kinds of programs: those conducted for the safety and well-being of students residing on campuses and occupying classroom spaces, and fire service training programs to improve the public fire departments.

An example of state university fire safety outreach is the smoke alarm initiative of the University of Alabama at the Birmingham Injury Control Research Center. This project seeks to have smoke alarms in every Alabama residence. The state fire marshal, state department of health, fire departments, and community action groups are cooperating in this program.

State fire training programs operated by colleges and universities are among the most effective in the fire service training field. By being housed on campuses, they have the advantage of the tie-in with academic programs and of offering participation in research work that might not otherwise be available. Foremost among the fire service training programs are those conducted by Louisiana State University, the University of Maryland, Oklahoma State University, and Texas A&M University. A number of colleges and universities offer 4-year and advanced-degree programs in fire protection. In some states, fire service training programs are operated as part of the state department of vocational education.

Judicial Branch

The judicial branch of state government has a bearing on fire prevention through legal processes in fire prevention code enforcement and arson control. Legal action in code enforcement is effective only with the backing and support of the judicial branch of government.

An Ohio Supreme Court case is an example. The court held that the use of adapters to connect hose to fire hydrants because of thread differences was not a violation of Ohio statute, since there was no support for the argument that the use of adapters created potential problems for the fire department. A local fire chief had alleged that since fire hydrant threads used by a county water system were dissimilar to those in his jurisdiction, necessitating the use of adapters, a noncompatibility situation was created.[8]

Legislative Branch

The legislative branch of state government must be mentioned because of the major role it plays in enacting legislation relating to fire prevention, as well as fire and arson control, and its role in enacting budgets for fire prevention and fire protection.

County Agencies

Fire safety activities at the county level (a county is referred to as a *parish* in Louisiana and a *borough* in Alaska) are quite varied from one county to another. An increasing number of county governments are operating fire departments for the protection of the entire county, or at least for portions outside incorporated cities. The Los Angeles County Fire Department protects unincorporated areas as well as a number of incorporated cities, with which it contracts to provide fire protection. Fire prevention services are also provided.

Although some metropolitan counties provide a full range of services and activities similar to those usually associated with municipal governments, the majority of the counties in the United States are not heavily involved in fire prevention programs.

Many counties provide fiscal support for volunteer fire department operations, and some maintain fire marshals' offices, communications centers, and training programs to assist local fire departments. In addition, an ever-expanding number of counties are enforcing planning and zoning requirements at the county level.

Municipal Government

Municipal government has a tremendous impact on fire prevention and fire control. There are few municipalities, regardless of area or population, that do not have some provisions for public fire protection in the form of an organized fire department, which may have career, volunteer, or both types of personnel. Details of the municipal government role in fire prevention are delineated in Chapter 7.

Support Agencies

Certain agencies operating at all levels of government have what may be called an indirect association with fire prevention. Government organizations responsible for taxation and assessment and for fiscal operation of government at all levels provide important support for effective fire prevention. Their part in ensuring funding of necessary programs cannot be taken lightly.

Agencies connected with emergency management, art commissions, transportation, law, and beautification, local and state environmental programs, historical preservation, and a number of other areas have some limited involvement with fire prevention and should not be overlooked for community cooperation.

Summary

Government, except at the municipal level, has not been as active in fire prevention and protection in the United States and Canada as it has been in other countries. However, because population growth and economic changes in the 21st century are concentrating the population in urban areas, the influence of governmental agencies in fire prevention efforts at all levels is steadily increasing. Indicative of this trend is the creation by Congress in 1973 of the National Fire Prevention and Control Administration, now known as the U.S. Fire Administration, a part of the Department of Homeland Security.

Among federal agencies, the Department of Agriculture has traditionally had the greatest impact on the general public with respect to fire prevention (mainly through the Forest Service). However, the National Institute of Standards and Technology and FEMA, an arm of the Department of Homeland Security, have made many contributions in the field.

Under the Occupational Safety and Health Act of 1970, the Department of Labor has an impact on fire prevention through the enforcement of the act, which applies to businesses, industries, institutions, farms, and other places of employment that engage in interstate commerce.

The Department of Transportation is an important force in fire prevention, as is the Department of Defense, which has maintained fire prevention efforts and training for members of the armed forces that carry over into civilian life.

Among state agencies, the office of the state fire marshal is paramount, having wide responsibilities in fire code enforcement and inspections and in investigations of suspicious fires, among other duties. The state fire marshal also lends aid and support to municipal fire departments.

At the municipal level, the community fire department is without question the major bulwark against the perils and ravages of fire in the United States. Practically every community has such a force, whether it is career, volunteer, or a combination of the two.

Review Questions

1. The Nationally Recognized Testing Labs Accreditation Program under OSHA serves what function?
   a. Conducts an award program for testing labs
   b. Issues licenses to testing labs
   c. Maintains an index of accredited labs
   d. Publishes national fire safety standards
2. The U.S. Army has contributed to the field of fire prevention through:
   a. development of fire-retardant materials.
   b. studies of hazardous materials.
   c. pipeline safety.
   d. all of the above.
3. The FBI can help in fire prevention by:
   a. arresting arsonists.
   b. taking fingerprints.
   c. examining the materials found at a fire scene.
   d. all of the above.
4. Personnel of the Bureau of Alcohol, Tobacco, Firearms, and Explosives can investigate:
   a. certain arson cases.
   b. explosions.
   c. use of firebombs.
   d. all of the above.

5. The U.S. Forest Service programs include:
   a. Smokey Bear.
   b. fire prevention.
   c. wildland–urban interface.
   d. all of the above.
6. The number of accidental wildfires caused by humans has been reduced by _____ percent since Smokey Bear arrived.
   a. 20
   b. 30
   c. 40
   d. 50
7. The Federal Motor Carrier Safety Administration:
   a. investigates all collisions.
   b. investigates accidents on federal highways.
   c. investigates accidents involving controlled carriers.
   d. all of the above.
8. The Department of Homeland Security was created in:
   a. 1999.
   b. 2001.
   c. 2002.
   d. 2003.
9. The National Fire Academy is located in:
   a. Washington, D.C.
   b. Chicago, Illinois.
   c. Denver, Colorado.
   d. Emmitsburg, Maryland.
10. State fire marshals carry out the following state-level functions except for:
   a. compiling state fire statistics.
   b. determining actual cause of death for fire victims.
   c. enforcing fire codes.
   d. investigating fires.

End Notes

1. U.S. National Commission on Fire Prevention and Control Report, *America Burning* (Washington, DC: U.S. Government Printing Office, 1973).
2. "Smokey Bear Turns 50," *Parade* (February 6, 1994).
3. Anthony P. Hamins, et al., *Reducing the Risk of Fire in Buildings and Communities: A Strategic Roadmap to Guide and Prioritize Research*, Special Publication 1130 (Gaithersburg, MD: National Institute of Standards and Technology, April 2012). Accessed June 17, 2013, at http://www.nist.gov/customcf/get_pdf.cfm?pub_id=909653
4. Ibid.
5. Personal communication to author Robertson from Director, National Institute of Science and Technology Building and Fire Research Laboratory, Gaithersburg, MD, 2003.
6. U.S. National Commission on Fire Prevention and Control Report, *America Burning* (Washington DC: U.S. Government Printing Office, 1973).
7. *Model Fire Marshal Law* (Quincy, MA: International Fire Marshals Association).
8. *Gamble et al. v. Dobrasky* (Supreme Court of Ohio, Case 99-1311, 2000).

CHAPTER

11 Fire Prevention Through Arson Suppression

Mike Love

KEY TERMS

arson, *p. 208*
corpus delicti, *p. 216*
Daubert v. Merrell Dow Pharmaceuticals, Inc., *p. 226*
F.I.R.E., *p. 228*
Michigan v. Clifford, *p. 226*
Michigan v. Tyler, *p. 226*
model arson law, *p. 211*
motive, *p. 213*
NFPA 921 Guide for Fire and Explosion Investigations, *p. 214*
NFPA 1033, *p. 216*
offender-based studies, *p. 213*
State of South Dakota v. Jorgensen, *p. 226*
wildland, *p. 213*

OBJECTIVES

After reading this chapter, you should be able to:

- Demonstrate an understanding of the characteristics of fire and the need for a thorough fire investigation.
- List the most commonly accepted motives for arson.
- Explain the model arson law, identify and explain the four degrees of arson, and describe how the model arson law relates to a modern state penal code.
- Explain the importance of preserving evidence and the potential conflicts that may arise between fire suppression efforts and fire investigations.
- Explain the delicate relationship between fire investigation and the media.
- Explain how the court cases *Michigan v. Tyler, Michigan v. Clifford, Daubert v. Merrell Dow Pharmaceuticals, Inc.*, and *The State of South Dakota v. Jorgensen* relate to fire investigation.
- List some of the conclusions of NFPA's Arson Control Symposium with regard to arson control needs.

FIGURE 11-1 Fire investigators spend hours digging through ruins to determine points of origin and causes.

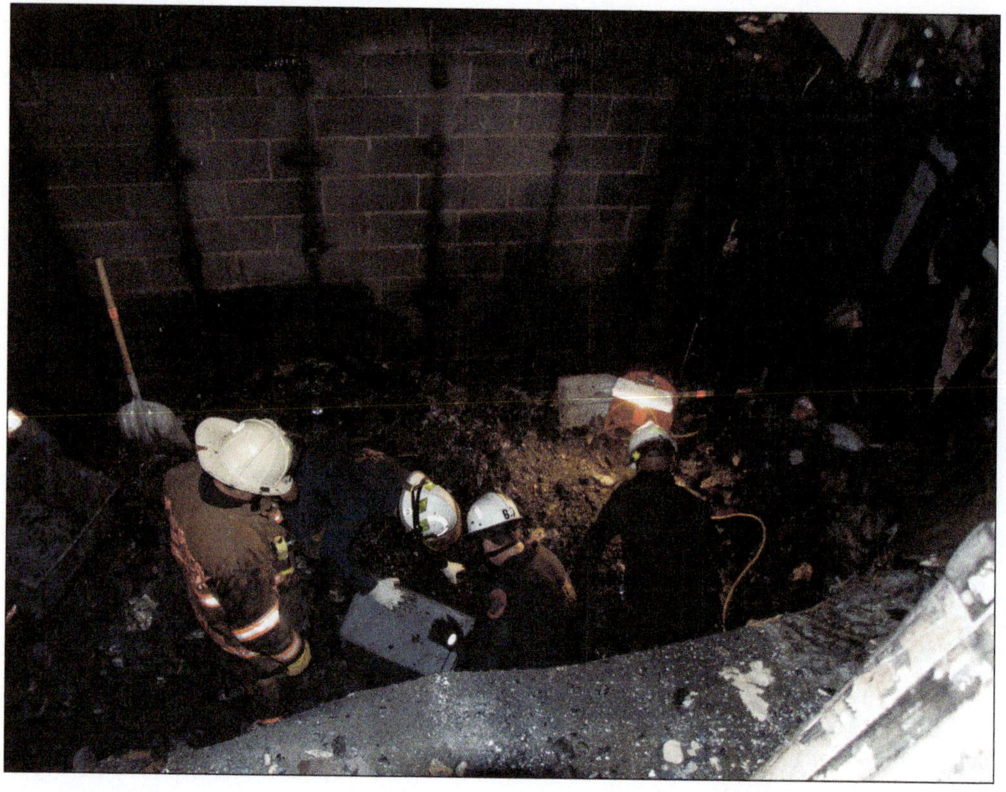

Mike Love

arson
■ the crime of maliciously and intentionally, or recklessly, starting a fire or causing an explosion.

Suppression of **arson**, "the crime of maliciously and intentionally, or recklessly, starting a fire or causing an explosion,"[1] is clearly a factor in fire prevention efforts. In many communities, arson is the cause of the majority of fires; therefore any steps that can be taken to prevent arson reflect favorably on the community's fire statistics.

Although law enforcement agencies often share with fire services the responsibility for suppressing arson, the fire service is usually the prime mover in arson control. Effective investigation and prosecution can discourage arson. Many communities have mounted successful programs aimed at preventing it (Figure 11-1).

The Crime of Arson

As a type of criminal activity, the maliciously set fire has historically been considered a serious crime. The first arson law in Maryland, enacted by the General Assembly meeting in St. Mary's City in 1638, classified arson as a felony. Punishment was death by hanging, loss of a hand, or being burned on the hand or forehead with a hot iron. The offender also forfeited ownership of all properties. If the offender escaped death for the first offense, there was a mandatory death sentence for the second offense.[2]

In Illinois, which obtained statehood in 1818, arson was considered so serious an offense that between 1819 and 1827 it was one of only four crimes punishable by death.[3]

Early western towns were likewise hostile toward arsonists. Vigilante justice occasionally superseded the established justice system. Closely built frame buildings, coupled with meager fire suppression capabilities, could spell doom for the entire town when an arsonist was at work. In Virginia City, Nevada, a vigilante band stormed the local jail in 1871, removing and lynching a suspected arsonist. Other members of the arson ring were sentenced to the state prison. After that time, arson did not occur frequently.[4]

Arson is one of the few criminal activities in which there is often no immediate recognition that a crime has been committed. A fire is considered an accident unless a thorough investigation shows evidence of arson. The fire accompanying civil disorder is a different matter. In that type of maliciously set fire, it is readily apparent in most cases that a crime has been committed. But the investigator must still prepare a tight case that examines as far as possible the ways in which the fire could have been started accidentally. Fires set by individuals for political reasons will be discussed in detail later in this chapter.

The crime of arson has another unusual feature. The weapon most often used, a match, is readily available at convenience stores, service stations, and restaurants. No program could be developed to control the acquisition of matches, nor would such a program have popular support.

ARSON STATISTICS

Uniform Crime Reports, published in 2012 by the Federal Bureau of Investigation (FBI) of the U.S. Department of Justice, provides interesting statistics on arson in the United States. The statistics for 2011 are used in this chapter.

For example, a total of 43,412 arson offenses were reported to the FBI's Uniform Crime Reporting (UCR) Program in 2011. The reporting program uses the following definition of arson: It is any willful or malicious burning or attempt to burn, with or without intent to defraud, a dwelling house, public building, motor vehicle or aircraft, or personal property of another. In addition, only fires determined through investigation to have been willfully or maliciously set are classified as arsons. Fires of undetermined origin are excluded from this statistical reporting program. NFPA 921 defines an *incendiary fire* as "a classification of the cause of a fire that is intentionally ignited under circumstances in which the person igniting the fire knows the fire should not be ignited."[5]

As shown in Table 11-1, the arson rates ranged from 44.2 per 100,000 inhabitants in cities with populations over 1 million to 28.2 per 100,000 rural county inhabitants. The suburban counties and all cities collectively recorded rates of 12.8 and 20.7 per 100,000 inhabitants, respectively. Overall, the 2011 national arson rate was 18.2 per 100,000 inhabitants.

The National Fire Protection Association (NFPA) bases its statistics for 2011 on surveys reported by fire departments. NFPA estimates there were 26,500 intentionally set (arson) structure fires in 2011, a decrease of 3.6 percent from 2010. Those intentionally set structure fires resulted in an estimated 190 civilian deaths, a decrease of 5.0 percent from the previous year. Monetary loss from the set structure fires resulted in $601 million in property loss, an increase of 2.7 percent over 2010 statistics.

Also during 2011, NFPA estimated there were 14,000 intentionally set vehicle fires, resulting in $88 million in property loss, a decrease of 1.1 percent over 2010.[6]

TABLE 11-1 Arson Rate by Population Group, 2011[1]

POPULATION GROUP	RATE
TOTAL ALL AGENCIES	18.2
TOTAL CITIES	20.7
Group I (250,000 and over)	32.8
1,000,000 and over (Group I subset)	28.2
500,000 to 999,999 (Group I subset)	31.8
250,000 to 499,999 (Group I subset)	38.6
Group II (100,000 to 249,999)	20.7
Group III (50,000 to 99,999)	16.9
Group IV (25,000 to 49,999)	14.9
Group V (10,000 to 24,999)	13.3
Group VI (under 10,000)	20.3
Metropolitan counties	13.2
Nonmetropolitan counties	12.2
Suburban areas[2]	12.8

1 13,045 agencies; 2011 estimated population 263,872,098; rate per 100,000 inhabitants.

2 Suburban areas include law enforcement agencies in cities with less than 50,000 inhabitants and county law enforcement agencies that are within a Metropolitan Statistical Area. Suburban areas exclude all metropolitan agencies associated with a principal city. The agencies associated with suburban areas also appear in other groups within this table.

Source: Federal Bureau of Investigation, Crime in the United States 2011: Uniform Crime Reports: Arson Rate by Population Group, 2011 [13,045 agencies; 2011 estimated population 263,872,098; rate per 100,000 inhabitants] (Washington, DC: U.S. Department of Justice, n.d.). Accessed December 5, 2013, at http://www.fbi.gov/about-us/cjis/ucr/crime-in-the-u.s/2011/crime-in-the-u.s.-2011/tables/arson-table-1

Vehicle fire estimates are probably very low, since many U.S. agencies do not report arson incidents (owing to political or administrative concerns). Many investigators think that the total of intentionally set structure fires is closer to 40 percent.[7]

The FBI's UCR Program also identifies arson by type of property burned. As shown in Table 11-2, in 2011, structural arsons (residential, commercial, industrial, etc.) were the most frequent type, representing 45.9 percent of those arsons. Mobile properties (motor vehicles, trailers, etc.) represented 23.9 percent of arsons, and other properties (crops, timber, etc.) accounted for 30.2 percent of arson offenses.

In 2011, the average dollar loss per arson offense was $13,196. The average loss from structural property arson was $23,918 loss; from mobile property arson, $7,016; and from arson of other property types, $1,813.[8]

For historical purposes, NFPA reported that intentionally set fires in structures resulted in 2,781 civilian deaths in 2001; 2,451 due to the events of September 11, 2001, and 330 in other set structure fires. Intentionally set structure fires resulted in $34.453 billion in property loss: $33.44 billion due to the events of September 11, 2001, and $1.013 billion in other set structure fires.[9]

TABLE 11-2 Arson by Type of Property, 2011[1]

PROPERTY CLASSIFICATION	NUMBER OF ARSON OFFENSES	PERCENT DISTRIBUTION[2]	PERCENT NOT IN USE	AVERAGE DAMAGE	TOTAL CLEARANCES	PERCENT OF ARSONS CLEARED[3]	PERCENT OF CLEARANCES UNDER 18
Total	43,412	100.0		$13,196	8,627	19.9	32.7
Total structure:	19,912	45.9	18.6	$23,918	4,784	24.0	30.7
Single occupancy residential	9,627	22.2	19.0	$24,990	2,180	22.6	21.0
Other residential	3,100	7.1	11.2	$22,690	814	26.3	23.8
Storage	1,402	3.2	18.2	$16,911	302	21.5	31.1
Industrial/ manufacturing	159	0.4	23.9	$68,349	29	18.2	34.5
Other commercial	1,581	3.6	18.2	$37,855	311	19.7	26.4
Community/public	1,899	4.4	20.1	$26,610	687	36.2	64.8
Other structure	2,144	4.9	26.7	$ 9,508	461	21.5	40.1
Total mobile:	10,381	23.9		$ 7,016	1,046	10.1	12.9
Motor vehicles	9,833	22.7		$ 6,580	946	9.6	11.5
Other mobile	548	1.3		$14,848	100	18.2	26.0
Other	13,119	30.2		$ 1,813	2,797	21.3	43.7

1 14,887 agencies; 2011 estimated population 253,539,067.

2 Because of rounding, the percentages may not add to 100.0.

3 Includes arsons cleared by arrest or exceptional means.

Source: Federal Bureau of Investigation, Crime in the United States 2011: Uniform Crime Reports: Arson by Type of Property, 2011 [14,887 agencies; 2011 estimated population 253,539,067] (Washington, DC: U.S. Department of Justice, n.d.). Accessed December 5, 2013, at http://www.fbi.gov/about-us/cjis/ucr/crime-in-the-u.s/2011/crime-in-the-u.s.-2011/tables/arson-table-2

Model Arson Laws

The types of arson and the seriousness of this crime are best discussed in terms of the model arson law. Developed by the International Fire Marshals Association, the **Model Arson Law** is a part of the laws of many states.[10] This law defines four degrees of arson offenses.

Model Arson Law
■ a law developed by the International Fire Marshals Association; it is a part of the laws of many states.

First-degree arson under the model law is the burning of a dwelling. In this category, any person who willfully and maliciously sets fire to, burns, or causes to be burned, or aids, counsels, or procures the burning of a dwelling is guilty of arson. The model law implies that the offense is the same whether the building is occupied, unoccupied, or vacant. It further indicates that the section applies whether the property is the property of the accused or of some other person. The section includes structures that are parts of the dwelling, such as kitchens or shops. The prison sentence for first-degree arson is generally from 1 or 2 years to 20 years.

Second-degree arson under the model law is burning buildings other than dwellings. This section also requires that the fire be willfully and maliciously set and provides that an individual is chargeable if he or she set the fire, caused it to be set, or aided, counseled, or procured the burning of the building. Again, the offense is applicable whether the property belongs to the person charged or to someone else. The statute provides for a possible prison sentence of 1–10 years.

Third-degree arson in the model law pertains to the burning of property owned by others. Individuals who willfully and maliciously set fire to, burn, or cause to be burned, or aid, counsel, or procure the burning of any property of any class, as long as the property has a value of $25 or more and is the property of another, are guilty and may be sentenced for a period of 1–3 years.

Fourth-degree arson in the model law relates to attempts to burn. This provides that a willfully and maliciously set fire may be chargeable as an attempt even if combustible and/or flammable materials have only been distributed and includes any acts preliminary to the setting of the fire. A separate offense category is included for the crime of burning with intent to defraud an insurer. This category likewise requires that the act be willful and malicious and that the property be insured at the time of the fire.

The phrase "willfully and maliciously" is often used in the Model Arson Law; a fire set accidentally is not chargeable under the Model Arson Law. To obtain a conviction of arson, a prosecuting attorney must be able to prove in court that the act was willfully and maliciously perpetrated. In fact, the law enforcement officer should not consider making an arrest on an arson charge if willful and malicious action cannot be proved. A distinction is made in the law between dwelling places and other occupancies because there is a greater chance of death in a fire maliciously set in a dwelling.

A number of states have adopted the Model Penal Code of the American Law Institute as first published in 1962.[11] This code covers the entire criminal law field and encompasses some concepts that are at variance with more traditional modes.

In the Model Penal Code, arson is covered in the following manner (for example, this is the first subsection of the code as taken from the Arkansas Criminal Code Annotated as of 2012):

A.C.A. § 5-38-301. Arson.

a. A person commits arson if he or she:
   1. Starts a fire or causes an explosion with the purpose of destroying or otherwise damaging:
      (A) An occupiable structure or motor vehicle that is the property of another person;
      (B) Any property, whether his or her own or property of another person, for the purpose of collecting any insurance for the property;
      (C) Any property, whether his or her own or property of another person, if the act thereby negligently creates a risk of death or serious physical injury to any person;
      (D) A vital public facility;
      (E) Any dedicated church property used as a place of worship exempt from taxes pursuant to § 26-3-301; or
      (F) Any public building or occupiable structure that is either owned or leased by the state or any political subdivision of the state.[12]

Most statutes patterned after the Model Penal Code use similar language. A comparison of the Model Arson Law and the arson sections of the Model Penal Code reveals advantages and disadvantages in each.

Arkansas and all other states also have statutes relating to **wildland** fires. In some states wildland arson is considered a felony; in others it is a misdemeanor.

As noted in Chapter 10, federal agencies have statutory responsibilities in arson suppression. The Bureau of Alcohol, Tobacco, Firearms, and Explosives is the primary agency with this responsibility.

wildland
▪ often a remote land area characterized by natural, sometimes dense vegetation minimally modified by human activity, such as woodlands, forests, brush, and meadows; wildland can contain dangerous buildups of fuel from vegetation that may or may not be a risk for fire; trends in migration and development currently bring together increased wildland interface with urban areas throughout the world.

Motives for Arson

Although not absolutely necessary, it is most helpful to the prosecution of a case if a motive for the setting of a fire can be established. **Motive** is an inner drive or impulse that causes a person to do something or act in a certain way. One popular motive is avarice, a great desire for wealth: The offender believes that an insurance company will settle a loss without realizing that the insured in fact set the fire to defraud the insurer. This type of arson occurs most frequently during times of poor economic conditions. However, there are exceptions.

motive
▪ an inner drive or impulse that is the cause, reason, or incentive that induces or prompts a specific behavior.

A jury often wants to have a verifiable motive considered with the evidence so that the jury members can justify the verdict to themselves, even though a verifiable motive is not a legal requirement. Although motive is not essential to establish the crime of arson and need not be demonstrated in court, the development of a motive frequently leads to the identity of the offender.

Establishing motive also provides the prosecution with a vital argument to be presented to the judge and jury during trial. It is thought that the motive in an arson case often holds together the elements of the crime.

Law enforcement studies on arson motives are **offender-based studies**; that is, they examine the relationship between crime scene characteristics and the behavioral characteristics of the offender as they relate to motive. One of the largest present-day offender-based studies was put together by the Prince George's County (Maryland) Fire Department (PGFD), Fire Investigations Division, and consists of 1,016 interviews with both juveniles and adults arrested for arson and fire-related crimes from 1980 through 1984. The offenses included 504 arrests for arson, 303 for malicious false alarms, 159 for violations of bombing/explosives/fireworks laws, and 50 for miscellaneous fire-related offenses.

offender-based studies
▪ examinations of the relationship between crime scene characteristics and the behavioral characteristics of the offender as they relate to motive.

The study was conducted primarily because fire and law enforcement professionals were given the opportunity to conduct an independent research study of violent incendiary crimes. The PGFD study determined that arrested and incarcerated arsonists most often cite the following motives:

- Vandalism
- Excitement
- Revenge
- Crime concealment
- Profit
- Extremist beliefs[13]

NFPA 921 Guide for Fire and Explosion Investigations
▪ a standard that identifies guidelines and recommendations for processes and procedures to conduct investigations or analysis of fire and explosions.

For the purposes of classification, FBI behavioral science research defines *motive* as an inner drive or impulse that is the cause, reason, or incentive that induces or prompts a specific behavior.[14] A motive-based method of analysis can be used to identify personal traits and characteristics exhibited by an unknown offender.[15] For legal purposes, the motive is often helpful in explaining why an offender committed his or her crime. However, motive is not usually a statutory element of a criminal offense.

The motivations discussed in this chapter are also outlined and described in **NFPA 921 Guide for Fire and Explosion Investigations.**[16]

Investigation of Suspected Arson

New approaches to investigative methodologies include the documentation of the facts relied upon by the lead investigator or an investigative team. These proper investigative methodologies include listing the documents, information, and data reviewed in the preparation of the report; qualifications and publications of the team members; and prior testimony in other cases. In some cases, the compensation received by the investigator may need to be revealed.

The basic tenets of investigator methodology come down to documentation. The investigator must collect and document sufficient evidence and data regarding the fire, document thorough examination of hypotheses based upon an exhaustive review, and correctly apply the methodology to arrive (if possible) at a final hypothesis. As in any profession, the background, education, training, certifications, and experience of a fire investigator often has a great bearing on the proper application of the methodology.

PROFESSIONAL STANDARDS OF CARE

The professional standards of care in conducting fire investigations rest upon two essential documents and several expert treatises in the field. Standards are viewed as the essentials of an industry, or, in other words, those tasks, materials, and activities that are mandatory. In fire investigation, the fire and life safety industry recognizes NFPA as the standard bearer, with NFPA 1033 as the standard and NFPA 921 as the way to professionally accomplish the standard. The methodology and opinions expressed in reports should rely primarily upon the 2014 edition of the NFPA 921, Guide for Fire and Explosion Investigations, approved and issued by the NFPA effective in January, 2014. Fire investigators should be familiar with the application of this standard.

The 2014 edition of NFPA 921 supersedes all previous editions and is considered the primary peer-reviewed standard of care for conducting fire and explosion investigations in the United States and throughout the world. The accepted standard of care under NFPA 921 is applicable to all forensic investigations, reports of investigations, expert opinions, and court testimony after that date, requiring them to be compliant with and evaluated under the 2014 edition.

NFPA 1033
▪ Standard for Professional Qualifications for Fire Investigator.

The second standard of care relies upon NFPA 921's companion publication, the 2014 edition of **NFPA 1033** Standard for Professional Qualifications for Fire Investigator,[17] which was approved by the NFPA Standards Council on May 28, 2013, with an effective date of June 17, 2014, and supersedes all previous editions. The 2014 edition of NFPA 1033 is a standard under the provisions of the NFPA and was also approved as an American National Standard on June 17, 2014. It is the universally recognized professional standard for fire investigators in both the public and private sectors, and it has been adopted by law in virtually every jurisdiction. Table 11-3 summarizes the professional levels of job performance required by NFPA 1033.

NFPA 921 is classified by the NFPA and is published as a guide. However, anyone who suggests that NFPA 921 is no more than a guide is simply uninformed.

TABLE 11-3 Professional Levels of Job Performance for Fire Investigators as Cited in NFPA 1033, 2014 Edition

General Requirements for a Fire Investigator	4.1.2 Employ all elements of the scientific method as the operating analytical process 4.1.3 Complete site safety assessments on all scenes 4.1.4 Maintain necessary liaison with other interested professionals and entities 4.1.5 Adhere to all applicable legal and regulatory requirements 4.1.6 Understand the organization and operation of the investigative team and incident management system
Scene Examination	4.2.1 Secure the fire ground 4.2.3 Conduct an interior survey 4.2.4 Interpret fire patterns 4.2.5 Interpret and analyze fire patterns 4.2.6 Examine and remove fire debris 4.2.7 Reconstruct the area of origin 4.2.8 Inspect the performance of building systems 4.2.9 Discriminate the effects of explosions from other types of damage
Documenting the Scene	4.3.1 Diagram the scene 4.3.2 Photographically document the scene 4.3.3 Construct investigative notes
Evidence Collection and Preservation	4.4.1 Utilize proper procedures for managing victims and fatalities 4.4.2 Locate, collect, and package evidence 4.4.3 Select evidence for analysis 4.4.4 Maintain a chain of custody 4.4.5 Dispose of evidence

(continued)

TABLE 11-3 Professional Levels of Job Performance for Fire Investigators as Cited in NFPA 1033, 2014 Edition (*continued*)

Interview	4.5.1 Develop an interview plan 4.5.2 Conduct interviews 4.5.3 Evaluate interview information
Post-Incident Investigation	4.6.1 Gather reports and records 4.6.2 Evaluate the investigative file 4.6.3 Coordinate expert resources 4.6.4 Establish evidence as to motive and/or opportunity 4.6.5 Formulate an opinion concerning origin, cause, or responsibility for the fire
Presentations	4.7.1 Prepare a written report 4.7.2 Express investigative findings verbally 4.7.3 Testify during legal proceedings

Source: NFPA 1033. (2014). NFPA 1033: Standard for Professional Qualifications for Fire Investigator. Quincy, MA: National Fire Protection Association.

The document is considered the leading peer-reviewed authoritative source for the standard in conducting fire and explosion investigations. NFPA provides a glossary of terms in most if not all its publications. NFPA describes a guide as being advisory or informative but generally nonmandatory. However, a guide sometimes becomes a standard and the accepted practice in an industry, discipline, or profession. In 2000 the U.S. Department of Justice issued a research report entitled "Fire and Arson Scene Evidence: A Guide for Public Safety Personnel," which stated that NFPA 921 "has become a benchmark for the training and expertise of everyone who purports to be an expert in the origin and cause determination of fires."[18]

corpus delicti ▪ the body of evidence that proves a crime has in fact been committed.

Successful investigation of maliciously set fires requires the establishment of a **corpus delicti**, the body of evidence that proves a crime has been committed. Every potential natural cause for such a fire must be eliminated. It is necessary to prove beyond a reasonable doubt that the fire could not have occurred accidentally or naturally. In defending the accused, the defense attorney in an arson case usually attempts to prove that the fire was accidental. The prosecuting attorney makes every effort to prove that the fire could not have started accidentally.

RESPONSIBILITY FOR INVESTIGATIONS

Fire investigation responsibilities are generally carried out in one of three ways:

- The entire job, including cause determination, investigation, and arrest, is carried out by fire service investigators.
- The entire job, including cause determination, investigation, and arrest, is carried out by police service investigators

- The responsibility is split, with fire service personnel being responsible for cause determination, followed by police investigation and arrest. In some communities, a team concept is employed. In others, fire service personnel merely notify the police when arson is suspected.

Many major cities, including New York, Los Angeles, Denver, Miami, and Houston, assign complete responsibility for arson suppression to the fire service. Personnel assigned must be thoroughly trained in law enforcement procedures and are empowered to make arrests. This method has an advantage because one agency is entirely responsible, and it eliminates shifting blame for poor performance and the jealousies that may arise when two agencies are investigating the same crime. Of course, communities using this method continue to rely on police support, primarily in intelligence and patrol activities. It can be helpful to provide fire company officers with awareness training of what they might see as they approach the scene of a fire. If an officer suspects that the fire is set, doing only enough to extinguish the fire but avoiding disturbing the scene, especially potential evidence (Figure 11-2), can have significant value.

In some communities, police departments are vested with the entire responsibility for arson detection and investigation. This system, to be successful, entails a great deal of training for police officers in examining fire scenes. To obtain any convictions, the officers must qualify as expert witnesses in examining fire scenes. A person with fire suppression experience is usually better qualified.

Responsibility assigned to a joint police–fire team has worked well in many communities; however, it has failed in others. A problem with this method is that

Mike Love

FIGURE 11-2 Fire investigator examines and determines the possibility of every ignition source for the cause of a fire.

work schedules in police and fire departments are generally not the same. Days off are not the same, and the result is lost time when one of the team members is not on duty. If one team member goes ahead with the investigation and successfully concludes it, the other may be unhappy.

Some communities have an arrangement whereby the fire suppression officer or fire marshal contacts the police department after determining that arson is the possible cause of a fire. The police department assumes all responsibility for the subsequent investigation and possible arrest. Both agencies must testify in court for a successful prosecution. Press publicity often gives credit to only one of the agencies, causing resentment in the unmentioned agency.

Smaller communities often depend on a state or county fire marshal's office for fire investigation services. Local fire officers or the state fire investigator may make the determination of arson. In either case, the state investigator carries out the investigation and makes an arrest where appropriate. This is a desirable arrangement, especially because it is difficult for those who have investigated very few fires to qualify as experts in court. Some smaller municipalities have developed an on-call system for mutual aid in fire investigations in an effort to overcome the experience problem.

Mississippi law requires each of the state's 82 counties to have at least one deputy sheriff trained in arson detection and investigation. Trained by the state's fire academy, the individuals are able to conduct a preliminary investigation and summon a state fire investigator when appropriate.

Charlotte, North Carolina, established a unique fire investigation task force, which includes fire and police department investigators; agents from the U.S. Bureau of Alcohol, Tobacco, Firearms, and Explosives; agents of the North Carolina State Bureau of Investigation; and a district attorney. The task force has maintained a case clearance rate above 30 percent since its formation in 1985.

State fire schools as well as the National Fire Academy provide training for fire investigators. In addition, the International Association of Arson Investigators holds instructional seminars at both the national and regional levels. Similar seminars are held in Canada.

Judges and prosecuting attorneys also need education in arson control. Seminars have proved to be quite successful in creating a greater awareness of the intricacies of prosecution of arson cases. One state raised its arson conviction rates from less than 1 percent to 50 percent in less than 3 years by conducting seminars and mock trials for judges and attorneys.

INSURANCE FRAUD

It is important to obtain a full inventory of materials and contents in fire-damaged structures, especially if insurers attempt to prove that a building was overinsured. Adjusters and other people familiar with evaluation techniques can be helpful in this process. The records of insurance carriers should be checked to determine if there has been any recent change in insurance coverage on the building. Such a change may be an indication of an attempt to defraud the insurance carrier. To prove fraud as a motive, first, there must have been a claim filed by the insured.

Insurance companies are concerned about soaring losses as a result of arson. In an effort to combat rising losses, which raise insurance rates for property owners, insurance companies have intensified their arson suppression programs.

One major insurance industry project is the Property Insurance Loss Register. This industrywide program is designed to record pertinent data about fires from adjusters. Computers store the names of insured tenants, owners, partners, and corporate officers; types of occupancy; cause, time, and date of losses; and insurers and amounts of coverage. This information is available for use in investigations.

Insurance carriers' denial of liability when fires are suspicious places the insured in the position of having to sue to recover the alleged losses. Carriers often employ their own investigators to assist in such cases. And the courts often uphold the denial of liability.

Some insurance companies assist communities in providing accelerant-detection canines. Companies also provide major funding for arson reward programs, in which tips leading to arrests are rewarded.

THE VALUE OF CONFESSIONS

Occasionally, in the investigation of an arson case, investigators obtain a confession to the crime. Such a confession is valid if properly obtained and can be of great assistance in the successful prosecution of the case. If a confession is to be considered valid, no threats may be used nor any promises made in taking a statement that turns out to be a confession. It is extremely important to forewarn the individual regarding his or her full legal rights as required by the U.S. Supreme Court's *Miranda* decision. This decision requires that interrogation not begin until the suspect has agreed to submit to the procedure. The individual must be warned, according to the decision, that a statement need not be made. The suspect should also be told that he or she may stop at any time and request the presence of an attorney; that in the event the suspect cannot afford a lawyer, one will be provided; and that any statement made can be used in a court of law.

In the case of arson or malicious burning, a confession alone may be used for convicting the defendant if there is corroborative evidence connecting the individual to a fire. This means that it is not always necessary for a witness to place the defendant at the scene of the fire at the time of its burning. It may be sufficient to have a witness testify that the defendant was seen around the area 15 or 20 minutes before the fire or around the fire scene shortly after the discovery of the fire. In using such a witness, the prosecutor should consider the need to report accurate times so the defendant can be placed at or near the scene within a short time before or after the fire.

FIRE SCENE EXAMINATION

Fire scene examination is a critical part of the investigation of fires. This examination cannot take place until the fire has been brought under control. However, the scene should not be left unguarded at any time. The fire ground should be securely maintained until the investigation has been completed and evidence or photographs at the scene are no longer needed.

In examining the fire scene, every effort should be made to locate sources of ignition that may have accidentally caused the fire, including electrical wiring, heating appliances, and carelessly discarded cigarettes (Figure 11-3). The investigator should be alert for any possible evidence pointing to the origin of the fire. Proof must also be established that a fire did actually occur, and the investigator must have the owner or owner's agent testify to the condition of the structure before the fire.

The firefighter plays a major role in the investigation of a maliciously set fire. Without the complete cooperation of the fire department, successful investigation and prosecution will be extremely difficult. Fire suppression officials responding to the alarm often have the first trained eyes on the scene. The fire department may also identify conditions relating to the spread of the fire, the use of an accelerant, and other factors.

Prefire conditions noted in the building during fire inspections and in planning surveys can be helpful in determining any significant changes in conditions at the time of the incendiary fire. The fire inspector may have recorded the original position of cabinets and other equipment that may have been moved by the person starting the fire.

On their approach, fire company officers can observe the general conditions at the scene, such as weather and road conditions. These factors may have a bearing on the subsequent investigation (through consideration of tire tracks in the snow, as an example).

The first fire department personnel to arrive should observe people and automobiles, as well as license numbers and descriptions of vehicles in the vicinity. This is not an easy task for a firefighter while fighting the fire or rescuing occupants. Any remembered detail, such as a license number, might be the key to tracing a suspected

FIGURE 11-3 Rarely are fire cause and point of origin this simple to determine.

Mike Love

arsonist. A suspect can sometimes be arrested on the scene, especially because whoever started the fire is apt to stay around long enough to see that the job was successful.

The first personnel on the scene should observe the size of the fire and the speed at which it is traveling. Again, like license plate numbers, these conditions are not easy to observe carefully under the pressure of fire control and rescue. Responders should particularly observe whether separate fires may be burning, along with the intensity and rapidity of spread. This information may be extremely critical at a later time when investigation indicates that the fire was, in fact, maliciously set.

Other indicators that may be of value in an investigation are odors, such as of gasoline or kerosene. Methods of extinguishing the fire might also have significance. The condition of windows and doors in the building should be observed for indications of forcible entry into the building, or for signs that material has been removed from the premises. Bystanders at the fire scene may offer some clues. Familiar faces in the crowd may have been noticed at other fires or emergencies in the community.

FIRE INVESTIGATOR

When the fire is under control and before the fire suppression forces depart, the fire suppression officer may decide that a fire investigator should be called to the scene. To preserve evidence, the scene should be guarded until the fire investigator arrives. Police officers, security guards, or other persons having a legal responsibility may need to stand by at the fire scene while awaiting the arrival of the fire investigator.

The fire suppression officer, while awaiting the arrival of the investigator, may look for evidence of arson intent. Among things to look for are unusual odors in the building (possibly indicating the use of flammable accelerants), multiple fires, undue charring, and uneven burning. A trained investigator can readily note significant fire pattern indicators that help determine the areas and points of the origin. The fire officer or investigator, or both, may look for penetrations that have burned through the walls or ceilings of the building. There is a growing tendency for fire setters to deactivate fire detection and sprinkler systems, block fire doors, and otherwise impede the operation of fire protection equipment. All such devices should be checked for evidence of tampering.

Tracks around the premises may be quite revealing. Fire personnel arriving early at the fire can best notice tracks. An observation may be made that there were people walking around the building before the fire. There is a good possibility that such evidence will not be useful later because fire service personnel carrying out fire suppression duties can unintentionally obliterate footprints.

The same is true of fingerprints. Although smoke and heat may obliterate most fingerprints, some might be found during thorough processing of the area. Most fire service personnel do not have training in fingerprint techniques and may overlook the possibilities of lifting prints.

Both NFPA 921 and NFPA 1033 require that the scientific method be used to perform a thorough and systematic examination of the fire scene to determine potential hypotheses for the fire's origin and cause. The scientific method is defined by NFPA 921 as "the systematic pursuit of knowledge involving the recognition and formulation of a problem, the collection of data through observation and experiment, and the formulation and testing of a hypothesis."[19]

Exhaustive examination of all potential ignition sources should be undertaken by the fire investigator. Heating equipment should be checked to see if gas valves were left open. Electric and gas stoves should be checked for the possibility that burners were turned on. Electrical equipment and appliances should also be examined.

Fatal fires deserve special attention. An autopsy should be conducted in all cases to ascertain the true cause of death. Where possible, the investigator should be present at the autopsy to ensure an adequate investigation of the entire body, including photographs.

A search should be made to detect any possible use of flammable liquids and for containers left behind at the fire scene that may have held flammable liquids. Ignition sources to look for are matches in the area, candles, or trailers (strips of combustible materials) placed between sources of ignition. Mechanical igniters or timing devices are not frequently found but have on occasion been used to start a fire. Not to be discounted is the possibility that the fire was started as a result of magnification of the rays of the sun by glass. The fire investigators should note excessive amounts of debris, rags, waste, and accumulations of papers or other combustibles in the area. All these factors should be carefully assessed and pointed out in the fire investigator's report.

PRESERVATION OF EVIDENCE

Every effort must be made to preserve and protect all evidence found of the cause of the fire. Extreme care is needed in using water around the fire scene where evidence may be involved. This precaution is not always easy to observe. Care should be taken in salvage and overhaul operations.

There is often a conflict between investigation and suppression activities. Overzealous cleanup activities sometimes result in the removal or destruction of necessary evidence. This outcome will decrease with adequate fire suppression training as it relates to fire scene examination and investigation. The owner or occupants of the building should under no conditions be permitted to return until the investigation has been completed. Visitors, including reporters, should not be allowed to enter the fire scene.

A major concern to fire investigators is preventing the destruction or alteration of evidence before it is submitted for thorough examination and analysis. Destruction or alteration of evidence, particularly when it will be the subject of pending or future litigation, is often referred to as *spoliation*.

Failing to prevent spoliation can result in sanctions, potential civil or criminal remedies, and disallowance of testimony.[20] Standards now establish practices for examining and testing items of evidence that may or may not be involved in product liability litigation. The American Society for Testing and Materials publishes a number of standards to follow to properly prepare and manage evidentiary materials. In laboratory examinations where the evidence will be altered or destroyed, all persons involved in present or potential cases should be given the opportunity to make their opinions known and to be present at the testing.

The use of barricades can be helpful if other means of restraining entry are ineffective. The fire service generally has the legal means of keeping unauthorized people from entering the structure while firefighting operations, including overhaul, are under way.

The fire officer should make a note of the time each piece of evidence is discovered. If at all possible, some other person should witness the discoveries and verify the conditions found. This verification may be a key factor in court.

There should be a mandatory procedure for handling evidence so that it is not damaged or distorted. The evidence should be maintained at its original location until collected and officially entered. Prior to its removal, all evidence should be photographed in place, and a sketch should be made to identify its exact location in the fire scene. All evidence must be marked or labeled for identification at the time of collection. Physical evidence should be stored in a secured location designed and designated for this purpose.

When collecting physical evidence for examination and testing, it is often necessary to collect comparison samples. This is especially important when collecting materials believed to contain liquid or solid accelerants. The laboratory can evaluate the possibility that flammable liquids were introduced to the scene and were not the normal fuels present.

All these factors should be carefully assessed and included in the fire investigator's report with photographs, videos, diagrams, and sketches. A recommendation for adequate documentation of a fire scene is found in NFPA 921, Guide for Fire and Explosion Investigations; DeHaan and Icove, *Kirk's Fire Investigation;* and Icove, DeHaan, and Haynes, *Forensic Fire Scene Reconstruction.*[21]

CONTROL OF STATEMENTS TO THE PRESS

Statements to the press regarding fire causes should be carefully controlled. Some fire service personnel have a tendency to make statements to the press regarding the details of fires without having specific knowledge of conditions. It is very easy to make comments to the press that later turn out to be incorrect. Statements about causes of the fire made by authorities who are not members of the fire service, including elected officials, may be damaging to the prosecution should an arson case be developed. Many investigations have been jeopardized because of unwise release of information to the news media. The best approach is to have one individual responsible for all press releases. Professional public information officers are well suited to assist in managing information and press releases.

PREVENTION OF ARSON

The U.S. Fire Administration has a "Coffee Break Training" program entitled "Safeguarding Homes from Arson," published as Issue No. FM-2013-1. The learning objective for the training program is to assist public safety officers in identifying the steps community residents can take to safeguard their homes from arson.[22]

Arson and Civil Unrest

Multiple arsons during periods of civil unrest during the 1960s and 1970s were the result of an arson motive that had not been prevalent for many years. Such fires impose a severe strain on routine investigative procedures.

HISTORICAL EXPERIENCE

Tulsa, Oklahoma, and several other cities experienced disorders in the 1920s. At least 27 fatalities occurred in the 1921 racial disturbances in Tulsa. Fires destroyed more than 1,000 buildings, primarily residences. Detroit also experienced riots during World War II.

In August 1908 a racial disturbance in Springfield, Illinois, resulted in arson against black neighborhoods, with rioters blocking fire wagons and cutting their hoses, triggered by an unconfirmed report of rape of a white woman by a black male.

In 1992, Los Angeles experienced the most expansive, violent, and costly epidemic of urban unrest in the 20th century. The rioting resulted in over 45,000 fire department responses with a total of 53 civilians killed.[23] Estimates on damage after the riot exceeded $1 billion.[24]

PROACTIVE MEASURES DURING CIVIL UNREST

Operations during times of civil unrest or disorder require planning. Investigative fire personnel cannot merely sit and wait for calls and expect to be successful once multiple fires start occurring. For example, it could be very useful to assign fire investigation personnel to patrol duties during times of disorder. The patrols could detect fires early and thereby increase chances for successful suppression; they could also aid in apprehending those responsible for setting the fire. The use of unmarked vehicles is probably desirable for this purpose. In a few cities, patrols of this kind have been equipped with portable fire extinguishers to enable them to control incipient fires. The possibility that such personnel might be lured into an area for physical attack must be kept in mind.

Fires that occur during periods of disorder should be as thoroughly investigated as those that occur during normal times. During periods of unrest, many people set fires for the first time. It is easy to lay aside investigation of minor fires during a disorder in deference to fires having a great monetary loss or loss of life. Failure to investigate the smaller fires may serve only to improve the confidence of their setters and to allay their fear of being apprehended.

In addition to a multitude of set fires, during periods of unrest there may be a number of false alarms and bomb threats. Those calls must receive attention because they could be genuine. Individuals have not hesitated to use explosive devices during periods of disorder. False alarms may be intended to send fire equipment to the other side of town to permit the unabated spread of planned fires.

INVESTIGATIONS DURING CIVIL UNREST

Fires must be investigated in such a manner as to provide adequate information in the event an arrest is made. The possibility that a property owner deliberately set fire to the premises during a period of unrest cannot be overlooked. Fires occurring during civil disorder may encourage individuals who are naturally inclined to set fires. An impulsive fire setter, for example, could be strongly influenced by television publicity given to fires during disorders.

Fire investigation during a civil disturbance involves a number of unusual problems. Guarding the scene of the fire for evidence preservation may be extremely difficult. Angry mobs may make it impossible to maintain effective safeguards. Also, the same building may be set on fire several times during the course of the night. Procedures usually followed in preserving evidence must be laid aside. Members of the National Guard may be asked to guard the fire scene until the investigator can safely return.

Some communities have investigators respond to fire scenes along with or just behind fire apparatus during periods of disorder. This policy gives the investigators an opportunity to locate and photograph evidence before it is removed. Fires set during disorders are usually rather crudely set.

Collecting and preserving evidence is just as important for fires set during times of civil disorder as for any other type of maliciously set fire. There is no reason fires cannot start from natural or accidental acts during times of disorder as well as during less turbulent times. The case will come to trial under quieter conditions. The investigator must be prepared to testify as to whether the fire could have been of natural or accidental origin. At least, it must be shown that an examination of heating appliances, electrical systems, stoves, flammable-liquid operations, and the like did not indicate them to be the cause of the fire.

If flammable liquids were used, the investigator should determine, if possible, where the flammable liquids were purchased and the means used to transfer them to the premises. Containers are, of course, necessary to transport flammable liquids.

When many fires occur in a short period of time, handling evidence generally becomes an issue. It is imperative that the evidence be adequately marked and properly stored. Materials gathered at the fire may present a problem, because portions of firebombs are often quite fragile. A scoop-type box is useful for the collection of firebomb material.

Arson Arrests

Many law enforcement and fire service investigators consider arson a difficult crime to solve with an arrest. Although arson convictions may require diligent efforts, the *Uniform Crime Reports* indicate that in 2011, law enforcement agencies in the nation cleared by arrest or exceptional means 19.9 percent of arson offenses. Almost one-third (32.7 percent) of arsons cleared in the nation in 2011 involved juvenile offenders (those under 18 years of age).[25]

JUVENILE FIRE SETTERS

As noted in the foregoing FBI Uniform Crime Reporting statistics, the juvenile fire setter is a major problem in the arson field today. The term *juvenile fire setter* includes children under the age of responsibility as well as children who can legally be held responsible for their acts. Some jurisdictions use the term *child fire setter* because the word *juvenile* may be associated with *juvenile delinquents* and therefore may cause parents to be reluctant to report their child's fire-setting activities to authorities.

Studies have indicated that a high percentage of fires are set by juveniles, many by boys under the age of 10, although fire setting by girls is increasing. A number of model programs address this problem. Most involve long- and short-term counseling, fire safety skill instruction, fire safety education, fire prevention projects, and parental awareness, as well as extensive networking.

In addition to the fire department, juvenile courts, school systems, state family and children's agencies, and a wide variety of other human service agencies and organizations are often involved in these model programs, which also address turning in false alarms and damaging fire safety equipment. The age of the involved child is, of course, a major consideration in determining the appropriate remedial action.

COURT DECISIONS

Two U.S. Supreme Court decisions have a bearing on the investigation of fires. The 1978 decision ***Michigan v. Tyler*** established guidelines for right of entry to investigate cause and origin and to subsequently gather evidence should a fire be determined incendiary. The court held that the original entry for fire suppression and subsequent investigation to determine cause and origin is permissible without a warrant; however, subsequent entries may necessitate obtaining consent to enter or a warrant.

Michigan v. Tyler
■ the U.S. Supreme Court decision that established guidelines for right of entry to investigate cause and origin and to subsequently gather evidence should a fire be determined to be incendiary.

Michigan v. Clifford
■ the U.S. Supreme Court decision that reaffirmed its findings in *Michigan v. Tyler* and recognized the right of the investigator to seize evidence in clear view at time of entry.

State of South Dakota v. Jorgensen
■ the court case that upheld use of evidence found during the state fire investigator's initial entry while the fire was still smoldering but upheld suppression of evidence found in later, warrantless entries.

Daubert v. Merrell Dow Pharmaceuticals, Inc.
■ the case that allows the Court to gauge whether expert testimony aligns with the facts of the case as presented.

In a later decision, ***Michigan v. Clifford***, the Court reaffirmed its findings in the earlier case and recognized the right of the investigator to seize evidence in clear view at time of entry. Because both cases directly relate to the investigation of fires, fire investigators should study details of these decisions.

Following the lead of the U.S. Supreme Court, state courts have made similar rulings. In a South Dakota Supreme Court case, ***State of South Dakota v. Jorgensen***, the court upheld use of evidence found during the state fire investigator's initial entry while the fire was still smoldering but upheld suppression of evidence found in later, warrantless entries.

The recent trend in U.S. Supreme Court decisions continues to define the admissibility of expert scientific and technical opinions, particularly as they relate to fire scene investigations. The decisions affect how expert testimony is accepted and interpreted.

A judge has the discretion to exclude testimony that is speculative or based on unreliable information. In the case ***Daubert v. Merrell Dow Pharmaceuticals, Inc.***, the Court placed on a trial judge the responsibility of ensuring that expert testimony was not only relevant but also reliable. The judge's role is to serve as a gatekeeper to determine the reliability of a particular scientific theory or technology. The Court defined criteria to be used by the gatekeeper to determine whether the expert's theory or underlying technology should be admitted. *Daubert* allows the Court to gauge whether the expert testimony aligns with the facts of the case as presented.

Since the *Daubert* decision, Federal Rule 702 on "Testimony by Expert Witnesses" in 2011 was amended to state:

> *Rule 702 Testimony by Expert Witnesses*
>
> A witness who is qualified as an expert by knowledge, skill, experience, training, or education may testify in the form of an opinion or otherwise if:
>
> a. The expert's scientific, technical, or other specialized knowledge will help the trier of fact to understand the evidence or to determine a fact in issue;
> b. The testimony is based on sufficient facts or data;

c. The testimony is the product of reliable principles and methods; and
d. The expert has reliably applied the principles and methods to the facts of the case.[26]

The four criteria listed above are often used by judges to determine whether the expert's theory or underlying technology should be admitted. Some states have adopted similar language for the admission of testimony in state courts.

Arson-Related Research Projects

The National Center for the Analysis of Violent Crime, located at the FBI Academy in Quantico, Virginia, conducted a comprehensive study of serial arsonists. *Serial arson* is defined as an offense committed by fire setters who set three or more fires with a significant cooling-off period between the fires. Researchers reviewed records of almost 1,000 incarcerated arsonists and conducted detailed interviews of 83 confirmed serial arsonists in their places of confinement. The research was included in *A Report of Essential Findings from a Study of Serial Arsonists.*[27]

In addition to the gender and racial breakdown revealed by arrest records, the study indicated that almost half of the serial arsonists interviewed had some type of tattoo, and almost one-fourth had some form of physical disfigurement. More than 60 percent had multiple felony arrest records for offenses other than arson, and almost two-thirds had multiple misdemeanor arrest records. More than half had been held in juvenile detention. One-fourth reported at least one suicide attempt.

A majority of the serial arsonists grew up in middle-class neighborhoods. Relationships with mothers were much closer than with fathers in a high percentage of the cases. Many started setting fires at a very young age. The 83 offenders interviewed had set a total of 2,611 fires, an average of 31.5 arsons per offender.

Of special interest to investigators was the fact that offenders reported they had been questioned by law enforcement officers an average of 3.7 times each before being arrested. The arsonists indicated that they had set an average of 25.3 fires without being questioned. They indicated that efforts by law enforcement investigators were responsible for 38 percent of the apprehensions. In 21 percent of the cases, the individual went to a law enforcement agency and confessed. Informants were responsible in 7 percent of the cases. Witnesses were the key element in 12 percent of the arrests.

Accomplices assisted in 20 percent of the fires. In a majority of cases (59 percent), material available at the scene was the accelerant of choice. Gasoline and other petroleum products were used in 28 percent of the fires. Matches were the predominant ignition device.

Of these serial arsonists, 31 percent remained at the scene of the fire; 28 percent went to another location. Slightly more than half returned to the scene at some time after the fire, usually within 24 hours of the arson.

Nearly half of the serial arsonists consumed alcohol before setting fires. More than half cited revenge as the motive for their arsons. The next most prevalent motive was "excitement." Nearly half entertained no thought of ever being arrested for their crimes. Almost half admitted that their acts were premeditated as opposed to impulsive.

A review of the findings indicates that serial arsonists generally lack the skills to deal with problems of life in general. They are failures in interpersonal relationships and in occupational activities. The report states that "for many, arson may be the only thing they have tried in their life that yields relative success."[28]

Another study illustrated the value of thorough fire investigations followed by vigorous prosecution. In this project, which was prepared for the Law Enforcement Assistance Administration, arson incidents, arrests, and convictions were analyzed in 108 cities over a 4-year period. Cities ranking in the upper third according to arson arrest rates had 22 percent less arson per 100,000 population than cities ranking in the bottom third; cities in the upper third, according to conviction rate, had 26 percent less arson.[29]

Arson Control Needs

The International Association of Arson Investigators and the National Fire Protection Association sponsored a symposium on Arson Control in the 1990s in Houston, Texas. Nationally recognized speakers for example from law enforcement, fire services, and the insurance industry participated in this program. Several of their conclusions follow:

- Automatic sprinklers, public fire safety education, smoke detectors, and code enforcement all play an important role in arson suppression.
- Training for fire investigators must be improved. In some cases, the arsonist is more sophisticated than the fire investigator.
- Safety and health issues for fire investigators need additional study.
- Key members of state legislatures need to be made aware of arson problems and solutions.
- Public awareness of the arson problem is essential to effective abatement. Accurate data are of fundamental value in this effort.
- Wildland arson problems are becoming closely associated with suburban and urban areas as the trend toward construction of homes in woodlands continues.
- Some aspects of programs that address serial arsonists are similar to those that relate to juvenile fire setters. Interconnection responsibilities should be explored.
- The sheer volume of motor vehicle fires creates a gap in terms of thorough investigation. This gap needs to be addressed.[30]

F.I.R.E.
▪ an acronym for Fire Investigation, Research, and Education; a facility funded by Congress that assists investigators in understanding and reconstructing the physical effects of fire in buildings.

In 1996, the Bureau of Alcohol, Tobacco, and Firearms received funding from Congress to construct a Fire Investigation, Research, and Education (**F.I.R.E.**) facility to assist investigators in understanding and reconstructing the physical effects of fire in buildings. The F.I.R.E. Center has become an internationally recognized laboratory that provides law enforcement agencies, as well as other fire investigators across the nation, access to a source of the scientific research and forensic support needed to determine the causes and characteristics of suspicious fires. The center has also become a repository for the collection of scientific facts, experimental results, physical property data of different materials, and other knowledge related to fire incident investigation, analysis, and reconstruction research.

Summary

The suppression of arson through the detection, investigation, and control of maliciously set fires has an important place in efforts directed toward prevention. Arson remains a serious problem and is a significant cause of fires in the United States. Many communities need a coordinated effort between police and fire agencies to investigate suspicious fires and to bring about prosecution of arsonists. A further need is more qualified fire investigators in the field and thorough training in investigative techniques to discover and preserve evidence of arson that will hold up in court. The crime of arson often goes unrecognized because of the lack of thorough investigation or because of failure to develop sufficient information to determine that arson has occurred.

The Model Arson Law, developed by the International Fire Marshals Association and adopted in the laws of many states, defines four degrees of arson, with corresponding penalties. The phrase "willfully and maliciously" is the key factor in these provisions. To obtain a conviction for arson, a prosecuting attorney must be able to prove in court that the fire was, in fact, willfully and maliciously perpetrated. Some states have adopted the Model Penal Code of the American Law Institute, which differs in some respects from the Model Arson Law.

It is helpful to the prosecution of a case if a motive can be established. Motives for arson range from insurance fraud, revenge, and cover-up of another crime to satisfaction of the abnormal urges of the vanity fire setter and the pathological fire setter.

A valid approach to the suppression of arson is a thorough investigation of all fires. No longer can the fire officer assume that a fire is accidental just because the location or the owner is well known in the community. Because motives for a deliberately set fire may not be readily apparent, no presumptions can safely be made.

A systematic examination of the fire scene is all-important in establishing the origin of a fire, whether accidental or not. The fire service personnel who arrive first should observe general conditions—the people and vehicles in the vicinity, the presence of distinctive odors, the intensity and travel speed of the blaze—all of which can be clues for the investigation. If arson is suspected, a fire investigator is usually called to the scene. A trained investigator looks for any traces of incendiary origin of the fire, such as the use of flammable liquids or accelerants and evidence of tampering with equipment, such as gas valves left open on heating equipment or stove burners turned on. Photographs may also be taken at the scene. All evidence should be delineated in the fire investigator's report, and all physical evidence should be carefully preserved for laboratory examination and possible use in criminal court proceedings.

Arson is especially prevalent during periods of civil and economic unrest. Difficulties of investigation, collection, and preservation of evidence are compounded in fires that occur during these periods because of multiple fire calls, false alarms, and confusion and mob intervention at the scene.

Statistics of the FBI *Uniform Crime Reports* show arson and malicious burning as a major crime problem in the country, and figures prepared by the National Fire Protection Association show that losses from incendiary and suspicious fires have increased. This is a serious matter for the fire preventionist. The only way to combat it is to persist in efforts to expose and suppress the crime of arson.

Review Questions

1. The first arson law in Maryland was enacted in:
   a. 1638.
   b. 1697.
   c. 1742.
   d. 1804.
2. In 2011 overall arson rates were just over 18 incidents per 100,000 population in U.S. communities. What two types of community areas had the highest rates nationally?
   a. Suburban and rural
   b. Cities over 1 million and rural
   c. Cities over 1 million and suburban
   d. None of the above
3. The Model Penal Code of the American Law Institute was first published in:
   a. 1953.
   b. 1955.
   c. 1958.
   d. 1962.
4. For successful prosecution of arson, it is helpful to establish:
   a. motive.
   b. corpus delicti.
   c. habeas corpus.
   d. tort.
5. Motivations for arson are described in NFPA:
   a. 1001.
   b. 1500.
   c. 1710.
   d. 921.
6. Fire investigation responsibilities may be assigned to:
   a. fire departments only.
   b. police departments only.
   c. shared fire and police departments.
   d. all of the above.
7. Possible sources of ignition include:
   a. electrical wiring.
   b. heating appliances.
   c. cigarettes.
   d. all of the above.
8. The 1992 riots in Los Angeles killed _____ civilians.
   a. 53
   b. 47
   c. 39
   d. 26
9. Approximately _____ of arson offenses are cleared by arrest or exceptional means.
   a. 25 percent
   b. 20 percent
   c. 16 percent
   d. 6 percent
10. The 1978 decision *Michigan v. Tyler:*
    a. established guidelines for entry to investigate cause and origins.
    b. allowed expert testimony for arson.
    c. set criminal sentencing guidelines.
    d. all of the above.

End Notes

1. National Fire Protection Association, *NFPA 921: Guide for Fire and Explosion Investigations* (Quincy, MA: Author, 2011).
2. Archives of Maryland Online, *Proceedings and Acts of the General Assembly, January 1637/8–September 1664*, Vol. 1 (Annapolis, MD: Maryland State Archives, n.d.), p. 72. Accessed December 2, 2013, at http://aomol.net/megafile/msa/speccol/sc2900/sc2908/000001/000001/html/am172.html
3. Betty Richardson and Dennis Henson, *Serving Together: 150 Years of Firefighting in Madison County, Illinois* (Collinsville, IL: Madison County Firemen's Association, 1984).
4. Steven R. Frady, *Red Shirts and Leather Helmets* (Reno: University of Nevada Press, 1984), pp. 144–145.

5. National Fire Protection Association, *NFPA 921: Guide for Fire and Explosion Investigations* (Quincy, MA: Author, 2011).
6. M. J. Karter, Jr., *Fire Loss in the United States During 2011* (Quincy MA: National Fire Protection Association, Fire Analysis and Research Division, 2012).
7. D. J. Icove, J. D. DeHaan, and G. A. Haynes, *Forensic Fire Scene Reconstruction,* 3rd ed. (Upper Saddle River, NJ: Pearson Prentice Hall, 2013).
8. Federal Bureau of Investigation, *Uniform Crime Reports: Crime in United States 2011,* (Washington, DC: U.S. Department of Justice), Section II, pp. 41–43.
9. Karter.
10. *Model Arson Law* (Quincy, MA: National Fire Protection Association, 1931).
11. *Model Penal Code* (Philadelphia: American Law Institute, 1962), p. 152.
12. Arkansas Code of 1987, Annotated, as amended 2012.
13. D. J. Icove and M. H. Estepp, "Motive-Based Offender Profiles of Arson and Fire-Related Crimes," *FBI Law Enforcement Bulletin* (April 1987).
14. A. O. Rider, "The Firesetter: A Psychological profile." *FBI Law Enforcement Bulletin* (June 1980), p.11. Accessed March 29, 2014, at https://www.ncjrs.gov/pdffiles1/Digitization/68373NCJRS.pdf
15. D. J. Icove, J. E. Douglas, G. Gary, T. G. Huff, and P. A. Smerick, "Arson," in J. E. Douglas, A. W. Burgess, A. G. Burgess, and R. K. Ressler, *Crime Classification Manual,* Ch. 2 (New York: Macmillan, 1992), pp. 165–166.
16. National Fire Protection Association, *NFPA 921: Guide for Fire and Explosion Investigations* (Quincy, MA: Author, 2011).
17. National Fire Protection Association, *NFPA 1033: Standard for Professional Qualifications for Fire Investigator* (Quincy, MA: Author, 2014).
18. U.S. Department of Justice, *Fire and Arson Scene Evidence: A Guide for Public Safety Personnel*, Research Report NCJ 181584 (Washington, DC: U.S. Department of Justice, June 2000), p. 6. Accessed December 2, 2013, at https://www.ncjrs.gov/pdffiles1/nij/181584.pdf
19. National Fire Protection Association, *NFPA 921: Guide for Fire and Explosion Investigations* (Quincy, MA: Author, 2011).
20. G. E. Burnette, Jr., "Spoliation of Evidence: A Fire Scene Dilemma" (*Interfire Online*, n.d.). Accessed December 2, 2013, at http://www.interfire.org/res_file/spoliatn.asp
21. J. D. DeHaan and D. J. Icove, *Kirk's Fire Investigation*, 7th ed.; and D. J. Icove, J. D. DeHaan, and G. A. Haynes, *Forensic Fire Scene Reconstruction*, 3rd ed. (Upper Saddle River, NJ: Pearson, 2012).
22. U.S. Fire Administration, *Coffee Break Training: Safeguarding Homes from Arson*, No. FM-2013-1 (Emmitsburg, MD: March 28, 2013). Accessed December 2, 2013, at http://www.usfa.fema.gov/downloads/pdf/coffee-break/fm/fm_2013_1.pdf
23. U.S. Fire Administration, *Report of the Joint Fire/Police Task Force on Civil Unrest: Recommendations for Organization and Operations During Civil Disturbance* (Emmitsburg, MD: Author, February 1994), p. 9. Accessed December 5, 2013, at http://www.usfa.fema.gov/downloads/pdf/publications/fa-142.pdf
24. Ibid, p. 9.
25. Federal Bureau of Investigation, *Uniform Crime Reports: Number of Offenses Cleared by Arrest or Exceptional Means, Percent of Clearances Involving Persons Under 18 Years of Age by Population Group, 2011: Table 28*. (Washington, DC: U.S. Department of Justice, n.d.). Accessed December 5, 2013, at http://www.fbi.gov/about-us/cjis/ucr/crime-in-the-u.s/2011/crime-in-the-u.s.-2011/tables/table-28

26. Legal Information Institute, "Rule 702. Testimony by Expert Witnesses" (Cornell Law School, Legal Information Institute, n.d.). Accessed November 16, 2013, at http://www.law.cornell.edu/rules/fre/rule_702
27. Allen D. Sapp, Timothy G. Huff, Gordon P. Gary, David J. Icove, and Philip Horbert, *A Report of Essential Findings from a Study of Serial Arsonists* (Quantico, VA: National Center for the Analysis of Violent Crime, 1994).
28. Ibid.
29. Aerospace Corp. Survey and Assessment of Arson and Arson Investigation, prepared for Law Enforcement Assistance Administration (El Segundo, CA, 1976).
30. *Report on Arson Control in the 1990s: A Symposium* (Houston, TX: International Association of Arson Investigators and National Fire Protection Association, 1991).

CHAPTER

12 Fire Prevention Research

Andrea Booher/FEMA

KEY TERMS

1-megawatt fires, *p. 241*
cone calorimeter, *p. 244*
fire dynamic simulator (FDS), *p. 239*
Fire Research Division of NIST, *p. 238*
flow path, *p. 239*
tenability, *p. 240*
Institute for Research in Construction, *p. 242*
platooning, *p. 241*
research, *p. 233*
research and development (R&D), *p. 234*
wildland, *p. 234*

OBJECTIVES

After reading this chapter, you should be able to:

- Define the term *research*.
- List recent research by the U.S. Forest Service.
- Identify the audiences to which the Vision 20/20 project plans to direct its marketing.
- Explain the significance of code enforcement research conducted by the Urban Institute, NFPA, and the U.S. Fire Administration.

The term **research** is defined as "careful, systematic study and investigation in some field of knowledge."[1] Until the 1970s very limited research was conducted in the field of structural fire safety and prevention, though some excellent studies were done in forest fire prevention. It was not until the last quarter of the 20th century that research increased in both structural and forest fire prevention.

research
■ careful, systematic study and investigation in some field of knowledge.

U.S. Forest Service Fire Prevention Research

The Forest Service of the U.S. Department of Agriculture has carried out a number of excellent fire prevention research projects. The Forest Service employs over 500 researchers and carefully plans priority projects that lead to the sustainability and management of the diverse U.S. forests and rangelands. Some of the research projects are conducted by Forest Service personnel. Others are conducted by university personnel under contract to the Forest Service. The resulting research continues to guide Forest Service procedures.

research and development (R&D)
▪ systematic work in organizations with the goal to produce a product, solve problems, create new processes, or acquire knowledge.

wildland
▪ often a remote land area characterized by natural, sometimes dense vegetation minimally modified by human activity, such as woodlands, forests, brush, and meadows; wildland can contain dangerous buildups of fuel from vegetation that may or may not be a risk for fire; trends in migration and development currently bring together increased wildland interface with urban areas throughout the world.

The U.S. Forest Service offers extensive information about its research programs on its website. A specific page dedicated to **research and development (R&D)** lists summaries of 76 different projects related to **wildland** fire and fuels. The service's wildland fire and fuels research and development programs have been working under a plan that includes three primary strategies:

- To advance the sciences that improve understanding of fire processes, which can lead to next-generation decisions for wildland fuel management
- To develop and deliver knowledge and tools to policymakers, wildland fire managers, and communities
- To provide federal leadership, responsiveness, coordination and collaboration, and forward-looking wildland fire-related research and development

While there are several goals within that R&D strategic plan, the most prominent one is to reduce risk of catastrophic wildland fires. Key areas of work are the improvement of prevention and suppression processes; the reduction of hazardous fuels; the restoration of fire-adapted ecosystems; and the promotion of community assistance.[2]

More recent research that supports prevention includes evaluation of the intensification of the results of climate change, such as drought, hurricanes, storm-induced flooding, and heat waves. The Forest Service wants to understand better how climate change will further affect the watershed and how well the forest and rangeland ecosystems will renew themselves in the face of those disturbances.[3]

Another important area of prevention related to R&D is the effect of invasive pests that can kill vegetation in the forests and rangeland. The result is a denser then usual fuel load that can increase the intensity of fire. The prevention or quick mitigation of these infestations can reduce the unnatural density of the fuel available for a fire. The difficulty is sometimes first detecting the pest's presence, then understanding the life cycle and all about the pest, so that no interventions have unintended consequences, such as harming other animals or insects that are native. Conditions such as weather and rainfall may increase density of the infestation, so many areas of research must come together to create a breakthrough for even one pest. The current study includes ongoing research into early detection and consequent reduction in invasive species.[4]

A key area of prevention R&D is the reduction of hazardous fuels. Nature would accomplish this through naturally occurring fires in forest and rangeland.

However, social changes in the United States since the arrival of European settlers have drastically altered the occurrence of natural fires. The result has been an ever-increasing build-up of biomass, or fuel, so the fires that do occur, from either natural or human-related means, are catastrophic fires that greatly impact the surrounding region.

Research into improved hazardous fuel reduction is one of the efforts of the U.S. Forest Service. One report issued for the benefit of forest managers and communities is the 2003 publication *A Strategic Assessment of Forest Biomass and Fuel Reduction Treatments in Western States.*[5] In helpful geographic information images, it provides information on predicting and finding the highest concentrations of fuel. Maps that offer data as visual interpretations of where fuel is massed can facilitate decisions and planning concerning the management of the fuels.

The 2003 report also offers a significant discussion on the cost–benefit of the various means available to manage fuel. For instance, it identifies prescribed burnings as the lowest cost ($35 per 300 acres) alternative to all other means of mitigating hazardous fuel, but this solution has many limitations, such as the possibility that a prescribed burn will become a hostile out-of-control fire (Figure 12-1). There are also limited time frames due to conditions that may limit where and when a prescribed fire will be safe. Other alternatives include the following:

- Chipping and mastication costs $100 per 1,000 acres; the key problem is all chipped material left in the woods with no purposeful use. Cutting, piling, and burning material costs $100 per 750 acres, but this alternative can produce significant smoke and results in no purposeful use of the cut material.
- Cut and skid-load material costs $30–$40 per bone-dry ton; it can have a negative effect on soil, but it offers the opportunity for productive use of the removed material, such as in energy development.

Andrea Booher/FEMA

FIGURE 12-1 Biomass thinning through fuel reduction has become a significant effort by the U.S. Forest Service to reduce fire risk. While it is the least expensive solution, prescribed fires to burn biomass can be dangerous and unpredictable.

- Cutting, skidding, and chipping is the most costly of all the alternatives ($34–$48 per bone-dry ton), but it does produce usable fiber material for energy development and other uses.

The 2003 report offers greater detail on these alternatives and a host of procedures for planning and implementing hazardous fuel management.

The cost-effectiveness of various means of wildland fire fuel management is affected by any number of factors. These factors may change the choice of processes, the potential valuation or devaluation of the materials managed, and the need to look for even more cost-effective measures. Since this study was concluded, the United States has experienced a difficult recession that has had deep effects on all areas of the economy. This text did not explore the current value of the biomass of vegetation for, say, conversion into alcohol, which has become an important commodity as an approach to energy. With this kind of consideration, the biomass may become valuable instead of just being waste material that has to be managed. As a result, the costs for chipping or mastication (the chewing, crushing, and grinding of materials to a pulp) and then shipping may fall into a range that makes this approach an efficient choice.

Some existing uses include the burning of the mass as a fuel to generate energy such as electricity and heat for industrial processes and use as an ingredient of biofuel (ethanol and biodiesel). The material may even be charred for use as a soil additive. These possibilities lead right back to the consideration of cost. Cost is a constantly changing major factor requiring research and development that will produce regular reports for the managers who make decisions both strategically and in the daily tactical management of operations.

Managing wildfires is most often seen in the significant efforts of suppression after a fire has started and in the management and reduction of fuels for defensible space and fire breaks. But public education that reduces the number of fires humans cause also may be significant. For example, note the campaigns to educate the public about environmental conditions, such as those offered by the Haines Index. The Haines Index provides a relative indication of dry, unstable air that could contribute to the severity of a fire. Its warnings are components of weather forecasts and are given as a range from 0 to 6, with 6 being a high potential that existing fires may become larger and exhibit erratic behavior; 5 is medium potential; and 4 and below indicate low potential. This information may be cause for actions such as prohibiting outside fires when the Haines Index is at level 6.

Smokey Bear has been an ongoing educational program by the U.S. Forest Service to make people aware of actions that can lead to or contribute to dangerous fires. In spite of the significant resources and effort in the area of Wildfire Prevention Education (WPE), little is known about how effective it actually is.

The 2010 study by USDA Forest Service Southern Research Station in Florida[6] was the first to document the effectiveness of education on the various human-caused fires, such as the escape of fire from the burning of debris and waste, fire play by children, campfires, and improperly discarded cigarettes. The findings showed that prevention education in Florida saves millions of dollars in firefighting costs and reduces damages from fires caused by people. Using a model to quantify the effects of education on fires identified as being preventable, researchers analyzed the impact of wildfire prevention activities on reducing fire occurrence, the area burned, and economic losses. Examples of educational efforts covered in the study were public service announcements, brochures, presentations, and home visits.

Researchers from the Southern Research Station also found that the time and place of education are important and that focused education during the highest risk times of the year (in regard to wildland fire cycles and the places of highest risk) is more effective than all educational resources evenly distributed across regions. A fact sheet produced by the researchers shows that concentrated effort at the beginning of and during the wildfire season is effective. Wildland fires with accidental causes occur mostly where a high concentration of people live and work and at the wildland–urban interface. Fire prevention education efforts in those areas tend to be effective and to enhance overall wildland fire prevention, according to the fact sheet.[7]

As an important component of strategic planning conducted by the U.S. Forest Service, research results in improvements to safety and property conservation in communities at high risk for wildfires. Direct involvement with communities and the transformation of science into usable prevention guidelines have been effective in some of the work of community action partnerships, such as Firewise Communities, which is an NFPA program that partners with the U.S. Forest Service and other federal, state, and community organizations, and Fire Adapted Communities, which is a broad coalition of federal, local, and private fire interests.

The Firewise Communities program, for example, lists and promotes tips that improve a home's and a community's defensible space by improving the fire resistance of the exterior structure and landscaping. Available guidance includes plant and landscaping lists and home construction and retrofit information specifically designed for increased fire defense. A 2011 news release from Firewise Communities[8] offers eight tips specifically focused on the wildland fire risk in Arizona. The list includes tips on reducing fuel, landscape and tree trimming suggestions, and cleanup of organic debris in rain gutters and other areas where it can collect and be ignited by flying embers from a wildland fire.

The fire defense tips promoted by Firewise Communities are examples of vital links developing among communities and groups that increase collaboration in reduction of the risk of fires. Research plays a critical role by formulating what fire scientists understand into guidelines for safer community interface with fire-prone wildlands.

Public Fire Education Research

It would seem that presenting fire safety to the public to increase awareness would be pretty straightforward, with little need for additional research, but this supposition could not be further from the truth. While much of what is promoted in the form of education comes from research into such areas as fire dynamics and human fire behavior, considerable research is available on how best to approach public fire education. Research on what people think of or know about fire, what behaviors have changed because of past fire education they have received, and how or where they turn to find information about fire behavior has been carefully reviewed.

Often, because of limited resources, local governments turn to fire safety research to determine the best means to reach more people with the least amount of effort. It is critical to deliver the most meaningful safety message. The Vision 20/20 project conducted extensive work in identifying and filling critical gaps in fire prevention efforts. Work completed under the strategy, "Prevention Marketing," focused on development of a unifying fire safety theme with subthemes that compelled people to accept and embrace a personal responsibility for their own fire safety. The themes were to be data-driven and simple to understand, especially to high-risk groups such as older adults, adults in lower income levels, and English-speaking Hispanic adults. The social marketing consultant Salter and Mitchell completed the work on this project. Their message-testing research led to the development of the "Fire Is Everyone's Fight™" theme implemented by the U.S. Fire Administration in 2013. The second round of grants funding the Prevention Marketing work helped to develop a series of messages, such as "Keep an eye on what you fry," that focuses action on the prevention of cooktop fires. The First Alert Corporation enabled additional social marketing research that created a simple yet powerful message inspiring people to maintain a working smoke alarm in their homes and that is expected to be made public in 2014.

National Institute of Standards and Technology Research

In 1974, steps were taken to increase fire prevention in the United States by means of the Federal Fire Prevention and Control Act of 1974. This act was a reaction to the landmark research report *America Burning*,[9] which described the state of U.S. fire safety and what was needed to begin to reduce the staggering fire loss, which at the time was triple our current (2014) average of around 3,000 deaths per year. Clearly, both the experts reporting on the fire problem in *America Burning* and the developers of the fire prevention law saw a critical need for research into all aspects of the fire problem.

The repeated use of the word *research* (59 uses) in the current version of the fire prevention law emphasizes its importance in the intended efforts. But one of those references to research was key in establishing a new, focused study of fire through a "Fire Research Center" under the authority of the U.S. Department of Commerce and its National Institute for Standards and Technology (NIST). The new research center's mission would be to perform and support research that focused on identifying scientific and technical knowledge "applicable to the prevention of and control of fires."[10] Some of the very early work of the Fire Research Center included new research on residential sprinklers and smoke alarms, including *Detector Sensitivity and Siting Requirements for Dwellings*.[11] The center's research led to early standards related to home smoke alarms.

Fire Research Division of NIST
▪ a division that conducts research in building materials; computer-integrated construction practices; fire science and fire safety engineering; and structural, mechanical, and environmental engineering.

NIST continues to be one of the most important sources of research in fire protection and fire prevention. The name of the Fire Research Center changed as the organization evolved, first to the most recent long-term name of Building Fire Research Lab (BFRL) and now the **Fire Research Division of NIST**. The division is a component of NIST's Engineering Laboratory in Gaithersburg, Maryland. A list of the current work groups under the Fire Research Division include Fire Fighting Technology, Engineered

Fire Safety, Flammability Reduction, Wildland–Urban Interface Fire, and National Fire Research Laboratory. The National Fire Research Lab is a new, state-of-the-art test facility capable of real-scale construction of structural experiments. All the groups continue to add depth to the science and understanding of the fire problem.

In the investigation of critical fires, carefully documenting a timeline of events is one of the important roles of the Fire Research Division. The process of investigating a fire is not unlike the process of research. In fact, standards in fire investigation call for the use of the scientific process to develop findings, so it is only the starting point of the research that is different. The Fire Research Division investigated notable fires over the years, including fires in the World Trade Center (2001), The Station nightclub (2003), the Cook County (Illinois) Administration Building (2003), and the Sofa Super Store in Charleston, South Carolina (2007). Those investigations represent a significant body of knowledge that helps us understand fire. To highlight some of the value, look at a fire occurring in a Washington, D.C., home that resulted in the horrific death of two firefighters and substantial burns to a third.

The investigation released by NIST in 2000[12] was one of the early uses of the NIST **fire dynamic simulator (FDS)**, a modeling application invented by NIST that predicts the thermal conditions in a compartment fire (fire in a room or enclosed space in a building). FDS provided a graphic representation of the flow path of hot fire gases that, without warning, encompassed the firefighters. The deadly gases hit the firefighters as they were about to descend the stairs to the basement, where the fire had started. Firefighters were entering from the basement level at the same time. A basement sliding door was opened, introducing an entry point for fresh air that completed the path from the basement, up the stairs, and out the front entrance. This was one of the first times fire scientists saw a visual data representation of what is now called **flow path**.

fire dynamic simulator (FDS)
■ a computer-modeling application that simulates fires and predicts fire conditions in a compartment.

flow path
■ a route of movement taken by hot gases between a fire and exhaust openings (windows, doors, cut vent holes) and the movement of air toward and feeding the fire.

FDS has helped the researchers understand better the effects on a fire of structural openings that suddenly introduce ventilation (Figure 12-2).

National Institute of Standards and Technology

FIGURE 12-2 NIST's fire dynamic simulator, when paired with animation software, predicted changing fire dynamics when a door was opened on a lower level of a home. Fire is demonstrated in red and can be seen flowing up the stairs just as firefighters prepare to go down the stairs.

This knowledge is now being presented to firefighters all over the United States to help them learn about fire dynamics and how their tactical activities can impact the fire and their safety. The visual data representation can add new meaning to many fire dynamics that were not understood. Simulations using computer models are becoming useful in building design with performance codes. Data can be entered to produce a simulation that shows how a fire protection device or construction feature may work, without actual real-life testing. This resource may lead to new methods that save construction costs and possibly increase safety in structural design.

In another high-profile project, NIST released a report[13] detailing the events of the fire that consumed the Sofa Super Store, where nine Charleston, South Carolina, firefighters were trapped and killed. The study combined the efforts of fire scientists who collected the data and analyzed them with computer simulations that showed the dynamics of the fire.

Computer simulation offers the possibility of changing fire conditions, structural and fuel contributions, and availability of ventilation to show how a fire will react. In the case of the fire in Charleston, the NIST team's development of a model to simulate that fire's conditions allowed them to evaluate conditions in the store, including fire spread, smoke movement, **tenability**, and the operation of passive and active fire protection systems. According to the final report, NIST found that the presence of fire sprinkler protection would have slowed the fire in the loading dock area and would have helped to maintain a safe environment for firefighters and escaping occupants.

tenability
▪ the fire environmental impact on human survival; used in discussion or presentation of fire protection and fire safety subjects to demonstrate when conditions are survivable or not.

The Fire Research Division's long history of fire research and investigation of disasters has opened the fire service to a better understanding of fire and its impact. Much of the fire research is available at a convenient portal on NIST's website,[14] where the reader can find an intuitive search function by author, title, and keyword. It includes over 4,500 documents. Most have been added since 1993, but many are from prior years of research. NIST has a very robust internal review process for its research reports. Its fire research offers the opportunity for other researchers to discover links and often next steps to further research, or findings among data and information that were captured in experiments but were not necessarily germane to that research.

University Research

University research is one of the many sources of ongoing work in the area of fire science that has had an impact on fire prevention. The conduct of research helps to tie a university to the various industries and stakeholders impacted by the research. Research opportunities and state-of-the-art laboratories often help sell a university to prospective students, as well as to people and groups who can help facilitate research through their philanthropic support. The research component also provides the opportunity for graduate academic work, experiential work for students, and jobs.

Examples of fire research schools include the University of California at Berkeley, Worcester Polytechnic Institute, and the University of Texas. Another,

Eastern Kentucky University (EKU), recently conducted experiments testing the typical materials in a home kitchen and their propensity to ignite and spread fire. It was funded by and supported efforts by Vision 20/20 and its Strategy Four Prevention Technology. The project evaluated the ability of several types of electric stovetop surfaces to ignite common kitchen items. Cooktop types included ceramic-glass, electric coil, electric coil with cast iron plate, and electric coil with temperature-limiting control sensor. EKU researchers selected nine commonly identified fuels listed as first fuels ignited in kitchen and cooking fires: oil-filled cook pans (corn, vegetable, and canola), a cardboard pizza box, a cotton dish towel, a roll of paper towels, a plastic storage container, a kitchen appliance, and a nylon spatula. Researchers evaluated the heat from 8- and 6-inch burners while energized to the low, medium, and high control settings. Researchers used various heat-sensing instruments in performing 54 tests on each type of cooktop. A report on the findings was published in October 2013.[15]

The University of Maryland is one of the many institutions that conduct research on a for-hire basis. It also sponsors academic research for students at the graduate level. Maryland has laboratory facilities suitable for different types of research: a lab set up to study sprinkler spray characteristics; a lab for medium-scale live-fire testing for up to **1-megawatt fires**; a computer lab that supports student fire modeling and simulations; and a lab for product testing.

1-megawatt fires
■ a significant heat release rate of a specific burning object; sometimes referred to as a minimum for producing a room flashover.

Engineering students working with the Maryland State Fire Marshal developed a test setup that enabled observation and documentation of hazards associated with so-called fire pots, the small ceramic burners used for ambient lighting. If refueled improperly, those devices can experience burn-back along the fill stream of the combustible gel used to fuel them and have caused explosive ignition, splattering burning fuel and exposing the user to injury. The Consumer Product Safety Commission has issued a recall of the fuel.

Other graduate research is going on at the University of Maryland all the time. For instance, graduate students Matthew Baker and Christopher Campbell completed research projects that analyzed NIST videos of fire drills in multiple buildings that had cameras placed in stairwells to capture evacuee behavior. Their project, *Human Behavior During Egress of High Rise Buildings,*[16] offered an opportunity to analyze data that had been collected but not yet interpreted and provided a better understanding of human movement patterns and egress times during evacuations. This one project allowed the students to study this behavior and provide individual conclusions in their own personal reports.

Matthew Baker's analysis resulted in a report titled *Observed Behavior of Platoon Dynamics During High-Rise Stairwell Evacuations.*[17] His research focused on the critical life safety process of evacuation to better understand factors influencing the amount of time required for people to reach safety. Specific to the research was the human pattern known as **platooning**, or grouping of evacuees. His research concluded that platooning behavior occurs frequently and that when platoons merge and grow, the overall flow of evacuees slows. Those kinds of observations can help improve input for evacuation modeling used in building design.

platooning
■ a grouping behavior of people during building evacuation.

Christopher Campbell's analysis of the videos resulted in a report titled *Occupant Merging Behavior During Egress from High Rise Buildings.*[18] The research focused on the behavior of occupant evacuees where they merge in building stairwells.

The merging involves two groups of evacuees: those already in the stairwell and those entering the stairwell from corridors. Campbell's most important conclusion was that nonqueued merging, or where there is only one choice for lining up to exit, is a very inefficient way to arrange exits if the intent is to speed up evacuation time. It was observed that evacuees lined up waiting in the corridor to merge into the flow of others already evacuating in the stairwell. There is a natural resistance here that slows the process down and could increase life safety risk to those closest to the fire, where it can be assumed that tenability is decreasing. Campbell suggests that future research should focus on the observed hesitation of people to merge at stairwells.

Findings from both studies can improve computer-modeling simulations used by building design professionals to better anticipate needs for new buildings.

Canadian Research Activities

Institute for Research in Construction
▪ a branch of the National Research Council of Canada established to serve the needs of Canada's construction industry.

The **Institute for Research in Construction** of the National Research Council of Canada was established to serve the needs of Canada's construction industry. It is Canada's largest industry, and an indicator of Canada's economic power. The Canadian construction industry employs 1.24 million people.[19] Established as a direct result of a major ship fire in the port of Toronto, the institute's National Fire Laboratory is responsible for conducting research aimed at reducing life and property losses by fire. Its priority is subjects related to the National Building and Fire Codes of Canada and to assistance to the building industry through research into fire reaction of building materials and assemblies. It also has a close working relationship with the Fire Research Division (NIST) in the United States.

The Canadian fire research facility located near Ottawa is probably the finest such facility in the world. It includes a 10-story tower in which high-rise fire safety studies can be conducted. It has been used to conduct evacuation research for such buildings. Other research activities include fire risk assessment, costs of fire safety including public protection, effectiveness of structural fire protection including fire barriers, and high-efficiency water-based fire suppression systems. Other aspects of fire protection are also studied.

Straw-built construction is on the rise in the United States and Canada. It has a very high energy-efficiency rating, which results in heating-cost savings. As an example of its work, the National Research Council of Canada tested tightly compacted bales of straw. Researchers found bales that are closely compacted greatly reduce the airflow potential, thereby preventing fire buildup. Another advantage of straw-built construction is the absence of wood studs separated by open space, which encourage upward fire spread. The compact straw bales preclude this possibility.[20]

United Kingdom Research

The London Fire Brigade, like most of the United Kingdom's fire service, has worked hard in recent years to reduce the risk of fire to residents. Its home fire safety visits have become a model to U.S. fire service efforts at Community

Risk Reduction. The London Fire Brigade realized that it needed to produce evidence that the fire safety work was achieving results, so it planned a data collection scheme that would be analyzed to determine if the performance lived up to the goals. A 2013 article[21] provides an overview of some of the value of collecting data.

The London Fire Brigade used a research prediction model that looked at a number of at-risk indicators and produced a target sample of over 700,000 homes with increased fire risk where safety visits were offered. The visits were voluntary by the residents, and the brigade captured data as to whether they participated or not. In their ongoing analysis of response and fire data, they compared the data of who participated and who did not. The findings to the research pointed to a significant decrease in the number of fires in homes where a home safety visit was conducted (2 fires per 10,000 people) compared to the number of fires for those who did not participate (36 fires per 10,000 people). This type of outcome performance would indicate that home fire safety visits are effective at reducing fire risk. The Chair of London Fire Authority's Strategy Committee, Councilor Sarah Hayward said:

> This new research is evidence that the Brigade's home fire safety visits is preventing thousands of fires and saving lives, particularly amongst some of the capital's most vulnerable people.[22]

Underwriters Laboratories Research

We see the UL (Underwriters Labs) label on many of our home and workplace electrical devices as a symbol that the device is listed to meet standards of electrical safety. This is only one of the many important functions performed by UL, which has expanded from its early leading work in electrical and fire safety and now, for example, includes the broader safety issues of environmental sustainability, food safety, and water quality in its scope of work. But another area that has been a core of the UL enterprise is research. UL's research work with the firefighting community has begun to shape understanding of the hazards of modern buildings and the impact of their contents on firefighter and occupant safety. Data collected through live burn experiments specifically investigated such building elements as floor beams that have been transformed from the legacy use of solid dimensional lumber into lightweight engineered components. The modern structural components meet normal daily loading needs but can fail early when exposed to fire and collapse just as firefighters enter a structure to search for victims and attack the fire. UL has also performed experiments to evaluate the changes to materials commonly found in residences suspected of increasing hazards to occupants and firefighters. The evolution from more natural legacy materials such as wood, cotton, wool, steel and leather to modern, predominantly synthetic materials may no longer be in sync with established design and testing criteria for fire protection essentials like smoke alarms. UL's work with the National Institute for Standards and Technology and firefighters across the United States in investigating and understanding fire dynamics in structures has led to the two organizations' presenting their findings and an introduction to tactical firefighting operations as suggestions to deal with the change in home fire severity at workshops and symposiums across the United States.

In 2006 UL and the Fire Protection Research Foundation conducted a Smoke Characterization Project[23] that enhanced the fire safety community's understanding of modern fire hazards. This first-of-a-kind study systematically investigated the characteristics of smoke and examined how a wide range of materials used in modern residential settings have affected the way fires behave in homes. The year-long project studied 27 synthetic and natural materials and various combinations of materials commonly found in homes.

In introducing the final report, the researchers described how earlier research conducted by NIST in 2004 identified that smoke alarms are working but that safe evacuation times had decreased from when initial smoke alarm studies were conducted. NIST also felt that smoke alarm performance could be more successfully studied if better data were available on the combustibility and smoke characteristics of a wider range of materials used in modern home construction. NIST emphasized that current smoke alarms perform within the standard criteria, but the experiments showed that time to activation was different for existing technology for flaming and nonflaming fires. The Smoke Characterization Final Report said the research was initiated to better understand flaming and nonflaming (smoldering) fires with the goal to achieve the following objectives:

- Develop smoke characterization analytical test protocols using nonflaming and flaming modes of combustion on selected materials found in residential settings
- Using materials from the analytical smoke program (data obtained on selected materials found in residential settings, where materials were classified chemically using the **cone calorimeter**), develop smoke-particle-size distribution data and smoke profiles in the UL 217/UL 268 Fire Test Room for both nonflaming and flaming modes of combustion.
- Provide data and analysis to the fire community for several possible initiatives:
  - Develop recommendations concerning the current residential smoke alarm standard (UL 217)
  - Develop new smoke-sensing technology
  - Provide data to the materials and additives industries to facilitate new smoke suppression technologies and improved end products

cone calorimeter
▪ a bench test instrument used primarily in fire safety engineering to evaluate small samples of material to obtain data measurements—for example, ignition time, heat release rate and mass loss.

In developing this project, UL's fire science experts were able to investigate the chemical and physical properties of smoke at a new level of sophistication. The results of this project showed that smoke can no longer be characterized just by color or thickness. The new technology allowed a much more detailed documentation of smoke particle size and identification of volume and makeup of the gases produced by fires. While we are talking differences in size smaller only by tenths of a micron, the finer differentiation may offer opportunities to introduce newer and faster technology in detecting and alarming when smoke is present. The documentation of more finite data allows further research and development to improve not only smoke alarm design but the tests used to ensure that new products will meet the requirements of any published standards.[24]

A second study by Underwriters Laboratory focused on the safety of firefighters in residential occupancies. The report *Analysis of Changing Residential Fire Dynamics*, published in 2012, helps the reader to better understand the critical time frames involving fire in modern residences that include larger and more open space, engineered construction materials and techniques, and contents primarily of synthetic materials. *Analysis of Changing Residential Fire Dynamics* helps to suggest new tactical approaches for firefighters. Suggestions for the alternative tactical approach result from increased understanding of fire dynamics resulting from conducting live burn experiments that simulated larger and more open homes with contents that ignited faster and burned with a significantly higher production of heat and toxic gases. Home size has more than doubled since the 1950s, with the National Association of Home Builders reporting new home size in the 1950s as 1,000 square feet or smaller compared to 2,265 square feet five decades later in 2000.[25] Larger and more open homes provide fire with a more favorable environment to grow because there is more availability of the air needed for combustion and less limitation on fire spreading. Additional details on the degree and reasons of change are outlined in research described in a report and workshop sponsored by the USFA: *The Changing Severity of Home Fires.*[26]

Residents of new homes may be even more at risk than firefighters because fires ignite faster and burn with more intensity. UL found evidence that on average, fires reach extremely high rates of heat production in just a few minutes with modern home contents compared to tests conducted in the 1970s, when home materials were made of natural products. One of the experiments included comparing tests of fires in two different room sizes with contents of modern materials (plastic, rayon, polyester, polyurethane) and legacy contents (wood, cotton, wool, steel, leather) to see how fast each would reach flashover. Flashover in a room greatly increases the rate of heat, smoke, and carbon monoxide developed by the fire. Room flashover conditions are severe enough to kill a person, and the high-energy expulsion of heat and gases can be so extreme as to allow the fire to easily spread and expose escaping occupants as well.

The tests used in *Analysis of Changing Residential Fire Dynamics* consisted rooms of two sizes: a smaller room 12 feet by 12 feet and 7 feet 10 inches high (3.7 meters by 3.7 meters and 2.4 meters high), for an area of 144 square feet (13.69 meters square), and a room 13 feet by 18 feet and again 7 feet 10 inches high (4.0 meters by 5.5 meters and 2.4 meters high) for an area of 234 square feet (22 meters square). There were six experiments using the same content quantity in total weight for each room. Three experiments used modern contents of synthetic materials (plastic, polyurethane, rayon, polystyrene), and three experiments used legacy contents of natural materials (cotton, wool, leather, wood, steel). The contents consisted of, for example, a sectional sofa, a coffee table, an end table, a television stand and bookcase, a flat-screen television, storage containers, toys, and a lamp with a shade. Modern materials were readily available at retail stores, and the legacy materials were obtained from a number of secondhand stores. Each experiment was ignited by laying a candle on the sofa.

The results of the room fire experiments confirmed the peril that residential occupants face in escaping a fire. The average time to reach flashover was less than 5 minutes in each of the experiments using the modern contents made of synthetic materials.

On average, room fire experiments for the legacy contents reached flashover in 29 minutes. One of the legacy experiments failed to sustain fire once it was ignited and never did reach flashover. Similar fires in a real residence could allow occupants to escape and even allow time for the fire department to arrive and search and remove victims if they are overcome. UL researchers used National Fire Incident Reporting System data to predict a reasonably accepted response time of just over 6 minutes. They added another 2 minutes for fire detection, a minute for calling the fire department, and a minute for dispatch and turn-out of the fire department, putting the fire department on scene in 10 minutes on average, still time enough for the fire department to rescue overcome occupants with an average time to flashover of 29 minutes, as found with the legacy content experiments.

It should be noted that experiments involving the room fires with legacy and modern contents were conducted with no limitation of air or ventilation to feed the fires. The reason this is mentioned is that under natural conditions, fires can have a range of impacts—from no ventilation and therefore no air being fed to the fire to unlimited ventilation. The amount of ventilation affects how fast or how slowly a fire builds toward flashover.

The Smoke Characterization Project with the Fire Protection Research Foundation and Analysis of Changing Residential Fire Dynamics are two research projects that document strong evidence of the significant hazards facing firefighters and occupants in the modern home. The fire service was aware for several years of a change in the fire environment. UL responded to its concerns by focusing research on those concerns.

Underwriters Laboratory and National Institute for Standards and Technology used findings from their combined research projects and began to deliver presentations to firefighters all over the United States, at conferences, workshops, and all sorts of gatherings to make firefighters aware of the fire changes that had been documented. These two examples of research are a small part of many hundreds of hours spent on hundreds of experiments conducted to help study fire dynamics in the modern home.

National Fire Academy Research

Students attending the Executive Fire Officer Program (EFOP) at the National Fire Academy (NFA) in Emmitsburg, Maryland, are required to prepare and submit applied research projects. EFOP is an initiative created to provide senior fire officers a "broader perspective"[27] in the varied and complex areas of the fire service. The purpose of the applied research is to build a professional standing for fire service leadership in the United States; to develop the knowledge, skill, and experience necessary for leaders to transform organizations; and to build a deep and diverse range of research that can be used when leaders search for solutions to problems in their communities. The research projects must be completed through standard research practices and guided by principles of the scientific method. Writing and documenting the projects follow the style described in the *Publication Manual of the American Psychological Association (APA)* to produce writing that is succinct, consistent, and straightforward. Student participants in the EFOP conduct applied research over 4 years, with the general subject area complementing the course they take in a given year. The research must address a need in the student's own community that corresponds with the course. There are hundreds if not thousands of projects germane to the various areas of fire prevention, life safety, fire investigation, and community risk reduction. While EFOP applied research focuses on the students' home-community needs, the problems they address are generally applied universally.

The following examples are chosen from the projects the NFA reviewers judged exemplary. Each project receives a final rating from 1 to 4, 4 being described by NFA staff as exemplary research. Out of all exemplary research projects, one is chosen from each year's course as Outstanding Research Project. A bibliography of all the exemplary research projects from the beginning of the EFOP through 2010 is posted online. There are approximately 1,000 exemplary projects, of which approximately 100 (10 percent) are projects related to subjects in the fire prevention field. The following are brief examples of some of the exemplary projects:

- Exemplary Project from Daniel Olson of South Kitsap, Washington, "Fire and Rescue"

  Completed in 2007, this research project was entitled *Protection System Selection and Implementation Strategies for Residential Stove Top Fires for South Kitsap Fire and Rescue*. The purpose was to develop an initiative to reduce loss from stovetop fires. Key to the initiative was the development of strategies selecting and implementing the protection. An analysis of fire response data identified statistical data confirming that stovetop fires were 16.5 percent of all fire causes in residential occupancies, with food or grease the first item ignited and cabinets and other structural combustibles contributing.

  The research identified three groups of residents at a higher fire risk: older adults, residents of multifamily residential sites, and occupants of lower income residential sites. As a result of the research, one of Olson's recommendations is the installation of stove fire extinguisher cans. The extinguishers operate by direct flame contact to an initiator that releases the extinguishing agent in a one-shot application. The extinguishers would be offered for installation in higher-risk residential areas.

  Another recommendation is to implement a community-wide education program addressing cooking hazards in the home. The author emphasized that implementation would include procedures to check visited homes for a working smoke alarm and install new ones or fresh batteries where needed.[28]

- Exemplary Project from John Webb, with the Derry, New Hampshire, Fire Department

  Completed in 2010, this research project was entitled *Improving Residential Fire and Life Safety Through Community Partnerships*. It sought to identify recommendations for a program of partnerships with and training of community groups that could go into homes of older adults and other higher fire-risk households. Partners visit homes and assess the resident's fire safety needs and then forward the information to the fire department for a follow-up safety visit. Fire service personnel then visit the home and offer to mitigate the observed risks. There would be three other visits to fix fire safety problems: a door-to-door one to homes in an area adjacent to recent fires, another to homes in targeted neighborhoods with higher risk, and finally to homes to which EMS or service calls have been made.

The most significant recommendation in terms of complexity and implementation was the formation of a coalition of community partners. The coalition would pair the fire department with nonprofits and service organizations that would help identify older adults and community members with disabilities who appear to be at increased risk of fires. The research described the potential of the partnership and home visits but also acknowledged limited resources and recommended planning and application for federal grant funding. The research again showed how an EFOP student performed the necessary analysis to determine a community's needs and conducted preliminary work to create an implementation plan to reduce the risk.[29]

- Exemplary Project from Fire Chief Brian Crawford of Plano, Texas, Fire Department

Completed while with the Shreveport, Louisiana, Fire Department, this project aimed at reducing fire risk to the city's poor. Chief Crawford's research described how fire deaths, injuries, and property loss had trended upward between 1999 and 2004. The goals in conducting the research were to identify the fire risk associated with the poor residents, to determine what impact fire was having on Shreveport communities, and to discover what actions could be taken to reduce the risk. The applied research project was started in 2004.

Shreveport, which is in western Louisiana, had a population of just over 200,000 (2000 U.S. Census). The city was over 50 percent African American (101,679 population per 2000 Census). The report indicated that this population made up 81 percent of Shreveport's fire deaths (17 of 21) in a recent 5.5-year period. Over 50 percent of the fires during that time in areas where household incomes fell below the poverty level. The fires in Shreveport appeared to disproportionately affect people living near or below the poverty level of impoverished sections of the city and most frequently impacted Shreveport's African American population.[30]

One of the more common factors reported by fire researchers is the role played by socioeconomic status in regard to fire risk. Organizations that perform fire research, such as NFPA and USFA, list poverty repeatedly as a predictor for high fire risk. Chief Crawford's research highlighted numerous examples from a literature review conducted as part of his research project that identify poverty as an indicator for fire risk. It is common knowledge in the fire and life safety industry that low socio-economic conditions relate to higher fire rates that can result in disproportionately higher rates of fire deaths and injuries. Research that included studies by Scheanman, Shainblatt, Hall, Swartz, and Karter for the Urban Institute and NFPA produced a significant body of research in the 1970s and 1980s. One published report from 1997 by Scheanman with Tri Data Corporation on behalf of the U.S. Fire Administration compiled literature of the earlier research on socio-economic impact on fire rates.[31]

Officials in Shreveport knew they had a problem, but it was the Executive Fire Officer Program EFOP research project that helped them understand all factors of why they were suffering such a high fire-death rate. Understanding was key to planning strategies. Crawford's report states that Shreveport Fire Department started efforts to reduce risk in the poor neighborhoods as early

2003. They increased public education efforts and instituted door-to-door fire safety visits to promote working smoke alarms. He indicated they knew it would be a long process to make an impact:

> Despite current public education efforts, the instances of fires and moreover the occurrence where a life or quality of life is cut short by the devastating consequences of fire, appear to be holding. Neighborhoods with the largest social issues related to poverty and depressed socioeconomic standing are continuing to suffer from a greater number of fires, fire deaths and injuries, and property loss as compared with other areas of the city.[32]

Chief Crawford saw from the hard work Shreveport fire officials had already invested that this problem would require more effort. Their efforts did pay off. They didn't just let the research and early attempts stand as a good try. Instead, they kept working at it. He later had an opportunity to lead with the ultimate authority to accomplish what he had envisioned in the research report as Shreveport's Fire Chief and then Assistant City Manager. His reflection on the efforts and outcomes of his research follows:

> The majority of the recommendations were championed heavily beginning in 2005 and in the following year of 2006 for the first time since records were kept, the city experienced no fire deaths. Remarkably, the historic feat [of zero fire deaths] was duplicated again in 2007.
>
> Upon further evaluation of the fire death numbers, it is noted that in the six years preceding 2005 (1999–2004), the average number of annual fire deaths was 3.33. In the six years following 2005 (2006–2011), the number of annual fire deaths dropped to 2.16, despite the fact of an anomaly year (2008) where seven deaths were experienced (a number of these were later attributed to crime and arson). The anomaly theory is supported by the fact that in the subsequent years from 2009–2011, the fire death averages again went back to record lows, averaging only 2 deaths over that three-year period (2009–2011).
>
> It should be noted that in that year of 2008, a number of the deaths were in apartment buildings with origins related to food on the stove. I also experienced several multi-alarm fires at apartment complexes with no deaths that same year. This moved us to further action and back to the research paper to implement Recommendations 3 and 11: first forming a Fire Prevention Task Force made up of community leaders and other internal and external stakeholders (sprinkler and construction industry experts, politicians, HOA leaders, apartment complex management companies, etc.).
>
> Through this Task Force we were able to gain multilateral support to pass local legislation that required all multifamily dwellings (namely apartment and condominium complexes) to provide overhead stove-top suppression systems. This alone drove our apartment fires down significantly and I believe was a factor in the fire death numbers returning back in-line with the six-year average of two deaths annually.
>
> This taught us a valuable lesson, in that, despite our best efforts and regardless of what programs and plans we had in place, fires and their devastating results can always buck a positive trend. The insight we gained through the 2008 experience was to not become complacent and stick to a dynamic and on-going fire prevention strategy. Be prepared to continuously reevaluate and reassess your community's fire risk and when trends shift (like what occurred to Shreveport in 2008) be prepared to respond.[33]

In Shreveport and other communities, implementation of EFOP applied research offered benefits of problem identification, understanding, and sometimes solutions. Based on the downward trend in fire deaths, it is likely that Shreveport's efforts were the right solutions.

Other Research Activities

The Fire Protection Research Foundation has become a vital asset in needed fire research. Its primary mission is to support the research needs of NFPA, but it completes a significant amount of work itself that benefits all areas of fire and life safety. Early research work of the foundation involved some of first analysis of how to replace Halon 1301 as a clean fire suppression system. At one point, Halon 1301 was the go-to agent for clean fire suppression. Halon 1301's manufacture was banned by the U.S. Environmental Protection Agency in 1994 because it was linked to the depletion of the ozone layer. The research foundation helped bring to light the reckless practice of testing procedures that allowed significant volumes of environmentally harmful Halon 1301 to escape. The foundation has created fire risk assessment tools that can assess product risk by opening consideration of alternative products as opposed to strictly considering combustibility and toxic nature. It also has a fire risk assessment that offers guidelines to evaluating fire protection alternatives available for big-box retail and large-space storage facilities.

In 2006, in affiliation with the foundation, NFPA developed a special program they referred to as a code fund initiative that introduced annual seed funding used to stimulate codes and standards research. Many of the projects funded to date have been small but have helped to build momentum and facilitate research for areas that may not attract much interest but need further investigation to improve current code or otherwise solve a problem. Some examples of projects are a residential sprinkler insulation study, fire pump data collection, fire safety in theaters, and public perception of high-rise building safety. Code fund project submittals have grown in number every year.[34]

One current project by the foundation has the potential for a very positive impact on America's number one home fire cause: cooking. The foundation is in the second phase of a project to better understand how fires can be prevented in the area of the stovetop. A third phase of the research is anticipated. The research focused on establishing the types of fire scenarios expected involving the stovetop cook area and a set of fire test performance goals associated with the scenarios. This work is intended to facilitate the development of the standard test methods needed to evaluate and certify the safety and effectiveness of any proposed products designed to prevent the stovetop fire. As mentioned, a third phase of research is needed to further solidify any data from tests and to refine areas related to the development of the performance goals and the testing standard.

The foundation has involved a diverse group of stakeholders including NIST, NFPA, Association of Home Appliance Manufacturers (AHAM), CPSC, Underwriters Lab, State Farm Insurance, and fire service representatives.

Summary

Research remains one of the essential elements of increasing fire safety. The very core inclusion of research in the Federal Fire Prevention Control Act of 1974, with its repeated mention and focus, foretold its importance to fire prevention. Fire research has played a vital role in the development of smoke alarms and residential fire sprinklers. Experiments with full-scale fires help to form the performance needed to create the standards for smoke alarms and residential sprinklers. Continued experimentation helps to make the technology even better and will help reduce the impact of cooking fires. Cooking is the number one cause of home fires in the United States.

Research programs carried out by the U.S. Forest Service direct efforts to understand all aspects of wildland fires and how humans influence and are impacted by them. Research helps us develop public warning procedures, such as the Haines Index, to provide a prediction of how weather conditions make it favorable or unfavorable for the development and control of fires.

Researchers have helped design public outreach programs focusing on wildland–urban interface areas to address preventable fires. With limited and sometimes severely restricted funding, researchers have helped to find the most economical means of managing biomass fuel in wildlands.

Vision 20/20's initiation of marketing research aimed at influencing the highest-risk audiences hopes to change indifference. Free home-fire safety visits in the United Kingdom appear to have been very successful in reducing the number of fires among that country's highest risk residents.

Advances in forensic fire research by NIST are helping us learn more and more about fire dynamics through computer models that predict fire behavior. Using state-of-the-art equipment on their home campuses, university research labs are offering students an opportunity to perform research that satisfies requirements at the graduate level while helping to better explain fire-related human behavior.

Review Questions

1. "Careful, systematic, patient study and investigation in some field of knowledge, undertaken to discover or establish facts or principles" is the definition of which one of the following terms?
   a. Discovery
   b. Science
   c. Physics
   d. Research
2. The top cause of home fires, as mentioned in this chapter, is:
   a. smoking.
   b. indifference.
   c. fire play.
   d. cooking.
3. What U.S. action emphasized the need for fire research?
   a. Federal fire prevention law
   b. Federal investigation
   c. Federal proclamation
   d. Federal tax
4. The most prominent goal of the wildland fires and fuel research and development is to:
   a. find a safe replacement for Halon 1301.
   b. reduce risk of catastrophic wildland fires.
   c. improve community fire evacuation routes.
   d. none of the above.

5. Prevention marketing under the Vision 20/20 project focused on the development of:
   a. a unifying fire safety theme.
   b. websites, Twitter and Facebook accounts, and online videos.
   c. a social indicators index of increased risk.
   d. mitigation plans for invasive pests.
6. The fire dynamic simulator developed by NIST simulates:
   a. effective prevention strategies.
   b. performance goals for standards making.
   c. fire conditions in compartments.
   d. building evacuation flow paths.
7. The National Fire Laboratory in Canada was established as a result of a _____ fire.
   a. hotel
   b. ship
   c. house
   d. school
8. London Fire Brigade officials claim fires have been reduced for those people receiving free:
   a. social media accounts.
   b. access to fire research.
   c. home fire safety visits.
   d. subscriptions to their newsletter.
9. The Fire Protection Research Foundation developed a code fund initiative to:
   a. stimulate codes and standards research.
   b. identify the scenarios for cooking fires.
   c. lobby for code changes.
   d. offer attendance stipends for standard making.

End Notes

1. *Merriam Webster Dictionary*, definition of the term *research* (Springfield, MA: Merriam Webster, Inc., 1995), p. 445.
2. U.S. Department of Agriculture Fire Service, *Wildland Fire and Fuels Research and Development Strategic Plan: Meeting the Needs of the Present, Anticipating the Needs of the Future*, FS-854 (Washington, DC: U.S. Department of Agriculture Fire Service, June 2006), pp. v–vi. Accessed August 5, 2013, at http://www.fs.fed.us/research/pdf/2006-10-20-wildland-book.pdf
3. U.S. Forest Service, *Extreme Event Effects* (Washington, DC: U.S. Department of Agriculture Fire Service, n.d.). Accessed August 5, 2013, at http://www.fs.fed.us/research/water-air-soil/extreme-effects.php
4. U.S. Forest Service, *Identifying and Preventing Invasive Species Threats* (Washington, DC: U.S. Department of Agriculture Fire Service, n.d.). Accessed August 5, 2013, at http://www.fs.fed.us/research/invasive-species/prevention/
5. U.S. Forest Service, *A Strategic Assessment of Forest Biomass and Fuel Reduction Treatments in Western States* (Washington, DC: U.S. Department of Agriculture Fire Service, April 2003), pp. 10–14. Accessed August 5, 2013, at http://www.fs.fed.us/research/pdf/Western_final.pdf
6. Jeffrey P. Prestemon, David T. Butry, Karen L. Abt, and Ronda Sutphen, "Net Benefits of Wildfire Prevention Education Efforts," *Forest Science,* Vol. 56, No. 2 (2010), pp. 181–192. Accessed August 14, 2013, at http://www.srs.fs.usda.gov/pubs/ja/ja_prestemon029.pdf
7. L. Annie Hermansen-Báez, Jeffrey P. Prestemon, David T. Butry, Karen L. Abt, and Ronda Sutphen, *The Economic Benefits of Wildfire Prevention Education* (U.S. Department of Agriculture, U.S. Forest Service, Centers for Urban and Interface Forestry–Interface South, n.d.). Accessed August 13, 2013, at http://www.srs.fs.usda.gov/pubs/ja/ja_hermansen002.pdf
8. National Fire Protection Association, *Arizonans Use Firewise Tips to Reduce Their Homes Risk from Wildfire: Wildfire Doesn't Have to Damage Your Home*, news release

(Quincy, MA: Author, June 6, 2011). Accessed August 6, 2013, at http://www.nfpa.org/press-room/news-releases/2011/arizonans-use-firewise-tips-to-reduce-their-homes-risk-from-wildfire

9. National Commission on Fire Prevention and Control, *America Burning* (Washington, DC: Author, 1973).

10. Report of the Committee of Conference on S. 1769, *Federal Fire Prevention and Control Act of 1974*, 93rd Congress, 2nd Session, 1974, S. Report No. 93-1088. Accessed August 7, 2013, at http://fire.nist.gov/bfrlpubs/fire75/PDF/f75006.pdf

11. R. W. Bukowski, T. E. Waterman, and W. J. Christian, *Detector Sensitivity and Siting Requirements for Dwellings,* NBS GCR 75-51 (Washington, DC: U.S. National Bureau of Standards, 1975). Accessed November 18, 2013, at http://fire.nist.gov/bfrlpubs/fire75/PDF/f75002.pdf

12. Daniel Madrzykowski and Robert L. Vettori, *Simulation of the Dynamics of the Fire at 3146 Cherry Road NE, Washington DC, May 30, 1999,* NISTIR 6510 (Gaithersburg, MD: National Institute of Standards and Technology, April 2000). Accessed November 18, 2013, at http://www.nist.gov/customcf/get_pdf.cfm?pub_id=908795

13. Nelson P. Bryner, Stephen P. Fuss, Bryan W. Klein, and Anthony D. Putorti, *Technical Study of the Sofa Super Store Fire—South Carolina, June 18, 2007,* NIST——SP 1118, Vol. I (Gaithersburg, MD: National Institute of Standards and Technology, March 2011). Accessed November 18, 2013, at http://nvlpubs.nist.gov/nistpubs/SpecialPublications/NIST.SP.1118v1.pdf

14. National Institute of Standards and Technology, *Building and Fire Research Portal—Overview* (Gaithersburg, MD: National Institute of Standards and Technology, n.d.). Accessed November 18, 2013, at http://www.nist.gov/building-and-fire-research-portal.cfm

15. Joshua B. Dinaburg, Daniel T. Gottuk, and Hughes and Associates, *Development of Standardized Cooking Fires for Evaluation of Prevention Technologies: Phase One Data Analysis: Final Report* (Quincy, MA: Fire Protection Research Foundation, October 2013). Accessed November 18, 2013, at http://www.nfpa.org/~/media/Files/Research/Research%20Foundation/Research%20Foundation%20reports/Other%20research%20topics/RFCookingFiresEvaluationPreventionTechnologies.pdf

16. Matthew Baker and Christopher Campbell, *Human Behavior During Egress of High Rise Buildings* (College Park, MD: A. James Clark School of Engineering, Department of Fire Protection Engineering, 2013). Accessed August 20, 2013, at http://www.enfp.umd.edu/research/projects/human-behavior-during-egress-high-rise-buildings

17. Matthew Daniel Baker, *Observed Behavior of Platoon Dynamics During High-Rise Stairwell Evacuations* (College Park: University of Maryland, 2012). Accessed August 20, 2013, at http://drum.lib.umd.edu/bitstream/1903/13657/1/Baker_umd_0117N_13896.pdf

18. Christopher Klly Campbell, *Occupant Merging Behavior During Egress from High Rise Buildings* (College Park: University of Maryland, 2012). Accessed August 20, 2013, at http://drum.lib.umd.edu/bitstream/1903/13562/1/Campbell_umd_0117N_13870.pdf

19. National Research Council, *Construction* (Ottawa, Canada, May 17, 2013). Accessed December 5, 2013, at http://www.nrc-cnrc.gc.ca/eng/rd/construction/index.html

20. Jonah Marc O'Neil, "Fire Resistance of Straw Bale Houses," *Canadian Fire Chief* (Ottawa, Ontario, Spring 2003), pp. 18–23.

21. "Brigade Visits Prevent Five Thousand Fires in London," *London Fire Brigade*, August 22, 2013. Accessed August 26, 2013, at http://www.london-fire.gov.uk/news/LatestNewsReleases_Visitsprevent5000fires.asp#.UhvqtRaDrzK

22. Ibid.

23. Thomas Z. Fabian, Pravinray D. Gandhi, and Underwriters Laboratories Inc., *Smoke*

Characterization Project (Quincy, MA: The Fire Protection Research Foundation, April 24, 2007). Accessed November 18, 2013, at http://www.nfpa.org/~/media/Files/Research/Research%20Foundation/Research%20Foundation%20reports/Detection%20and%20signaling/smokecharacterization.pdf

24. "Groundbreaking Study of Smoke Characterization Could Change the Face of Fire Safety," news release by Underwriters Laboratories, Northbrook, IL, April 17, 2007. Accessed December 3, 2013, at http://www.ul.com/asiaonthemark/as-en/2007-Issue22/page8.htm
25. National Association of Home Builders, *America's Housing 1900–2000: A Century of Progress* (Washington, DC: Author, April 2003). Accessed November 23, 2013, at http://www.nahb.org/assets/docs/files/v5_513200312545PM.pdf
26. U.S. Fire Administration/National Fire Data Center, *Changing Severity of Home Fires Workshop Report.* (Washington, DC: FEMA, 2012). Accessed March 31, 2014, at http://www.usfa.fema.gov/downloads/pdf/publications/severity_home_fires_workshop.pdf
27. U.S. Fire Administration/National Fire Academy, *Executive Fire Officer Program: Celebrating 25 Years of Excellence in Fire/Emergency Services Executive Education: 1985–2010* (Emmitsburg, MD: Author, n.d.). Accessed December 3, 2013, at http://www.usfa.fema.gov/downloads/pdf/publications/fa-279.pdf
28. Daniel N. Olson, *Protection System Selection and Implementation Strategies for Residential Stove Top Fires for South Kitsap Fire and Rescue* (Emmitsburg, MD: National Fire Academy, July 2007). Accessed August 16, 2013, at http://www.usfa.fema.gov/pdf/efop/efo40937.pdf
29. John Q. Webb, *Improving Residential Fire and Life Safety Through Community Partnerships* (Emmitsburg, MD: National Fire Academy, October 2010). Accessed August 16, 2013, at http://www.usfa.fema.gov/pdf/efop/efo45029.pdf
30. Brian A. Crawford, *Fire and the Poor: Identifying and Assessing Community Risk and Intervention Strategies* (Emmitsburg, MD: National Fire Academy, July 2004). Accessed August 17, 2013, at http://www.usfa.fema.gov/pdf/efop/efo37322.pdf
31. U.S. Fire Administration and Tri Data Corporation, *Socioeconomic Factors and the Incidence of Fire* (Washington, DC: FEMA, June 1997). Accessed March 29, 2014, at http://www.usfa.fema.gov/downloads/pdf/statistics/socio.pdfM.
32. Crawford.
33. Personal communication to author Love from Brian A. Crawford, August 28, 2013.
34. Fire Protection Research Foundation, *Code Fund.* (Quincy, MA: National Fire Protection Association, n.d.). Accessed August 22, 2013, at http://www.nfpa.org/research/fire-protection-research-foundation/code-fund

Billy Morris Director of Fire Science
Cisco College

CHAPTER 13

Proving Fire Prevention Works

KEY TERMS

Built for Life Fire Department, *p. 268*
casualty, *p. 262*
Cigarette Safety Act of 1984, *p. 270*
marketing, *p. 260*
National Fire Data Center, *p. 257*
National Fire Incident Reporting System (NFIRS), *p. 264*
performance measurement, *p. 258*
subrogation, *p. 262*

OBJECTIVES

After reading this chapter, you should be able to:

- Describe the importance of validating the effectiveness of fire prevention.
- List the evidence presented that supports the effectiveness of residential fire sprinklers.
- Explain how data can be used in annual reports, blogs, Facebook, Twitter, and other community venues to market positive fire prevention services.
- Describe how fire data are collected and compiled in the National Fire Incident Reporting System.
- Describe how one out of the three examples (residential fire sprinklers, changes in mattress flammability, and testing of cigarettes) can reduce the impact of unwanted fires in the United States.
- Explain how the U.S. Congress was involved in reduction of fires caused by smoking materials.

Fire prevention works. However, this fact must be proved year after year. It is said that in the 1920s, Chief D. W. Brosnan of Albany, Georgia, spoke on this subject at the first annual meeting of the Southeastern Association of Fire Chiefs. With the change of a few words, the talk would be as appropriate today as it was then:

> Any person who is at all conversant with fire safety knows that at least 85% of fires could be prevented. It is the duty of the Fire Chief to assume leadership and point out the way for the protection of life and the conservation of property of our citizens. The modern Fire Chief knows that he must be up and doing and prevent fires from starting, if he is to be successful in reducing the loss.[1]

It must be recognized that a tremendous amount of progress has been made in fire safety and prevention. The popularity of smoke alarms and fire sprinklers is an example. Public fire and life safety education is an accepted program of most fire departments. Even so, thousands of citizens of all ages are still not fortunate enough to have the benefit of this important information.

In recent years, economic hard times have resulted in the loss of many fire prevention programs, and in some ways, that loss can be attributed to stakeholders' lack of awareness of what services exist and why they are important.

A strategic plan for Georgia's fire service recognizes marketing and promoting fire prevention and risk reduction. The plan states:

> Interviews indicated the importance of the work conducted by the state fire marshal's office has been, for the most part, unknown to key stakeholders. The reduction in force in the fire marshal's office by more than 50% will have serious repercussions for many years to come. A program to reinstate many of the functions that have been lost will be necessary, and fire service support for those efforts will be needed. Stakeholders and legislators in particular need to be educated regarding the importance of code development, code refinement, and code enforcement. In addition, the need for additional investigators and inspectors will have to be brought to the attention of those who develop state budgets.[2]

Offsetting some of the progress in fire safety is the reduction of the staffing of fire prevention bureaus in many cities as an economy move. The economic challenges beginning in 2008 required state and local governments to make significant cuts to services. Significant cuts have been made to fire service delivery throughout the United States, but recent survey evidence has shown that fire prevention programs, especially fire inspections and public education, have taken a disproportionate reduction. The survey data are preliminary results of the Vision 20/20 Fire Prevention Cuts survey administered by Michael Donahue and Raymond O'Brocki.[3] In many cases, data from the survey indicated whole fire prevention programs have been eliminated with no remedial plans for the deferral of services or resuming programs when the economy improves.

Fire service leaders recognize that the fire department alone cannot successfully impart the message of fire prevention and safety without strong community involvement. In many inner-city neighborhoods and low-income rural areas, individuals who are known and respected by the residents may be quite effective in generating an interest in fire prevention.

Fire prevention chiefs must be able to adjust their programs to meet new challenges in the face of drastically reduced funding. Bureaus incapable of change will not survive (Figure 13-1).

Billy Morris Director of Fire Science Cisco College

FIGURE 13-1 Fire safety destinations, also known as safety villages and fire safety zones, have become popular ways to demonstrate fire safety to residents.

Measurement of Fire Prevention Effectiveness

Measuring the success of fire prevention efforts is subject to many variables. As an example, a nationally circulated magazine attempted a comparative evaluation of U.S. fire departments. Cities across the country were rated on their fire departments' efficiency according to the criterion of per capita fire losses. The fire department protecting a bedroom community had a definite advantage over a community in which there was a good deal of heavy industry with little residential area. Fire death totals have also been used as a means of measuring the effectiveness of fire prevention activities. Although such a comparison may have some validity, there are weaknesses in it as well.

Research has shown that socioeconomic factors have a bearing on fire safety measures and that a fair comparison of individual procedures in a community should take them into account. As an example, most cities experience a higher per capita fire death rate in low-income, densely populated areas than in high-income, detached-home areas.

Fire incidents are starting to show a decline, as is per capita dollar loss, according to a 2001 publication of the **National Fire Data Center** of the U.S. Fire Administration. Civilian and firefighter fatality and injury rates are down as well. The publication gives credit to the following factors: smoke alarms, sprinklers, strengthened fire codes, improved construction techniques and materials, public fire safety education, firefighter equipment, and training. The statement concludes, "If we could understand the relative importance of these factors to lessening the fire problem, resources could be better targeted to have the most impact."[4]

National Fire Data Center
▪ a division of the U.S. Fire Administration, it collects, analyzes, publishes, disseminates, and markets information about the U.S. fire problem.

Measuring the value and efficiency of fire prevention activities has been and remains inconsistent if not altogether absent, and while departments are beginning to evaluate their programs, measurement is still the exception rather than the rule. Failing to conduct **performance measurement** and to quantify fire prevention services could hurt opportunities for funding and community support. People are more concerned about government performance and accountability and are less likely to support programs without evidence or at least predictability of success. Promoting a new fire prevention program to one's community, or defending a program that may be on the chopping block, may depend on outcome information that results from evaluating one's program.

performance measurement
■ a system management process that quantifies value, effectiveness, and efficiency so services and programs such as fire prevention and safety can be better understood by stakeholders, community partners, residents, government management, and elected officials.

A chief fire officer described in a personal conversation[5] a negative experience: A fire prevention program was attacked in a public speech. The chief, a strong fire preventionist, was in the audience of a Rotary Club meeting. The CEO of a Fortune 100 company addressed the members (many of whom were the elite business leaders of his city), saying that the fire department, specifically the fire prevention bureau, was a barrier to development and innovation in the city, that the bureau was hampering economic growth and costing the community jobs and revenue. His perception was that the fire department was applying standards that were different from those of other jurisdictions, and that the time to review development and construction projects was excessively long, costing developers money. His final comment was that the fire prevention bureau added no value to the community through enhanced safety. True or not true, the damage was done, and the chief realized that successful people such as the speaker did not always get to that level by being nice. But how was he to go back and perform damage control? How was he to prove the CEO wrong and to spread the word that the fire prevention activities in that city did add value and enhance safety?

It may very well be that some of the images painted of fire prevention efforts are a result of customer service issues. Customer service is very important and should be a top priority of any fire prevention bureau large or small.

Another example of how fire prevention efforts have been badly portrayed is the comments of political opponents to the current federal Assistance to Firefighters Grants. More specifically, the Fire Prevention and Safety grants have been attacked as being a complete waste and offering no value. One pundit said:

> Grants had no impact on the occurrence of firefighter deaths, firefighter injuries, civilian deaths, or civilian injuries. Without receiving fire grants, comparison fire departments were just as successful at preventing fire casualties as grant-funded fire departments.[6]

The point was that Congress should eliminate the funding for federal fire grants. Again, in building advocacy for fire prevention programs, elected officials up to and including the president of the United States should be considered for receiving proof that programs work and offer measurable value. These attacks demonstrate that the lack of promotion of fire prevention program value may give others the opportunity to fill that void with negatives. So it is critical to sell programs regularly to ensure the community gets a positive vision of its services.

Fire safety in a community is a reflection of the culture of that community. Many communities are very interested in safety and do everything possible to promote fire safety and reduce risk. For example, the City of Albany, Georgia, has a unique record of fire safety going back to 1918, when fire safety began to really emerge as an industry. At the heart of Albany's success was a hard-driving fire chief who set the tone and expectation for safety in that community. Fire Chief D. W. Brosnan was a leader, not only in his city but throughout the United States. He was well known as a trainer and active member of the International Association of Fire Chiefs, where he served on committees and was the president in 1931. Chief Brosnan's motto concerning fire was "Stop 'em before they start."

Chief Brosnan used data and evidence management in government long before this approach became mainstream. He recognized that an aggressive approach to fire loss reduction benefited the city economically. He knew fire prevention could work to reduce risk in Albany, and he did everything possible to prove it. One way of creating a compelling program to promote the city's fire safety culture was to take part in some of the emerging contests that gave municipalities the opportunity to showcase their efforts. Albany was an active player in those promotional activities.

As early as 1908, fire safety was on the minds of the organization of U.S. governors. Their Conference of Governors partnered with the U.S. Chamber of Commerce to increase and promote local-level fire safety activity. The Chamber of Commerce formed an internal focus on fire safety it called the Fire Waste Council. The council facilitated safety competitions called Fire Waste Contests for member communities, offering awards and honors for the highest achieving communities. Albany, Georgia, became a regular honoree in the early years of the contest.

During the early years of the Fire Waste Contest, Albany was a city of approximately 18,000 population. In 1926 *The Quarterly*, a regular journal of the NFPA,[7] described Albany's accomplishments in fire safety. Albany's residents were proud of their average $1.73 per capita fire loss over the 5-year period of 1920 to 1924. In addition, the community had not registered a single fire death in over 15 years. Between 1916 and 1925 Albany had been able to reduce the number of annual total fires from 135 to 127, while the rates of fires in most other cities were increasing. Albany's rate of fire decline showed a positive trend and resulted in the lowest fire insurance rates of all cities in Georgia. So it is proved that fire prevention not only works but also pays high dividends when it does.

In 1923, Albany adopted a new building code with provisions that prohibited wood shingles. There was evidence of the need for the limitations on the shingles because the data showed that 45 of the 127 fires in 1925 had been started by sparks or burning embers that landed on wood shingles not yet replaced to conform with the new building code. The citizens of Albany made the data work for them and saw that they were very likely preventing other fires in many of the city's buildings with noncombustible shingles.

The Albany fire department was actively involved in searching out and eliminating fires before they started or could get beyond reasonable control. Uniformed firefighters made rounds every 3 hours looking for any indications of fires, with a keen focus on the cotton industry, which offered a unique fire hazard in town. During one bad drought, the cotton warehouses were under very close supervision by the firefighters, who removed numerous individual cotton bales at different times. In processing the cotton, the cotton gins could be a source of ignition when foreign materials, such as rocks and metal objects that had been picked up in harvesting, sparked, starting the cotton to smolder. Removal of the burning bales and in some cases the burning material from the individual bales resulted in an averted loss of the material that was one of the town's biggest commodities. Inspection rounds throughout the night in the business district led to the detection of many small fires that were extinguished by the firefighters and therefore resulted in minimal loss.

It does not much matter that the activities and fire prevention leadership of Chief Brosnan took place in the 1920s, 1930s, and 1940s. What matters is that Chief Brosnan recognized fire prevention as a way of life that managed risk and reduced losses to the city. He showed that it was more a mind-set and a process than specific activities. He also showed that when one pays attention to a concern, has goals and expectations, and keeps track of the results, others will pay attention and adopt that mind-set, too. This is what measuring performance is all about. Proving that fire prevention works is nothing more then measuring performance.

marketing
■ the action of promoting and selling products and services, such as programs associated with fire safety and fire prevention.

There are many ways to keep the positive efforts of fire prevention activities in the eyes of the community. **Marketing** fire prevention activities can include regular reporting by the news media, email forums, Facebook and Twitter pages, and traditional publications such as annual reports.

Many fire departments prepare an annual report to give community residents and the official governing body an accurate picture of the fire department's operation. The annual report should be simplified so members of the general public can readily understand it. It should be attractive and easy to read, with photographs and charts to help improve its appearance. A tight budget should not preclude the issuance of a report. Even a photocopied report can be successful in advising the public of fire department activities.

The Montgomery County, Maryland, government uses a system of measures to evaluate performance and outcomes. The fire and rescue service there has been recording the number of fire deaths since the government became chartered in the 1940s. In the late 1990s, fire officials began to track a trend of increasing fire deaths. The increase was proportionally greater than the county population growth. What was found was a disproportionate number of fire deaths among adults over age 65. A specific task force was assigned to study the problem and make recommendations, and as a result, the county began to notice another trend change with fire deaths going down from 2005, with 15 deaths, to 2008, with 10. The firefighters ramped up their door-to-door fire safety visits with a focus on areas where older adults made up a high percentage of the demographics. In 2009, fire deaths dropped below a total of 5 in a year's time, or about 0.5 deaths per 100,000 population.

Tracking the fire deaths and providing regular quarterly updates of performance keeps residents aware of the Montgomery County firefighters' efforts to keep fire deaths as low as possible. A trend in performance over the 5 years from 2005 through 2009 was a good indication the county was having a positive safety impact compared to the performance of just a decade before (Figure 13-2).

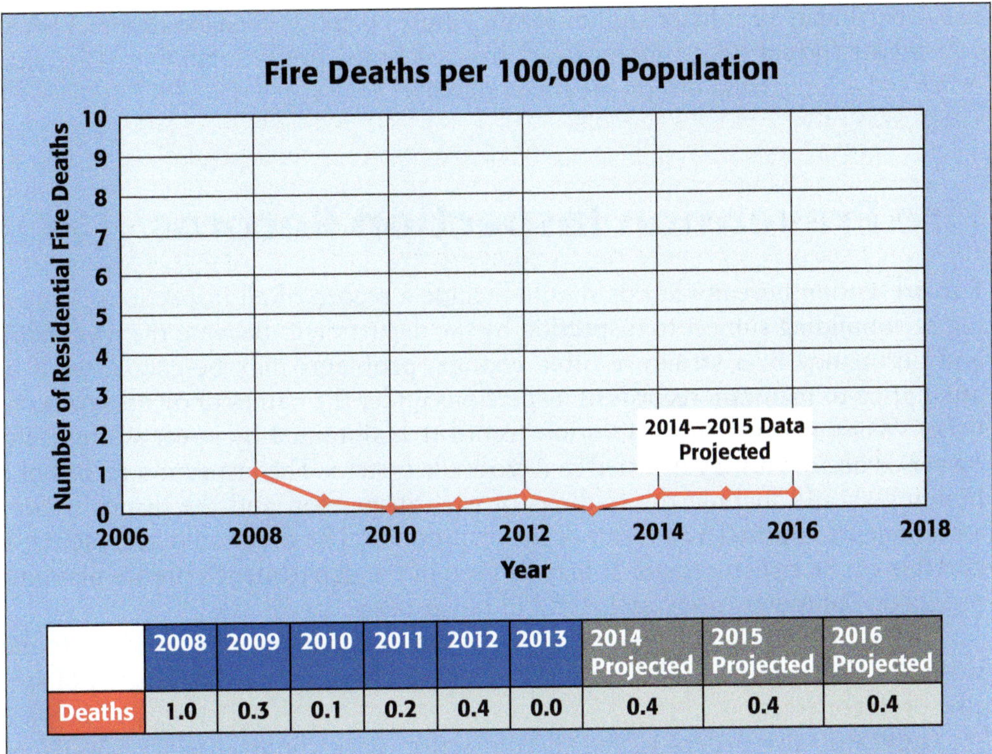

	2008	2009	2010	2011	2012	2013	2014 Projected	2015 Projected	2016 Projected
Deaths	1.0	0.3	0.1	0.2	0.4	0.0	0.4	0.4	0.4

Mike Love

FIGURE 13-2 Fire data recorded over a period of 5 years and plotted as a trend line can provide a useful visual indicator of fire safety performance. Maryland's Montgomery County evaluation of its fire death statistics indicates an average of only 0.2 fire deaths per 100,000 population between 2009 and 2013.

Recording Fire Safety Activities

The fire-reporting system should offer information regarding activities of the fire prevention bureau, including demonstrations, lectures, other public education programs, plan reviews, and all other activities that contribute to the goal of reducing fire losses. Any public contact has a bearing on the official function of the organization and should be included. Having a guest log in every fire station that nicely documents every walk-up contact and provides for data collection can show how surprisingly often the contact was influenced by the department's services.

Fire departments are also taking advantage of all their emergency runs to both educate the public and take care of life safety problems such as nonworking smoke alarms. This type of activity can make a difference and requires little extra effort since the crews are already on the scene of an incident. Equipping fire stations and apparatus with digital cameras is a good way to capture events that can then be used in all promotional activities.

The training division of the fire department should maintain complete records on time devoted to fire prevention training. The training division may call on the fire prevention bureau for assistance in such instruction.

A fire department record system should include information relating to all responses made by the department. Those responses should be classified and enumerated in a manner that gives an accurate picture of services performed by the fire department. Simplicity and uniformity must be major considerations. A separate report should be maintained on all inspections, prefire planning visits, and other activities carried out by the fire company.

Fire Prevention Inspection Records

Fire prevention bureau records should include a record of all inspections, including each building subject to inspection by the department. Because the ownership and occupancy of a structure often change, problems may be encountered in attempting to maintain records of inspections under the names of owners or occupants. When the primary inspection record is maintained by street address, the owner or occupant may be listed in a cross-file system. The inspection record of a building should include the findings of each inspection and the disposition of requirements imposed as a result of the inspections. The file should also include a diagram of the structure, which may be used in the department's prefire planning program. Computers are very helpful in this process.

The file should include photographs of the building and of conditions found within it. Photographs and diagrams can be used to pinpoint the location of specific problem areas.

Copies of any orders issued in connection with the occupancy should be included in the file. Newspaper clippings regarding the occupancy can also be helpful. A news item might indicate policy changes by the operators of the structure and give other information that can be of great value in the event of a fire, such as changes in business conditions that relate to the cause of a fire.

Fire prevention bureau records, with the exception of open arson cases, are considered public records in most states. **Subrogation** cases, arson cases, and other cases in litigation in which details of fires must be recounted make the maintenance of adequate and complete records mandatory. It is necessary to be able to immediately recount inspections conducted and to have information available regarding fire prevention education programs carried out by personnel.

subrogation
■ a legal process often used by insurance companies to recover damages or obtain a monetary reimbursement from a person or organization responsible for property damage or injury.

Recording Fire Deaths and Injuries

Fire department records should include facts about injuries sustained in fires, as well as facts about deaths incurred as a result of fires. Many communities do not have complete records because, in most cases, they record deaths and injuries only when fire department response was involved. Medical reports are usually not public records. The National Fire Incident Reporting System (NFIRS) has a **casualty** form that should be completed for each injury and death directly related to a fire. Most of the fire department data collection systems have a casualty form for collecting and storing the information.

casualty
■ a person who is killed or injured as a result of being directly exposed to a fire; a term used for both civilians and firefighters, though the two are recorded as distinctly separate data.

The fire department may not know about a death or injury by fire if the fire department was not called and the victim was taken directly to the hospital or mortician. This is often the case for burn injuries in which the victim is transported for medical treatment by a relative or neighbor and no fire department emergency medical or fire response is involved. Fire departments need to maintain close ties with hospital emergency departments and the medical examiner's or coroner's office. The hospital should be able to provide data that include demographic information on fire casualties, both injuries and deaths, because the hospital collects such information with a key data point: the type of treatment needed. It can become a little trickier in areas where patients are transported across state lines, such as to trauma or burn centers; their deaths would be recorded in the states in which they died.

Statistical reporting of deaths and injuries related to fire has another inherent obstacle. There are many borderline cases in which it is difficult to determine whether the death or injury should be attributed to fire or something else. For example, the question of assignment arises in traffic collisions in which the vehicles burn.

According to the National Fire Protection Association, the risk of dying in an unintentional building fire has fallen from more than 20 deaths for every 200,000 people to just 2 deaths for every 200,000 (between 1896 and 2006).[8] Those fire fatality decreases can be attributed to many factors, which are described in earlier chapters in this book.

Approximately 17 percent of the nation's total fire deaths occur in vehicles. These incidents are not easily abated by the usual fire service activities. Well over half of vehicular fire deaths occur as a result of a collision, an occurrence that relates to many factors: vehicle construction, driver ability, road configuration, and a myriad of other factors, none of which are under the control of fire prevention agencies.[9]

Certainly, the elimination of fire fatalities is the major goal of public fire service prevention activities. Merely recording rote figures on fatalities is not enough. An analysis of fire death trends by cause is quite helpful in determining proper avenues for fire safety education and enforcement activities.

Recording Loss Statistics

Widespread involvement of the fire service in the provision of emergency medical services has necessitated the development of data systems to record responses. Many of the data systems function at the local and state levels. Motor vehicle injuries and fatalities, falls, cardiac arrests, drownings, and other non-fire-related incidents are recorded in connection with fire service response.

The availability of statistics for non-fire-related responses has been the driving force behind the expanding involvement of fire services in educational programs aimed at abating injuries and fatalities. Fire departments may be joining with other community departments and organizations in disseminating information designed to lower the frequency of such incidents. Examples include CPR training and programs promoting seat belt usage, pool safety, and fall prevention.

The fire department record system should include information on fire losses in the community. Included should be data relating to causes of fires, buildings and occupancies in which fires occur, and information related to the losses incurred in fires. Also helpful is a community risk assessment that provides a comprehensive evaluation of fire risk, including demographics; economic indicators; description of residential, business, and industrial makeup; and so on. A risk assessment could help the community understand some of the concerns and help build support for efforts to solve fire safety problems in the highest risk areas.

Obtaining loss statistics for all fires is a problem because there are so many variables in developing such information. Many departments use loss statistics developed by fire department personnel. Others use those developed by insurance loss adjusters. In some cases, the departments count only losses not covered by insurance carriers. These departments believe that any dollar loss covered by insurance is not actually a fire loss and therefore does not count in their statistical report. Some fire departments classify losses as moderate, medium, high, or in other terms of a general nature. This "broad range" procedure makes it easy for personnel to complete their reports on the fire ground. However, accuracy is not served. Personnel assigned to estimating losses should have training in evaluation methods.

Records and reports are quite helpful in determining the value of fire protection devices such as fire sprinklers and smoke alarms. Statistics have authenticated the value of automatic sprinklers. However, smoke alarms are rapidly coming of age in this respect. Many communities are recording all smoke alarm operations in an effort to prove the value of the device. Statistics on performance coupled with examples of incidents in which smoke alarms and sprinklers were successful in saving lives can be of great assistance in promotional campaigns by fire departments.

National Fire Incident Reporting System

National Fire Incident Reporting System (NFIRS) ▪ a system operated by the National Fire Data Center that collects and analyzes fire incident and casualty data on a nationwide basis.

The **National Fire Incident Reporting System (NFIRS)** is operated by the National Fire Data Center, a part of the U.S. Fire Administration. The system is designed to collect and analyze fire incident and casualty data on a nationwide basis. NFPA 901, Standard Classifications for Incident Reporting and Fire Protection Data, is used as the basis for the system.[10]

Fifty states and the District of Columbia now report data. The U.S. Fire Administration uses the data to produce national estimates and to develop a database for answering information requests from government, industry, and the public. The NFIRS data center is feeding national summaries back to the states for their use in evaluating and improving their fire protection programs.

Data for the NFIRS, including fire incidents and casualty reports, are collected at the state level from local fire departments. Statewide summaries are prepared, as are feedbacks to local fire departments. Statewide data are sent to the NFIRS system, which is the largest national annual database of fire incident information in the world.

It is imperative that fire incident data, whether gathered on a local, state, provincial, or national level, be of value to the agency submitting the initial report. Fire department personnel, whether career or volunteer, rapidly lose interest in supplying accurate information to a system that provides their department with few or no returns.

Use of Computers

Computers and other forms of technology are now widely used in fire prevention as a means of expediting decisions in fire safety code enforcement. With computers, the fire prevention chief is able to immediately review inspection criteria as well as results of code enforcement activities.

Computers and other forms of technology are also widely used in the fire service. Routine correspondence, newsletters, document sharing, calendars and appointments, messages regarding hydrant and street closings, duty reassignments, inspection assignments, and other routine messages can be handled in this manner. Records of inspections, hydrant flow test data, fire protection systems maintenance, and building information such as scanned images of construction plans can be stored and retrieved as needed. With wireless data networks, community maintained and private Wi-Fi, cell phone, and fiber optic networks, any information can be transmitted easily and quickly.

Computers have enabled fire service managers to immediately ascertain trends and directions that may be significant when planning for fire safety within a community. Likewise, they play a major role in fire protection research. A computer can give the researcher accurate information on probabilities that may be difficult to formulate in the laboratory.

In many fire departments, all firefighters are proficient in the use of computers. Operational requirements demand that this degree of proficiency be reached if inspection programs, company response reports, company budget maintenance, and other essential activities are routinely processed by computer.

Computer-aided dispatch has become routine and usually includes methods of tracking the status of all firefighting resources, occupancy data for hazardous and target locations within the city, and information on fire ground operational requirements. Terminals for receipt of necessary data may be provided on fire apparatus and in command vehicles.

An example of the use of computers for fire code enforcement is the program through which fire departments may report inspections and fires. Each station is given a printout with information on the status of all fire inspections. The jurisdiction receives copies of the reports and is able to see the number and types of deficiencies within the community on receipt of each printout. This system enables the fire prevention officer to keep track of inspections and to make appropriate personnel assignments from month to month. Many cities in the United States use a similar system for their fire inspection reporting and inspection frequency control.

Other routine activities of the fire prevention bureau can be handled similarly by computers. Handheld computers allow the tracking of whether the person involved in a fire has received fire safety training.

Computer applications in fire safety management, plan reviews, statistical packages, fire modeling programs, and the use of geographic information systems in determining necessary development planning information are other examples of the widespread use of this equipment. Fire modeling is being used in the reconstruction of fires in connection with cause and origin investigations as well as in planning for fire protection.

Fire departments throughout the United States and many other countries are using technology to more efficiently handle the high workload of fire inspectors and code enforcement officers. The following information has been provided by the fire marshal in Montgomery County, Maryland, to show some of the ways computers are being used in the daily operations of the department:

- *Personal computers.* Desktop computers or laptops that dock at the inspector's desk provide all the usual office administrative tools from word processing, database, and spreadsheets, to graphic programs. This is the primary portal for the inspector to access the department's data, including permit and electronic-based plan reviews and other critical records for tracking occupancy, licensing, construction, and permit work. Ideally, departments responsible for enforcement of codes and regulations all use integrated systems so that they can access all of the information on every address or building plan.

  A proprietary county system for electronic plan review and permit management provides inspectors access from their desks to any plan, including the notes and comments from the plan reviewer, which document requirements specific to that project. Copies of the files can also be made available to the inspector in the mobile environment thus making wasteful trips to the office for additional information unnecessary.

  Other essential office applications include financial systems for fee and permit management, a calendar for scheduling inspector appointments, task and to-do list management, and e-mail. In January 2007, the Montgomery County Fire Marshal's office implemented a new fee-based, full-cost-recovery process. The first phase of this new system uses an "off the shelf" software program that is compatible with the county's fire data application.
- *Computers in the mobile environment.* Computers of all sorts, including laptops, tablets, and smart phones, have become an essential tool for fire inspectors because they can be used remotely to access most code references and to conduct fire and life safety activities. Frankly, there is little difference in the content or processes that can be undertaken; only the physical configuration of the device differs. Nearly all devices integrate communication, business processing, and Internet access into one unit. What form of computer is used is a matter of choice. With the right software, computers can be used to enter inspection data and manage notice of violations (NOV). Once an inspection or witnessed fire equipment test is complete, the department staff member can start the billing process and print out a copy of the NOV or permit certificate for the customer. Inspectors carry portable printers to use when needed or can complete transactions electronically by sending files with e-mail by way of cellular phone and wireless messaging.
- *Smart phones.* Smart phones are sophisticated computers that have evolved from the integration of personal digital assistants and cellular phones to provide the user with access to telephone service, e-mail, Internet, and an up-to-the-minute calendar, as well as cameras to capture still photos and videos and schedule changes. Some fire code enforcement agencies are also using smart phones as their inspection data collection tool.

- *Digital scanning of paper records.* Inspections still create a significant number of paper records, such as letters, memos, agreements, and equipment spec sheets. In an effort to eliminate unnecessary paper files, Montgomery County, Maryland, scanned its files for occupancy inspections that occurred in the past and created digital files to replace them. Software programs are available to create indexes that identify each scanned item and other management tools so that files and individual images can be easily found for future reference.[11]

Systems Analysis

Systems analysis uses information technology and specially designed software programs and powerful databases to methodically process data to help managers and decision makers solve problems. Developed by the U.S. Fire Administration in cooperation with Commission on Fire Accreditation International, Inc., the occupancy vulnerability assessment evaluation called Risk, Hazard and Value Evaluation (RHAVE) is an example of a computer-based system for analyzing and classifying occupancies under the purview of the using fire service agency. The RHAVE system stores information on seven major factors relating to a given structure:

- *Premises:* Assessed valuation, area, revenue benefit, description
- *Building construction:* Type of construction, access, size
- *Life safety:* Occupancy load, mobility, warning alarms, exiting system
- *Risk:* Regulations in effect, human activity, loss experience
- *Water demand:* Fire flow required and available, sprinkler protection
- *Value:* Personal, family business, impact to community
- *Summary:* A compilation of the hazard involved based upon all factors above with resulting score of maximum, significant, moderate, and low

Approaches to the Fire Problem

By now it should be obvious that the subject of fire prevention is not limited to the prevention of an unwanted fire. It is also about technology that can stop or at least limit the growth of a fire, ultimately reducing the risk to people or property exposed to the fire. One of the most effective technologies for containing and extinguishing fires is the fire sprinkler.

Residential fire sprinklers have been slow to catch on outside the fire service and the fire and life safety industry. In fact, there has been a significant effort by opponents to keep them out of new residential construction. A number of states have taken such drastic steps as to use legislation to prohibit them from being adopted or even considered at both the state and local levels. But this opposition has not stopped the positive results that prove residential sprinklers work well to reduce fire deaths.

The Northern Illinois Fire Sprinkler Advisory Board (NIFSAB) has worked very hard at the local level to promote residential sprinklers and help authorities

Built for Life Fire Department ▪ a program initiated and supported by the Home Fire Sprinkler Coalition, recognizing fire departments in the United States that focus on promoting residential fire sprinklers in their outreach programs.

implement local sprinkler requirements. The board uses resources from the Home Fire Sprinkler Coalition, especially the **Built for Life Fire Department**, to establish and sustain the local support, awareness, and networking needed to spread the word about sprinklers. The Built for Life program assists fire departments in educating residents about residential sprinklers and offers the opportunity to simply add the message to existing outreach materials. It also offers even more comprehensive support with free materials: a fire and sprinkler burn demonstration kit, educational outreach materials, a certificate of participation, and Built for Life Fire Department window decals and baseball caps.

The NIFSAB started in 1999 when there were only 2 municipalities with residential sprinkler ordinances requiring system design and installation following the NFPA 13D, Standard for the Installation of Sprinkler Systems in One-and Two-Family Dwellings and Manufactured Homes. There now are 76 such municipalities. The NIFSAB offers a number of services at no cost, which include:

- Consultation and research on local fire and building code changes and fire sprinkler plan review questions
- Liaison with sprinkler contractors during review and construction
- Fire sprinkler training for fire officials
- Public education and information materials for fire and elected officials
- Fire sprinkler demonstration trailers and customized live burn side-by-side demonstrations (see the explanation below)[12]

The NIFSAB reports that it has been successful in its awareness programs of how quickly fire can grow and spread in a home through its side-by-side live fire demonstrations. The demonstrations use typical home furnishings and construction and show how quickly the fire grows when unchecked by a fire sprinkler and how quickly a fire can be controlled when sprinklers are present. NIFSAB has conducted 363 side-by-side demonstrations for over 200 fire departments and now assists in developing the programs on a self-service basis so fire departments can conduct their own demos and reach more residents with awareness of the effectiveness of residential fire sprinklers.

The NIFSAB believes that the growth of communities adopting sprinkler ordinances is evidence of the positive effects of its program efforts. At least 87 municipalities and fire protection districts in Northern Illinois have adopted residential fire sprinkler ordinances for single-family homes. Approximately 20 more municipalities have requirements as well, but a minimum-square-foot area may be a threshold before the sprinkler requirement kicks in. The numbers of municipalities adopting residential fire sprinkler ordinances in Illinois is changing rapidly and for the foreseeable future will be a moving target. A list of the state's municipalities that have adopted residential fire sprinkler ordinances can be viewed on the NIFSAB web site. NIFSAB's web site is built to be a go-to resource to promote residential fire sprinklers in Illinois. One table listing the adopting municipalities includes information on whether all new homes apply (listed as zero square foot minimum size) as well as a range of minimums starting at 500 square feet to over 10,000 square feet.[13]

The strong evidence presented that fire deaths are nearly nonexistent in homes protected by residential sprinklers, as well as the number of whole regions adopting ordinances at the local level, such as in Illinois, California, and Maryland, has a positive impact on fire safety. In all three states, the local efforts resulted from heavy involvement by the local fire departments' strong and near constant education and outreach work.

It would be difficult to prove that a safety technology like sprinklers is of value with evidence from only a small area of implementation. But the information available on the performance of residential fire sprinklers is impressive. The bigger the sample becomes, the more broad is the evidence that can be applied to any jurisdiction, along with a prediction of the occurrence of fires and casualties that may be expected.

The first evidence of residential fire sprinkler performance was offered in a report about the first 15 years after Scottsdale, Arizona, implemented a requirement for residential sprinklers in all new homes. As of 2001 more than 50 percent of homes (41,408) had sprinklers. During that time, there were fewer than 600 home fires, 49 of them in homes with residential sprinklers. In the home fires with sprinklers, there were no fire deaths, while there were 13 fire deaths in homes without sprinklers.[14]

Additional evidence has come from Bucks County, Pennsylvania. A recent report on six municipalities that have implemented residential sprinkler requirements showed that between 1988 and 2010, there were 90 fire deaths in unsprinklered one- and two-family homes in all of Bucks County, Pennsylvania, while there were no fire deaths in dwellings protected by sprinklers. U.S. Census Bureau Quick Facts information on-line lists an estimated Bucks County population of 926,976 for 2013.[15]

The report also compared fire dollar loss in the six municipalities studied for individual fires between 2005 and 2010 with an average loss of $14,000 in homes with sprinklers and $179,896 in homes without sprinklers.[16]

The municipalities in Bucks County indicated that 88 percent of all fire deaths occur in dwellings. In addition, a report on residential sprinkler performance in Prince Georges County, Maryland, between 1992 and 2007 showed that 89 percent of their fire deaths occurred in dwellings. Of the 2,855 fire deaths reported by NFPA for the U.S. in 2012 just about 8 out of 10 fire deaths (83 percent) occurred in homes,[17] consistent with the 2010 and 2011 NFPA reports.[18,19] However, note that Prince Georges County also reported no fire deaths in homes with residential sprinklers during the period studied from 1992 to 2007, while 101 people died in unsprinklered homes.[20]

As described previously in the focused analysis of Scottsdale, Arizona; Prince Georges County, Maryland; and Bucks County, Pennsylvania, residential sprinklers have a proven record of reducing the risk to people in their homes. Despite the opposition to residential fire sprinklers and the passage of laws prohibiting adoption of sprinkler requirements by state and local government, sprinkler use will grow. The evidence of sprinklers' ability to preserve life is clear. The fact that all of the model codes for building and fire safety in the United States have included the requirement for residential sprinklers makes their use a standard of care. In anticipation of what could happen in the future in regard to the continued opposition to sprinklers, there is the potential for it to become a subject of litigation. But the fire and life safety industry has and will continue to measure the performance and life-saving capability of residential sprinklers and will push steadily for adoption nationally, because they work.

In more good news for fire safety, NFPA has reported that in 2010 fire deaths related to smoking materials dropped to the lowest point in 30 years. The NFPA attributes the gradual reduction of fire deaths to a decline in smoking as well as to stricter standards for mattresses and upholstered furniture. It points to engineering developments brought about by legislation requiring the so-called fire safe cigarettes as the likely reason for the improvement.[21]

The final implementation of the Federal *Standard for the Flammability (Open Flame) of Mattress Sets* was published in the Congressional Register for an effective date of July 1, 2007, and applied to all mattresses manufactured, imported, or renovated after that date in the United States. The Consumer Product Safety Commission believes that the mattress flammability standard could reduce the size of mattress fires and potentially result in 240 to 270 deaths and 1,150 to 1,330 injuries eliminated annually. The research work leading up to the change in mattress design and the resulting changes in industry standards directed by the federal government are examples of important fire safety engineering improvements. Another is the culmination of several decades of research and development that resulted from federal legislation designed to eliminate fires caused by cigarettes. According to NFPA, sleeping was a contributing factor to ignition with smoking materials in one-third (32 percent) of fire deaths in the home.[22]

As with residential fire sprinklers, addressing the problem of deaths resulting from fires caused by smoking materials can have an important impact in fire safety. In the case of smoking materials and fire death statistics, it may take some time to sort out exactly what has had the most impact, but prevention strategy remains the right direction to pursue. It took nearly 20 years from the first legislative work in the U.S. Congress with the **Cigarette Safety Act of 1984** to achieve the model standard for reducing the ignition propensity of cigarettes. Fire deaths from smoking materials had reached a 30-year high from 1980 to 2010, the highest year being 1982, when just under 2,000 people died in smoking-material-related fires.

Cigarette Safety Act of 1984
■ federal legislation that formed a technical study group to establish the technical and economic feasibility of making a fire-resistant cigarette, leading to follow-up legislation known as the Fire-Safe Cigarette Act of 1990, which directed the development of a standard test for cigarettes that measures their propensity to ignite other materials.

New York State had the first legislated regulation in 2000 for using a standard test method to measure the ignition strength of cigarettes. In 2002, New York adopted and provided the implementation leadership with ASTM E2187-02b, Standard Test Method for Measuring the Ignition Strength of Cigarettes. New York's original regulations required that cigarettes be less fire-prone, and with the ASTM test standard and the New York implementation as a model by 2012, the United States and Canada had both adopted the model approach. Most recent statistics from NFPA showed a reduction of smoking-related fire deaths to 540 in 2010 about one-quarter the number of deaths seen a little over 25 years earlier.[23]

The data reflecting the decline in smoking-related fire deaths in homes indicate that engineering and the regulatory process can improve fire safety and that fire prevention works. Neither of the two examples of bedding standards and test standards for cigarette ignition strength could have been accomplished without the will of government to overcome strong industrial influence, the same influence that can be seen in the opposition to residential sprinklers. There is a strong push to limit regulatory influence over business and industry as the United States recovers from the difficult economic recession that hit in 2008. A continuing refinement of measurement and evaluation of the many facets of fire safety is needed to continue reducing fire risk.

The U.S. Fire Administration and National Fire Protection Association continue to look at the fire problem from the wider perspective given by statistical information and national trends. The measure of success, however, comes from overlaying the efforts at the state and local levels with the many examples of meaningful proof. There are countless examples throughout this book of the careful examination of local fire problems and detailed planning, implementation, and evaluation of results that when added up show progress of how fire prevention works. The aggressive programs under way—in Morgantown, West Virginia; Scottsdale, Arizona; Prince George's County, Maryland; New York City; and Plano, Texas—exemplify the direction communities with a genuine interest in addressing the fire problem need to follow. Only through community-wide interest, spearheaded by the fire service, can the nation's fire problems be seriously abated.

CHAPTER REVIEW

Summary

Whether fire prevention works or not may be seen as something in the eye of the beholder: the stakeholders in the case of public prevention programs. If they are unaware of the good services delivered by the fire prevention bureau, they are likely to take those services for granted and be misled about their importance. The loss of services in recent difficult economic times in many communities may lead to a future slide in community safety. So it is important to regularly tell the good stories related to evidence of fire prevention value. To tell these stories, it is necessary to acquire and analyze data and then communicate the fire experience to the community.

Records are essential for the proper administration of any fire prevention program. Accurate and complete information about fire prevention and fire control activities should be available for a thorough job to be done. Analysis of statistical records can reveal specific problem areas toward which fire prevention education programs should be directed. Analysis of records can likewise point to a need for changes in fire prevention codes, modification of statutory requirements, and other changes in regulations. Note that examples taken from records of the local fire department are far more meaningful than are reports of fire conditions taken from communities hundreds or thousands of miles away.

Statutory requirements make it mandatory to maintain records of the information needed in subrogation cases, arson cases, and other legal proceedings. Court cases in which fire departments have been sued make it imperative to keep records of all fire inspection activities and fire prevention education programs.

Records and reports of both fire suppression and fire prevention activities are closely allied, and it is sometimes difficult to distinguish between the two. The types of reports and records the fire department should generally maintain are a complete record of fire department responses; statistical records of fire deaths, injuries, and losses; records of fire prevention and training activities; and reports for public information.

A complete record of all fire prevention inspections is all-important in the work of the fire prevention bureau to ensure proper follow-up of requirements imposed or changes necessary to conform to fire code regulations. The file should be kept up to date with any new information acquired.

From all records, many fire departments compile an annual report that gives community residents, as well as the official governing body, an accurate picture of the fire department's operation and accomplishments.

Review Questions

1. One way fire departments have quantified effectiveness is by:
   a. reporting the number of fire deaths.
   b. attending service club meetings.
   c. being able to obtain federal grants.
   d. focusing on densely populated residential areas.
2. New challenges needing the attention of fire prevention officials include:
   a. fire watch.
   b. water rescue.
   c. urban–wildland interface problems.
   d. urban search and rescue.
3. Most cities experience a higher per capita fire death rate in:
   a. high-rise buildings.
   b. detached houses in high-income areas.

c. low-income housing.
d. hotels and motels.

4. The National Fire Data Center is a federal government function in what agency?
   a. U.S. Product Safety Commission
   b. National Institute for Standards and Technology
   c. National Fire Protection Association
   d. U.S. Fire Administration
5. Opportunities for funding and community support of fire prevention services can be enhanced through use of what management process?
   a. Integrated business process
   b. Manufacturing management process
   c. Performance measurement
   d. Process architecture
6. Albany, Georgia, was honored by what program of the U.S. Chamber of Commerce?
   a. Safe Cotton Storage Initiative
   b. Fire Waste Contest
   c. Built for Life Community Project
   d. Firewise Community Coalition
7. A Built for Life Fire Department promotes fire safety through use of what technology?
   a. Sprinklers
   b. Preplanning
   c. Plan review
   d. Computers
8. A person killed or injured as a result of being directly exposed to fire is considered:
   a. unlucky.
   b. a victim.
   c. a casualty.
   d. a statistic.
9. Keep positive efforts of fire prevention activities in the eyes of the community through use of:
   a. news media.
   b. social media.
   c. annual reports.
   d. all of the above.
10. All of the model codes for building and fire safety in the United States have included the requirement for residential sprinklers.
    a. True
    b. False

End Notes

1. Betty Rehberg, "Vintage Albany: AFD Chief D.W. Brosnan," *Albany Journal: Southwest Georgia's Local Online Newspaper*, April 29, 2012. Accessed December 3, 2013, at http://thealbanyjournal.com/2012/04/vintage-albany-afd-chief-d-w-brosnan/#sthash.e5KhKeYH.dpuf
2. Georgia Fire Service, *Shaping the Future: Strategic Plan* (Atlanta: Office of Insurance and Safety Fire Commissioner, 2012), pp. 18–19. Accessed October 4, 2013, at http://www.oci.ga.gov/ExternalResources/Documents/Safety%20Fire/FireServicesStrategicPlan.pdf
3. Author Love's notes from Mid-Atlantic Life Safety Conference, Laurel, MD, September 2013.
4. Federal Emergency Management Agency, U.S. Fire Administration, and National Fire Data Center, *Fire in the United States: 1989–1998*, 12th ed., FA-216 (Arlington, VA: TriData Corp., August 2001), p. 16.
5. Personal communication to author Love from Jim Tidwell, Tidwell Code Consulting, September 2013.
6. Eugene K. Chow, "HSNW conversation with David Muhlhausen: Many DHS Grants Ineffective, Lack Proper Oversight," *Homeland Security News Wire*, September 1, 2011. Accessed September 1, 2013, at http://www.homelandsecuritynewswire.com/many-dhs-grants-ineffective-lack-proper-oversight
7. National Fire Protection Association, "A Small City's Achievement," *Quarterly of the National Fire Protection Association*, April 1926, p. 311.
8. Personal communication to the author Robertson from Dr. John Hall, National Fire Protection Association, 2008.
9. Federal Emergency Management Agency, U.S. Fire Administration, and National Fire Data Center, *Fire in the United States: 1989–1998*,

12th ed., FA-216 (Arlington, VA: TriData Corp., 2001), p. 9.

10. *NFPA 901: Standard Classifications for Incident Reporting and Fire Protection Data* (Quincy, MA: National Fire Protection Association, 2011).
11. Personal correspondence from author Love, Fire Marshal, Montgomery County, MD, September 2007.
12. Northern Illinois Fire Sprinkler Advisory Board (2010). Accessed September 30, 2013, at http://firesprinklerassoc.org
13. Northern Illinois Fire Sprinkler Advisory Board (2013). Accessed March 31, 2014, at http://firesprinklerassoc.org/pdfs/city%20square%20foot%202012-7.pdf
14. Home Fire Sprinkler Coalition, *Highlights: Scottsdale 15-Year Data (Power Point),* (Frankfort, IL: Home Fire Sprinkler Coalition). Accessed on March 31, 2014, at http://www.homefiresprinkler.org/images/stories/download/HFSC_15_year.ppt
15. U.S. Census, *State and County Quick Facts.* Accessed March 31, 2014, at http://quickfacts.census.gov/qfd/states/42/42017.html
16. Home Fire Sprinkler Coalition, *Communities with Home Fire Sprinklers: The Experience in Bucks County, Pennsylvania* (Frankfort, IL: Home Fire Sprinkler Coalition, 2012). Accessed September 17, 2013, at http://www.homefiresprinkler.org/index.php/fire-department-bucks-county-report
17. Michael J. Karter, Jr., *Fire Loss in the United States During 2012* (Quincy, MA: National Fire Protection Association, *September,* 2013), p. iv. Accessed November 26, 2013, at http://www.nfpa.org/~/media/Files/Research/NFPA%20reports/Overall%20Fire%20Statistics/osfireloss.pdf
18. Michael J. Karter, Jr., *Fire Loss in the United States During 2011* (Quincy, MA: National Fire Protection Association, *September,* 2012), p. ii. Accessed November 26, 2013, at http://www.nfpa.org/~/media/Files/Research/NFPA%20reports/Overall%20Fire%20Statistics/FireLoss2012.pdf
19. Michael J. Karter, Jr., *Fire Loss in the United States During 2010* (Quincy, MA: National Fire Protection Association, *September,* 2011), p. ii. Accessed November 26, 2013, at http://www.nfpa.org/~/media/Files/Research/NFPA%20reports/Overall%20Fire%20Statistics/OSfireloss2010.pdf
20. Home Fire Sprinkler Coalition, *Benefits of Residential Fire Sprinklers: Prince George's County 15-Year History with Its Single-Family Residential Dwelling Fire Sprinkler Ordinance* (Frankfort, IL: Home Fire Sprinkler Coalition, 2012). Accessed September 17, 2013, at http://www.homefiresprinkler.org/index.php/fire-department-prince-george-county-report
21. International Fire Marshals Association, "Smoking-Materials Fire Deaths Drop to 30 Year Low," *Fire Marshals Quarterly,* Summer 2012, p. 20.
22. Ibid.
23. John R. Hall, Jr., National Fire Protection Association, *The Smoking-Material Fire Problem* (Quincy, MA: National Fire Protection Association, July 2013). Accessed November 26, 2013, at http://www.nfpa.org/~/media/Files/Research/NFPA%20reports/Major%20Causes/ossmoking.pdf

ANSWERS TO REVIEW QUESTIONS

Chapter 1

1. b
2. d
3. a
4. b
5. d
6. d
7. b
8. d
9. a
10. d
11. c
12. a

Chapter 2

1. c
2. d
3. c
4. c
5. d
6. b
7. b
8. b
9. b
10. a
11. c
12. d
13. b

Chapter 3

1. d
2. d
3. c
4. a
5. d
6. b
7. c
8. b
9. b
10. b
11. a
12. d
13. d

Chapter 4

1. b
2. a
3. c
4. d
5. d
6. a
7. d
8. a
9. d
10. c

Chapter 5

1. a
2. d
3. d
4. a
5. b
6. a
7. b
8. c
9. b
10. d

Chapter 6

1. a
2. b
3. d
4. d
5. d
6. b
7. b
8. d
9. c
10. d

Chapter 7

1. c
2. a
3. d
4. d
5. d
6. b
7. a
8. d
9. b
10. b

Chapter 8

1. c
2. a
3. c
4. d
5. d
6. b
7. d
8. d

Chapter 9

1. b
2. b
3. d
4. b
5. d
6. d
7. d
8. c
9. a
10. d

Chapter 10

1. c
2. a
3. c
4. d
5. b
6. d
7. c
8. c
9. d
10. b

Chapter 11

1. a
2. b
3. d
4. a
5. d
6. d
7. d
8. a
9. b
10. a

Chapter 12

1. a
2. d
3. a
4. b
5. a
6. c
7. b
8. c
9. a

Chapter 13

1. a
2. c
3. c
4. d
5. c
6. b
7. a
8. c
9. d
10. a

GLOSSARY

1-megawatt fires—a significant heat release rate of a specific burning object; sometimes referred to as a minimum for producing a room flashover

A

America Burning—the report of the National Commission on Fire Prevention and Control that identified America's growing loss of life and property from fire, along with the parallel culture of indifference to fire, including comprehensive recommendations on how to begin to turn around the problem

American National Standards Institute (ANSI)—a clearinghouse for voluntary safety, engineering, and industrial standards nationally

arson—the crime of maliciously and intentionally, or recklessly, starting a fire or causing an explosion

ASTM International—a voluntary consensus standards organization founded in 1898 that publishes standard fire tests of building construction and materials and recommends practices for preparing fire test standards; formerly known as American Society for Testing of Materials

B

Built for Life Fire Department—a program initiated and supported by the Home Fire Sprinkler Coalition, recognizing fire departments in the United States that focus on promoting residential fire sprinklers in their outreach programs

Bureau of Alcohol, Tobacco, Firearms, and Explosives (ATF)—a law enforcement agency in the U.S. Department of Justice that regularly inspects and licenses explosives dealers, manufacturers, and users when any aspect of the operation may involve interstate commerce; investigates certain cases of arson, explosions, use of firebombs, and careless activities relating to explosives

C

casualty—a person who is killed or injured as a result of being directly exposed to a fire; a term used for both civilians and firefighters, though the two are recorded as distinctly separate data

Cigarette Safety Act of 1984—federal legislation that formed a technical study group to establish the technical and economic feasibility of making a fire-resistant cigarette, leading to follow-up legislation known as the Fire-Safe Cigarette Act of 1990, an act directing the development of a standard test for cigarettes that measures their propensity to ignite other materials

common fire hazards—hazards found in practically all occupancies, such as those related to normal use of electricity and heating equipment, as well as to the existence of everyday combustibles

community analysis—a process that identifies fire and life safety problems and the demographic characteristics of those at risk in a community

community partnership—a person, group, or organization willing to join forces with others and address a community risk

Community Risk Reduction (CRR)—local-level commitment to develop internal and external partnerships with the community to implement programs, initiatives, and services that promote an integrated approach to the risks of injury and loss through education, engineering, enforcement, economic incentives, and emergency response

Community Wildfire Protection Plan (CWPP)—based on the needs of people living in wildfire-threatened areas, a local collaborative planning process that addresses specific issues such as wildfire response, hazard mitigation, community preparedness, and structure protection

compliance—conformance with either applicable codes or a notice of violation issued or served by an authorized code official

cone calorimeter—a bench test instrument used primarily in fire safety engineering to evaluate small samples of material to obtain data measurements—for example, ignition time, heat release rate and mass loss

conflagration—a destructive fire covering a large area, often able to cross natural barriers and frequently causing major life and property loss

Congressional Fire Services Institute—an organization that provides a liaison between the Congressional Fire Caucus and the fire protection community

Consumer Product Safety Commission—the agency that endeavors to eliminate hazards associated with consumer products, including toys, hazardous substances, and flammable fabrics

control of occupancy—an enforcement method that controls occupancy load through determination of the adequacy of egress and through provision of fire safety personnel on scene to ensure an orderly evacuation in the event of fire

corpus delicti—the body of evidence that proves a crime has in fact been committed

curfew—an official action intended to restrict some activity; derived from the Anglo-Norman word *covre-feu,* which means "cover fire"

D

Daubert v. Merrell Dow Pharmaceuticals, Inc.—the case that allows a court to gauge whether expert testimony aligns with the facts of the case as presented

E

economic incentives—an intervention that promotes a variety of savings to developers, builders, and building owners in exchange for employing desired but often not required fire protection systems, materials, or construction practices

education—actions that teach, promote awareness, present information, and conduct activities intended to change behaviors that increase people's safety from fire

emergency response—the systematic and immediate response of fire suppression or other emergency services with the goal of stopping the loss of life or property or otherwise reducing the risk to community by the deployment of personnel and materials

enforcement—a systematic approach to adopting community codes and certifying officials with authority to inspect the extent of code compliance and to implement rules and regulations to clarify the application of codes

engineering—actions that employ physical science to find solutions for reducing the impact or eliminating the risk of fire and that may involve the development of codes, standards, equipment, and systems and construction techniques

exit interview—the closing portion of a fire safety inspection, during which a review is conducted with a person of authority or a responsible management representative to ensure that corrections are understood and will be made

exposures—areas of a structure where its proximity to other structures or combustibles could endanger the building in question if involved in a fire

F

Federal Aviation Administration—the agency under the U.S. Department of Transportation that is responsible for the control of all aspects of aviation, including safety features incorporated within aircraft

Federal Bureau of Investigation (FBI)—the top law enforcement agency in the U.S. Department of Justice, offering intelligence services and threat-focused national security; supports state and local law enforcement on crimes such as arson by facilitating access to national crime data and specialized forensic management of evidence

Federal Emergency Management Agency (FEMA)—the agency under the U.S. Department of Homeland Security responsible for emergency management and other coordinating programs relating to major disasters

Federal Housing Administration—an arm of the U.S. Department of Housing and Urban Development that has set certain standards for structures insured under the National Housing Act and other federal laws, which include minimum fire safety standards

F.I.R.E.—an acronym for Fire Investigation, Research, and Education; a facility funded by Congress that assists investigators in understanding and reconstructing the physical effects of fire in buildings

Fire Adapted Communities (FAC)—communities that have formed partnerships with homeowners, local agencies, and other organizations to reduce the potential for loss of life and property by providing information and expertise on strategies and actions to mitigate wildfire risks

fire and life safety education—planned activities focused on promoting, presenting, and making available information intended to change behavior and reduce the risk of fire and other injury-causing events in the community

fire dynamic simulator (FDS)—a computer-modeling application that simulates fires and predicts fire conditions in a compartment

fire exits—exits specifically identified and maintained as a means of egress, often a door or other opening that provides occupants a safe way out of a building or other structure

fire ground—the area where fire department emergency operations are conducted at the scene of a fire

fire investigation—a scientific inquiry conducted to determine the cause of fire so as to prevent a recurrence of the same scenario

fire prevention—a level of effort to decrease the chances of unwanted fire ignition; the philosophy and practice of reducing the hazards and risk of fire with the goal of decreasing the loss of life and property

fire prevention bureau—the organizational element that coordinates all fire prevention activities within the fire department and makes important contributions to fire department public relations through its inspection and fire safety education programs

fireproof—a term used, usually in the construction of buildings, to describe materials that do not demonstrate the characteristic of combustibility; a term found to serve little value since the structure was often filled with other combustible materials that still presented a hazard to the occupants of the building

fire protection master plan—a comprehensive analysis and descriptive report about a community's fire problem, which includes subsequent preparation and implementation of a plan that addresses the fire protection needs of the jurisdiction, with an economic analysis of the costs associated with the recommendations

Fire Protection Research Foundation—a nonprofit organization that provides fire and life safety research supporting NFPA's mission and provides specific guidance to NFPA technical committees

fire reaction—fire-related human behavior that must be planned for in designing fire prevention training and education that may include, for example, leaving by the same door a person entered from, taking in cues of what is going on before deciding to evacuate, helping others escape a fire, and having a narrowed focus while under stress

Fire Research Division of NIST—a division that conducts research in building materials, computer-integrated construction practices, fire science and fire safety engineering, and structural, mechanical, and environmental engineering

fire-resistive—a term currently used for some buildings that can resist and even prevent the spread of fire and products of combustion

fire safety—the concept of actions planned and taken that reduce the risk of human exposure to fire

fire safety manual—a document that forms the basis of fire drill procedures and is tailored to a specific occupancy

Firewise Communities Program—a wildfire planning and mitigation process cosponsored by the USDA Forest Service, the U.S. Department of the Interior, the National Association of State Foresters, and the National Fire Protection Association; it encourages solutions and action at the local level with community, government, and business participation

flashover—the point when a room fire becomes fully developed with all contents completely involved in fire; people intimate to a fire when it reaches flashover are not likely to survive

flow path—a route of movement taken by hot gases between a fire and exhaust openings (windows, doors, cut vent holes) and the movement of air toward and feeding the fire

FM Global Group—a significant leader in research to reduce property loss, specializing in insuring major corporations' commercial property

I

incapable of self-preservation—a condition associated, for example, with reduced cognitive ability, inability to ambulate independently, illness, and medical treatment that makes a person unable to take appropriate actions when faced with life-threatening hazards

Institute for Research in Construction—a branch of the National Research Council of Canada established to serve the needs of Canada's construction industry

Institution for Fire Engineers—an international organization with the mission to encourage and improve all areas of the science and practice involved in fire safety

International Association of Arson Investigators—an organization whose primary objective is to improve the professional development of fire and explosion investigators through its availability as a global resource for investigation, technology, and resources

International Association of Fire Chiefs—a worldwide organization with a primary objective of leadership and advocacy that provides positions on key issues important to the fire service and, in particular, fire chiefs; its Fire and Life Safety Section leads efforts in fire prevention and is active in codes and standards development

International Association of Fire Fighters (IAFF)—an organized-labor representative for the majority of the career firefighters in the United States and Canada, active in fire prevention through participation in standards development, code change, and fire and life safety forums

International Building Code (IBC)—a model building code published by the International Code Council; the most common building code used in the United States

International Fire Service Training Association (IFSTA)—an organization that identifies needed areas of training materials and supports those needs through the development and validation of the training materials for the fire service and other related services

interventions—fire prevention or Community Risk Reduction (CRR) actions to prevent or reduce loss; can include changing unsafe behavior, separating building occupants and combustible products, and installing fire protection or smoke detection

M

marketing—the action of promoting and selling products and services, such as programs associated with fire safety and fire prevention

mechanical codes—building codes that cover heating and air conditioning

Michigan v. Clifford—the U.S. Supreme Court decision that reaffirmed its findings in *Michigan v. Tyler* and recognized the right of the investigator to seize evidence in clear view at time of entry

Michigan v. Tyler—the U.S. Supreme Court decision that established guidelines for right of entry to investigate cause and origin and to subsequently gather evidence should a fire be determined to be incendiary

Mine Safety and Health Administration (MSHA)—an agency of the U.S. Department of Labor that has direct responsibility for fire prevention and safety within coal, metallic, and nonmetallic mines, often sharing this responsibility with a state bureau of mines

mini–maxi codes—codes that are adopted at the state level and cannot be amended at the local level

Model Arson Law—a law developed by the International Fire Marshals Association; forms part of the laws of many states

motive—an inner drive or impulse that is the cause, reason, or incentive that induces or prompts a specific behavior

N

National Board of Fire Service Professional Qualifications (NBFSPQ)—an organization, commonly referred to as the Pro Board, established to oversee a means of acknowledging firefighter and other related fields of achievement, primarily through the accreditation of organizations that certify members of public fire departments

National Fire Data Center—a division of the U.S. Fire Administration; collects, analyzes, publishes, disseminates, and markets information about the U.S. fire problem

National Fire Incident Reporting System (NFIRS)—a system operated by the National Fire Data Center that collects and analyzes fire incident and casualty data on a nationwide basis

National Fire Protection Association (NFPA)—an organization that aims to reduce the risk of fire and other hazards by offering consensus codes and standards, educational and training materials, and fire safety advocacy

National Park Service—a division of the U.S. Department of the Interior that promotes conservation and environmental awareness in the national parks—including countless historical structures and icons—through administration of an aggressive fire prevention program

National Safety Council—an organization whose major impact is in the field of safety, including fire prevention as an overall safety objective

National Volunteer Fire Council—an organization that represents the interests of the nation's volunteer fire services

NFPA 921 Guide for Fire and Explosion Investigations—a standard that identifies guidelines and recommendations for processes and procedures to conduct investigations or analysis of fire and explosions

NFPA 1031 Professional Qualifications for Fire Inspector—a standard that identifies professional performance qualifications for fire inspectors and plan examiners

NFPA 1033 Standard for Professional Qualifications for Fire Investigator—a standard that identifies professional performance qualifications for fire investigators

NFPA 1035 Professional Qualifications for Public Fire and Life Safety Educator—a standard that identifies professional performance qualifications for public fire and life safety educators, public information officers, and juvenile-fire-setting interventionists

O

Occupational Safety and Health Administration (OSHA)—a division of the U.S. Department of Labor that is charged with developing safe and healthy working conditions through standards and regulations, including those covering fire protection

offender-based studies—examinations of the relationship between crime scene characteristics and the behavioral characteristics of the offender as they relate to motive

P

performance-based design—an engineering design that incorporates fire safety solutions by means of using alternative materials and techniques to achieve specific goals

performance measurement—a system management process that quantifies value, effectiveness, and efficiency so services and programs such as fire prevention and safety can be better understood by stakeholders, community partners, residents, government management, and elected officials

plan review—prior to new construction or building modifications, an evaluation by certified officials that ensures compliance of architectural or engineering plans with building and fire codes

plan-review program—a review by the authority having jurisdiction over the plans and specifications for buildings to be constructed within that jurisdiction

platooning—a grouping behavior of people during building evacuation

prefire planning surveys—on-site visits to buildings and occupancies for the purpose of gathering and documenting information that will ultimately be used in prefire plans and building information files

public assembly—a type of area where at least 50 people tend to congregate, such as theaters, churches, auditoriums, dance halls, nightclubs, and restaurants

public fire education planning—efforts involved in the consideration and implementation of a program to raise awareness or teach new fire safety skills, including analyzing problems, developing partnerships, creating strategies to address priorities, implementing plans, and evaluating progress

R

research—careful, systematic study and investigation in some field of knowledge

research and development (R&D)—systematic work in organizations with the goal to produce a product, solve problems, create new processes, or acquire knowledge

S

SARA Title III—a federal program that requires industry and related occupancies to maintain records on hazardous materials and their characteristics

See v. City of Seattle*, and *Camera v. City and County of San Francisco—significant court cases holding that inspectors must not enter nonpublic areas of a business unless permission is granted; these decisions expanded the boundaries of the Fourth Amendment of the U.S. Constitution beyond just dwellings

Society of Fire Protection Engineers (SFPE)—a professional society representing those practicing in the field of fire protection engineering

special fire hazards—hazards a fire inspector associates with conditions unique to an occupancy, such as combustible liquids or gases in an auto repair shop

State of South Dakota v. Jorgensen—the court case that upheld use of evidence found during the state fire investigator's initial entry while the fire was still smoldering but upheld suppression of evidence found in later, warrantless entries

structural control—a means of enforcement effected through the plan-review function coupled with inspection before occupancy to be sure that all fire safety requirements have been fulfilled

subrogation—a legal process often used by insurance companies to recover damages or obtain a monetary reimbursement from a person or organization responsible for property damage or injury

T

target groups—members of a population who are the focus of a specific fire safety message

tenability—the fire environmental impact on human survival; used in discussion or presentation of fire protection and fire safety subjects to demonstrate when conditions are survivable or not

U

Underwriters Laboratories Inc. (UL)—an independent not-for-profit organization that has a comprehensive program of testing, listing, and reexamining equipment, devices, materials, and assemblies as they relate to fire protection and safety

U.S. Coast Guard—one of five U.S. armed forces branches and the only military branch within the U.S. Department of Homeland Security; safeguards U.S. maritime and environmental interests and has a long history of legal responsibility in the field of fire prevention, both aboard vessels and in U.S. ports

U.S. Department of Health and Human Services—formed in 1953; has had a major impact on fire safety through a number of agencies with direct interests in fire prevention

U.S. Department of Homeland Security—established as a result of the attacks on September 11, 2001; provides funds through matching grants to local fire departments and public safety agencies

U.S. Department of Housing and Urban Development—devotes considerable attention to fire protection for lower-income housing units; has funded several research projects in the fire protection field; also enforces requirements for manufactured housing

U.S. Forest Service—a division of the U.S. Department of Agriculture that maintains a widespread fire suppression force and conducts an extensive nationwide fire prevention promotional program

V

Vision 20/20 Project—sponsored by the Institution of Fire Engineers, a grant-funded fire safety project that includes five key prevention strategic initiatives—advocacy, marketing, culture, technology, and codes and standards—and a sixth overarching strategy providing local-level evidence of positive prevention results

W

wildland—often a remote land area characterized by natural, sometimes dense vegetation minimally modified by human activity, such as woodlands, forests, brush, and meadows; wildland can contain dangerous buildups of fuel from vegetation that may or may not be a risk for fire; trends in migration and development currently bring together increased wildland interface with urban areas throughout the world

Z

zoning codes—a land-use planning process that involves the limitation of various types of occupancies to given sections of the community, including, for example, the land's intended use, building size and height, and lot size

INDEX

D

E

F

G

H

I

J

K

L

M

N

S

T

U